EXPLORING MATTER AND ENERGY

PHYSICAL SCIENCE

David R. Kiefer
Assistant Principal for Physical Science
Midwood High School at Brooklyn College
Brooklyn, New York

Project Coordinator
Carl M. Raab

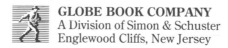

GLOBE BOOK COMPANY
A Division of Simon & Schuster
Englewood Cliffs, New Jersey

Other programs in this series
Exploring Earth and Space: Earth Science
Exploring Living Systems: Life Science

Project Coordinator
Carl M. Raab
Assistant Principal, Supervision: Science Department
Fort Hamilton High School
Brooklyn, New York

Reviewers
Estella V. Abel
Science Coordinator
Mt. Vernon High School
Mt. Vernon, New York

Judah Landa
Science Teacher
Midwood High School at Brooklyn College
Brooklyn, New York

Joel Berger
Professor, Science Education
College of Staten Island, CUNY
Staten Island, New York

Charlotte Smith
Math and Science Coordinator
Watertown City School District
Watertown, New York

John W. Kominski
Assistant Director of Science
New York City Board of Education
New York, New York

Dale Wescott
Science-Health Supervisor
Guilderland Central School District
Guilderland, New York

Developmental assistance by The Mazer Corporation.

Cover photo credits: Upper left (laser beams), Chuck O'Rear/WestLight; Upper right (flasks), Peticolas/Megna/Fundamental Photographs/Photo Researchers; Lower left (neon abstract), FourByFive; Lower right (microprocessor chip), IBM. Cover Photographs researched by Rhoda Sidney.

Illustration credits: Computer Generated by Hyper Graphics and The Mazer Corporation

GLOBE BOOK COMPANY
A Division of Simon & Schuster, Inc.
Englewood Cliffs, New Jersey 07632

CONTENTS

CONTENTS

5

CHAPTER 1

INTRODUCTION TO PHYSICAL SCIENCE

Would you throw away a lamp because its light goes out? Probably not. Figure 1-1 suggests some ideas. You could try a new bulb. If that did not work, you might test the electric outlet with another lamp. Maybe a fuse has blown. You would keep asking questions and making tests until you solve the problem. You may not realize it, but when you ask questions and apply tests, you are acting like a scientist.

Scientists ask questions about all sorts of things. Some of their questions relate to problems, like a lamp that mysteriously goes out. Other questions, such as "What is the world made of?" or, "How long does it take water to turn to ice?" come from simple curiosity.

Scientists can explore questions like these through experiments. For example, the scientist curious about water turning to ice could place water in a freezer and observe the water every 15 minutes for six hours. Observing the water is like the process you use to find out why your lamp stopped working. You often do things in a scientific way.

1-1 The Nature of Physical Science

■ *Objectives*

☐ *Describe the nature of science.*

☐ *Define physical science.*

☐ *List four topics within chemistry and physics.*

Throughout history, people have collected facts and ideas about how the world works. **Science** is more than just a collection of facts and ideas. Scientists use special ways

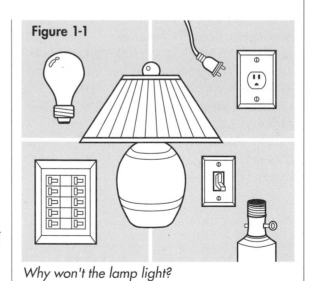

Figure 1-1

Why won't the lamp light?

of gathering these facts and ideas. They ask questions, do experiments, and ask more questions. Scientists also use logical thinking.

Science can be used to learn about most things in life. Different areas of life are studied by different areas of science. This book is about physical science.

The area of **physical science** involves the study of the materials and forces in the world around us. Physical science explores almost everything we do and see. Imagine you are a swimmer. Here are a few physical science questions you might ask about swimming.

- What is the water made of?
- Why do I float and not sink?
- What determines how fast I can swim?

These big questions lead to smaller ones. The questions physical scientists ask can be grouped into two areas of study.

One branch of physical science is **chemistry.** Chemistry explores what things are made of and how they can be changed. Scientists who study and work in chemistry are known as chemists. Chemists investigate subjects like the composition of soil or air, the way plants use sunlight, and ways chemicals can be changed to grow better crops. Chemists and chemistry students ask and try to answer questions about the nature of basic materials. For example, when cooking a pot of rice, a chemist might ask, "How does the rice change after water is added and heat is applied? Do the rice and water combine? How?"

Another branch of physical science is **physics.** Scientists use physics to explore the forces on objects and motions that objects make. Scientists who work in physics are called physicists.

If you throw a ball, as a physicist you would ask questions about the force of your arm on the ball, the ball's speed and distance, and the changes in energy that occur. Physicists investigate subjects like electricity, light, and heat.

Every day, chemists and physicists make new discoveries. Some discoveries, such as new ways of cleaning up pollution, can change the world. It is important, therefore, for you to understand the basic ideas of physical science.

▬▬ Section 1-1 Review ▬▬

Write the definitions for the following terms in your own words.

1. **science**   2. **chemistry**
3. **physical science** 4. **physics**

Answer these questions.

5. What do chemistry and physics examine?
6. List two topics a chemist might study and two topics a physicist might study.

1-2 The Method of Science

■ *Objectives*
☐ *List the steps of the scientific method.*
☐ *Describe a controlled experiment.*

Scientists, like detectives, use an organized approach when they want to answer questions or solve problems. This approach is called the **scientific method**. The scientific method is the process researchers use in stating problems and seeking solutions. The scientific method can be broken down into the steps described in this section.

Identify a Problem

Which of the objects in Figure 1-2 falls faster, a shoe or a pencil? Does an object's weight influence how fast it falls? As an object falls, does the air slow it down? Could an object's shape also be important to its falling speed?

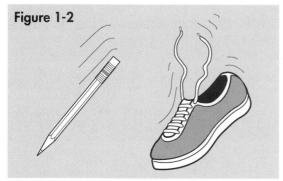

Figure 1-2

Which falls faster, the pencil or the shoe?

Now that you have asked questions, you can restate the problem in another question: How do weight and shape affect an object's speed of falling?

State a Hypothesis

Once you have clearly stated your problem or question, you can begin looking for a solution or an answer. Let's continue with the above example.

You can try to explain an object's falling speed in terms of its weight and shape by suggesting a possible solution, which scientists call a **hypothesis** (hy-PAHTH-uh-sis). Sometimes a hypothesis is called an educated guess because it is a guess that is based on related experience. Here are some possible hypotheses about falling things.

- A heavy object falls faster than a light object because gravity pulls harder on the heavy object.
- A smaller object falls faster because there is less for the air to slow down.
- Everything falls at the same speed.

Not all of these hypotheses can be correct. How would you find out which hypothesis is correct? You need to do what a scientist would do in this situation, test your hypothesis. One way to test all of your hypotheses at once is to drop a pencil and a shoe to see which hits the floor first.

Anyone can make a hypothesis, but it must be tested. A hypothesis may turn out to be wrong. Disproving a hypothesis is as important as finding evidence that a hypothesis is correct. This is because eliminating an incorrect hypothesis helps you focus on the hypothesis that may be correct. Disproving a hypothesis can also help you restate other hypotheses more clearly.

Design and Conduct an Experiment

Scientists carefully plan their research experiments. Equipment and procedures must be carefully chosen in order to test the hypothesis accurately. Figure 1-3 is an example of a procedure that does *not* accurately test a hypothesis.

Evan wants to compare the falling speed of the pencil and the shoe. He drops the

Figure 1-3 A Poor Experiment

Figure 1-4 A Controlled Experiment

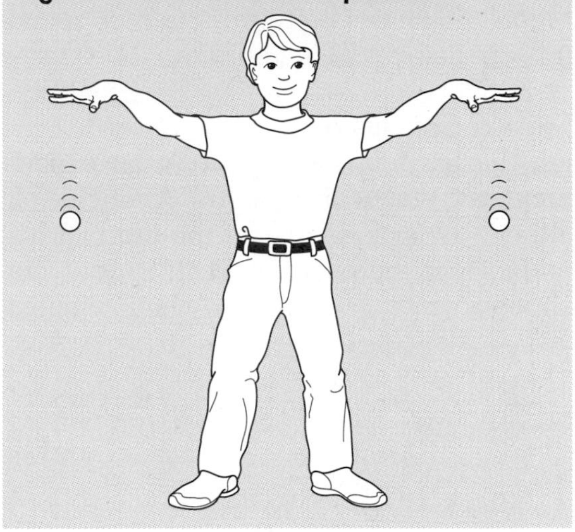

shoe from a lower point than he drops the pencil.

What is wrong with this procedure? You cannot compare the falling speed of objects by dropping them from different heights because it takes longer to drop from a higher point.

Good planning produces a **controlled experiment**. In a controlled experiment, you test one thing at a time and keep all other aspects, or conditions, of the experiment the same. Each aspect of an experiment that can affect the results is called a **variable** (VER-ee-uh-bul). For example, if you are testing the variable of weight by dropping objects, you use objects of different weights. You make sure all other variables, such as shape, size, height of the drop, and time of release, are exactly the same for each object dropped. In this way you can compare the effects of different weights precisely.

Look at Figure 1-4 and compare it with Figure 1-3. Instead of dropping a shoe and pencil, the experimenter in Figure 1-4 is dropping two balls of the same size, from the same height, at the same time. However, one ball is a heavy metal paperweight, and the other is a light ping-pong ball. If the metal ball hits the floor first, the experimenter may conclude that weight does indeed affect the falling speed of objects. If both balls hit the floor at the same time, he may conclude that weight does not affect falling speed.

Gather and Record Data

The observations and measurements taken during an experiment are known as **data**. Scientists accurately gather and record data in a neat, organized way. The final results of an experiment can be written in a report or in tables, graphs, and diagrams. It is important to plan the presentation before you begin experimenting, so that you can record data in an organized way. You will learn more about tables and graphs in Chapter 3. Figure 1-5 shows three ways to record and present data on falling objects.

State a Conclusion

On the basis of the data from an experiment, you state **conclusions**. A conclusion either supports or rejects the original hypothesis of the experiment. If it does not support the hypothesis, the scientist states that forces always come in pairs. When you push an object, it pushes you back. This idea is so well established and so important to every branch of science that it is called a law.

State New Questions

Finding an answer to one question usually leads to new questions. For example, if air causes different objects to fall at different

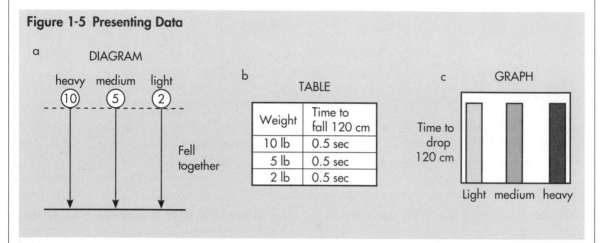

Figure 1-5 Presenting Data

usually proposes a new hypothesis and conducts another experiment. In the experiments about falling objects, for example, one conclusion might be that shape can affect how fast something falls.

Scientists use a **theory** to explain the results of many experiments. An example is the theory that all material is made of atoms. No one has ever seen an atom but this theory explains many experimental results in chemistry.

A major conclusion, supported by thousands of experiments and disproved by none, results in a **law of science**. For example, the **law of action and reaction** speeds, what would happen if there was no air? In the absence of air, or in a **vacuum** (VAK-yu-um), would all objects fall at the same speed? Scientists can test this hypothesis in the laboratory.

Scientists usually try to show that the conclusions of their experiments fit the laws of science. If a conclusion does not fit a law, the scientists will ask whether the conclusion is wrong or the law is wrong. The researcher may design a new experiment and draw new conclusions to try to bring the results within the laws of science. The laws of science have been modified, but very rarely.

How Do Different Objects Fall?

Process Skills *gathering and recording data; stating conclusions*

Materials paper, identical boxes, ball, various metal cubes if available

Procedures
Work with a partner so that one of you can experiment while the other observes. Conduct each experiment below twice, switching roles so that you and your partner can each be an observer.

1. Drop a crumpled piece of paper and a ball of about the same size from the same height at the same moment. Have your observer tell you when your arms are at the same height. Write down what you both observed. Which object hit the floor first? Did the objects move differently as they fell? What else did you observe?
2. Prepare two identical pieces of paper.

Keep one flat and crumple the other into a ball. Drop each piece from the same height. Write down what you observed.

3. From the same height, drop objects that have the same shape but different weights, like a ping-pong ball and a golf ball. Other possibilities include an empty coffee can and a full coffee can, or similar-sized cubes of various metals. You can even drop the ball of paper with metal objects of about the same size.

Conclusions
4. Which of the following hypotheses do your experiments support?

- A heavy object falls faster than a lighter one.
- A smaller object falls faster than a larger one.
- Everything falls together at the same speed.

■■■ Section 1-2 Review ■■■

Write the definitions for the following terms in your own words.

1. **hypothesis**
2. **controlled experiment**
3. **variable**
4. **data**
5. **theory**
6. **law of science**

Answer these questions.

7. List the six steps of the scientific method.
8. Lisel wants to learn whether she could ride her bicycle faster by adjusting the height of the seat or changing the air pressure in the tires. Describe a controlled experiment she could conduct to find out whether either of these variables make a difference.

1-3 Science Process Skills

■ *Objective*
☐ *List and illustrate science process skills.*

As you know, you can use the scientific method in solving problems. It will be easier for you to use the scientific method, however, if you master certain skills. The skills used to carry out the scientific method are **science process skills.** You will find special features in each chapter of this book that will give you practice in using science-process skills. One feature is titled *Skill Builder.* The other is *Activity.*

Observing

When you notice heat or a sweet smell, you are observing. When you **observe,** you use your five senses—sight, smell, touch, taste, and hearing—to gather information about your surroundings.

Some observations are **qualitative**, which means they describe how an object or event appears to the five senses. For example, sugar dissolves into water when you stir it and vinegar has a sharp odor.

Other observations are **quantitative**, which means they use numbers to measure and describe an object or event. You see nine geese flying in a group; you use a thermometer to take your temperature; you take a friend's pulse and find a rate of 70 beats per minute. In all three of these examples, you are making quantitative observations.

Measuring

When you **measure**, you use an instrument to make an observation. Look at lines A and B in Figure 1-6. Is one line longer? Use a ruler to learn the truth.

When scientists make observations with their senses, the observations are double-checked with instruments. For example, a thermometer can give the exact temperature more accurately than a person can.

Estimating

Many times it is impossible to make exact measurements. Suppose you wanted to know the number of students in your school. There is a quick way to get a reasonable answer. If there are approximately 30 students in a classroom, and

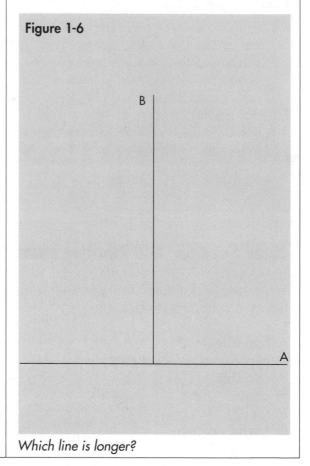

Figure 1-6

Which line is longer?

there are 20 classrooms per floor, there are approximately 600 students per floor. If there are four floors, your school contains approximately 2400 students. This is an **estimate**, which is an educated guess about a quantity.

Calculating

Calculating means working with numbers to obtain an answer. Usually this means adding, subtracting, multiplying, and dividing. Often, scientists need to perform calculations on a group of measurements when they report scientific results. Scientists use formulas that contain instructions for performing calculations. For example, how far will you run in one quarter of an hour if your speed is 12 miles per hour?

Distance = Speed × time
Distance = 12 miles/hour ×
$\qquad$ **¼ hour**
Distance = 3 miles

Organizing

Before beginning an experiment, scientists carefully organize equipment and plan procedures. During an experiment, they work neatly and according to a plan. After finishing an experiment, scientists present their data and conclusions in the clearest, most organized way so that others can understand the results. Organization helps scientists make fewer mistakes in experiments.

Figure 1-7 shows some ways that scientists present their results. They organize data into tables. Often, they create graphs

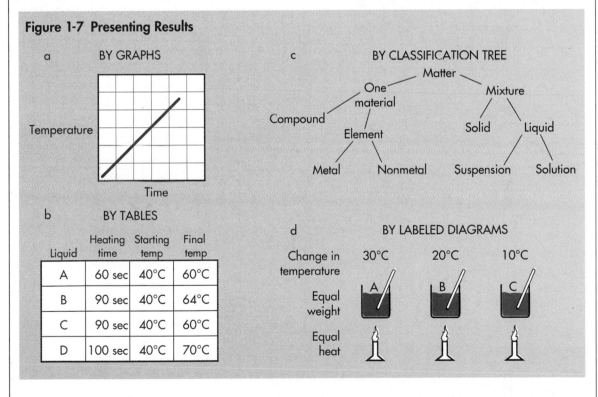

Figure 1-7 Presenting Results

a   BY GRAPHS

Temperature / Time

b   BY TABLES

Liquid	Heating time	Starting temp	Final temp
A	60 sec	40°C	60°C
B	90 sec	40°C	64°C
C	90 sec	40°C	60°C
D	100 sec	40°C	70°C

c   BY CLASSIFICATION TREE

Matter — One material / Mixture
Compound
Element — Metal, Nonmetal
Mixture — Solid, Liquid
Liquid — Suspension, Solution

d   BY LABELED DIAGRAMS

Change in temperature: 30°C   20°C   10°C
Equal weight: A   B   C
Equal heat

that show relationships among data. A labeled diagram clearly shows the parts of an object or the steps in a process. A classification tree helps readers follow the scientist's ideas.

Analyzing

When you **analyze,** you look for patterns in the observations and data of an experiment. For example, look at Figure 1-8, which contains some data on the cost of sailboats of various lengths.

By analyzing these numbers, you can spot some patterns. For instance, the longer the boat, the more it costs. You can also observe that cost increases much more rapidly than length. A ten-foot boat costs $1000, but if the length is doubled to 20 feet, the cost is far more than just

doubled. It is $10,000, or ten times larger.

One way to illustrate this rapid increase would be to analyze the data on a graph. Look at Figure 1-9 and compare it with Figure 1-8. The graph makes it obvious how cost and length each increase.

Making a Model

Scientists sometimes use a representation or description, called a **model,** to help other scientists understand their results. You are already familiar with models that are miniature copies of an object. Perhaps you build model cars, planes, trains, or buildings. Such models are useful for understanding and planning the real object.

Other types of models include diagrams of something that cannot be seen, such as an **atom,** a basic particle of matter.

Figure 1-8

Length of sailboat, in feet	10	15	20	25	30
Cost, in thousands of $	1	3	10	20	35

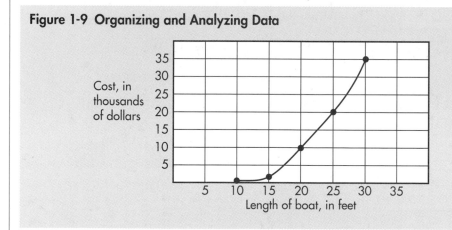

Figure 1-9 **Organizing and Analyzing Data**

Analyzing Data

Here are the ages of students in one class, not listed in any order.

14, 14, 13, 16, 15, 14, 13, 14, 15, 14, 13, 14, 13, 15, 15, 12, 13, 14, 15, 16, 15, 15, 13, 15, 14, 14, 14, 15, 14, 14

These data are disorganized and therefore are difficult to interpret.

Organize the data by completing a frequency table like the one shown in Figure 1-10. Frequency tables show how often something occurs. Use a separate sheet of paper.

1. You can use your completed table to obtain information quickly about this class. How many students are in the class? This should equal the total in the table above.
2. What age occurs most often in this class?
3. How many years difference are there between the youngest and oldest in the class?
4. These same data can be presented in a graph. Look at the graph in Figure 1-11. Which presents the information more clearly to you, the table or the graph? Why?

Figure 1-10	
Age	**Number of students of this age (the frequency)**
12	
13	
14	
15	
16	
	Total =

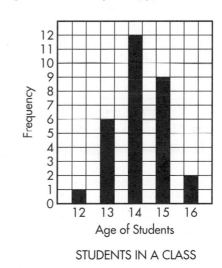

Figure 1-11 Sample Graph

STUDENTS IN A CLASS

Scientists use a model of an atom in studying chemistry. Look at the model in Figure 1-12. As you will learn, this picture of an atom helps explain the make-up of all things.

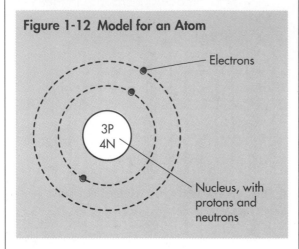

Figure 1-12 Model for an Atom

Electrons

3P
4N

Nucleus, with protons and neutrons

Predicting

When you predict something, you say what you think is going to happen. You can often do this by following the pattern of the data or the trend of a graph. Look at Figure 1-13 and predict the possible cost of a 35-foot sailboat.

Sometimes you make predictions based on the data of certain experiments or on the laws of science. For example, your knowledge of falling objects lets you predict that if a marble rolls toward the edge of a table, it will fall to the floor unless it is stopped.

Evaluating

When you evaluate, you think about what you have done. A scientist evaluates an experiment by asking questions like the following:

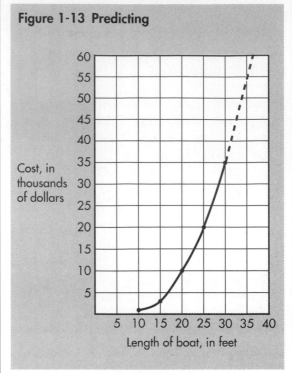

Figure 1-13 Predicting

Cost, in thousands of dollars

Length of boat, in feet

- Was the design of the experiment good?
- Should any procedures be repeated?
- Were the observations and measurements made accurately?
- Were the data carefully analyzed?
- Are there other ways to check the results?
- How do the results relate to other experiments and with the theories and laws of science?

Before announcing and publishing their results, scientists carefully evaluate the work they performed. Other scientists will try to reproduce the same results in order to evaluate the conclusions that have been announced. Science carefully moves forward through evaluation and reevaluation.

Write the definitions for the following terms in your own words.

1. **observe**
2. **qualitative observation**
3. **quantitative observation**
4. **analyze**
5. **model**

Answer these questions.

6. Choose five of the science process skills developed in this section and give your own example of each.
7. Compare measuring and estimating. Describe each of the skills, and use a single example to compare them.
8. Explain the difference between a quantitative and a qualitative observation.
9. How is organizing used before, during, and after an experiment?

1-4 Science, Technology, and Society

■ *Objectives*
☐ *Define technology.*
☐ *Illustrate how science, technology, and society interact and change one another.*
☐ *Contrast the benefits and burdens, and the short-term and long-term consequences of technology.*
☐ *Give examples of decision making concerning technology.*

Perhaps you or a friend owns a computer. Many students use computers for games, word processing, keeping records, and writing programs. You can find computers in businesses, homes, and schools.

Did you know that the computer is a recent invention? In the 1950s, only laboratories and government institutions had computers, and the machines were so large they took up entire rooms. In the 1970s, people learned to make computers smaller and faster, with enough memory to become useful to businesses. This is an example of how **technology** can change society. Technology is the use of scientific discoveries to develop devices and processes that improve the quality of human life.

Technology

Computers were developed through discoveries in the sciences of electronics, mathematics, and chemistry. Airplanes, nylon, television, processed foods, heart pacemakers, and medicines are a few other examples of products of technology that have come from scientific discoveries.

Technology is essential to our society. It continues to improve our lives. However, technology is not always helpful. Some of the side effects of technology have been harmful. Cars and power plants produce air pollution as they burn fuel. Industry causes pollution. You can see that technology offers benefits, but there is a price to pay.

Science and Technology

Science is a field in which people seek new knowledge about the world. Technology is a field in which people use this knowledge to develop helpful devices and processes.

Figure 1-14 shows a table that compares the questions that science and technology try to answer.

Figure 1-14	
Science	**Technology**
What happens to the way fibers burn if chemicals are added?	How can we make fabrics fire resistant? What is needed to weave such material into safe clothing?
How do very cold temperatures allow electricity to flow freely in wires?	How can very cold temperatures be used to produce magnets strong enough to make new types of trains possible?
How do sound waves act on surfaces?	Can sound be used to clean teeth or dishes?

Each chapter in this book ends with a Science, Technology, and Society (STS) feature. The purpose of the STS feature is to help you understand how science, technology, and society interact. Science and technology cannot, of course, solve all of society's problems. You have seen that they can even create problems. The STS features will let you look more closely at the way society faces problems caused by technology, and how technology can be used to solve some of society's problems.

The following sections will give you an overview of some of the STS themes that appear throughout the book. As you read this section and the STS features in each chapter, think about the issues they present. How do they affect your life? What is your responsibility in dealing with the issues, now and in the future?

Science and Technology Advance Each Other Science and technology help each other make progress. Scientific discoveries allow engineers to invent or improve technological devices. Also, new technologies help scientists carry out research faster and more accurately. The tables in Figures 1-15 and 1-16 illustrate these interactions.

Figure 1-15 Science Advances Technology	
Science	**Technology**
Study of energy stored in atoms	→ Energy from atoms can be used to fight cancer and provide electricity.
Research on silicon, a basic substance, and electricity, a form of energy	→ Silicon computer chips help make computers small and powerful.
Discovery of new fuels	→ New car engines use the newer, cleaner-burning fuels.

Figure 1-16 Technology Advances Science	
Technology	**Science**
Computers	→ Used to rapidly collect and analyze data
Electron microscopes	→ Allow scientists to see parts of cells and structure of solid materials
Radiation detectors	→ Permit scientists to track paths of unseen particles to better understand matter

Technology Affects Everyone Think of several products of technology you use every day, such as your watch, radio, and refrigerator. Technology clearly affects the way you spend your time. Think about the invention of the light bulb. Light bulbs permit you to work, play, and be active long after the sun goes down.

Every form of technology also has some affect on the physical environment around you. Some of the effects harm the environment. Detergents have contributed to water pollution. Mining for metals and other resources destroys the land and reduces farm production.

Both science and technology have caused major changes in society. For example, cars, trains, and planes have increased people's ability to travel. This has led to growth of suburbs around cities. The growth of suburbs has, in turn, resulted in a population shift that has led to bigger problems in some central cities.

Science and technology have resulted in other changes to business and industry. New products, new procedures, and new jobs have been made possible. Each new technological product creates an industry to produce the product. For example, you know that computers have replaced old filing systems. As a result, many thousands of people now work in making and repairing computers.

Society's Effects on Science Society certainly has a strong effect on science and technology. Sometimes society demands that scientists and engineers develop devices to meet particular needs. For example, the increase in cancer deaths has led society to demand that research and engineering find a cure. Society also controls scientific research through laws, government committees, and public opinion.

Global Effects of Technology The development of technological products sometimes have an effect worldwide, a **global effect.** For example, because nuclear radiation and pollution are carried by the wind, they can affect people and countries thousands of miles away. Should there be an accident at a nuclear power plant, nuclear radiation and pollution would likely spread far beyond the site of the accident.

Many technologies bring the world closer together. They include communications satellites, weather satellites, and international telephone services. Some of the global effects of communication satellites are shown in Figure 1-17. Technologies with global effects require that nations cooperate with one another.

Figure 1-17 Communicating Via Satellites

Communication satellites

Transmission dish

Receiving dish

Communication satellite

Benefits and Burdens
Every technology offers **benefits and burdens.** These can also be called advantages and disadvantages, or pros and cons. Figure 1-18 illustrates the two sides to several products of technology.

Short-Term and Long-Term Consequences
When engineers develop a new technology, they must collect information about possible short-term and long-term effects on people and the environment. Data collection is necessary in order to protect people

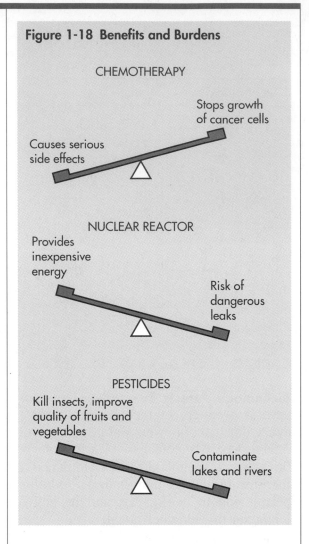

Figure 1-18 Benefits and Burdens

CHEMOTHERAPY

Stops growth of cancer cells

Causes serious side effects

NUCLEAR REACTOR

Provides inexpensive energy

Risk of dangerous leaks

PESTICIDES

Kill insects, improve quality of fruits and vegetables

Contaminate lakes and rivers

and the environment as much as possible. For example, it is known that pesticides have a positive short-term consequence. They kill insects and weeds that destroy farm crops. Pesticides result, therefore, in bigger and better harvests. However, pesticides are eventually washed into lakes and rivers where they can harm the fish and the water, which, in turn, directly harms humans. Contaminated fish and water are important long-term consequences of pesticides.

Decision Making and Trade-offs Should the use of one particular technology, such as that of pesticides, be widespread or limited to severely affected areas? Do the disadvantages of some technologies outweigh their advantages? Individuals, community groups, and government agencies often must deal with such questions. Data must be gathered in order to make wise decisions. Often, people must make a compromise, or a **trade-off**.

Returnable bottles and cans create a short-term inconvenience. They must be stored at home, returned to a store, and shipped to recycling plants. Reusing cans and bottles, however, cuts down on the use of energy and natural resources to make new bottles and cans. It also cuts down the amount of garbage. For this reason, some states have decided to use deposits to encourage people to return bottles and cans.

Some technology raises moral questions. For instance, organ transplants can save lives, but spare organs are in short supply. If there are not enough organs for people who need transplants, who should get the available organs?

The Role of Government
Federal and state governments sometimes pass laws to control science and technology. These laws are meant to protect individuals and society. Some examples of federal and state laws that control science and technology include mandatory seat belt laws and laws that control gas fumes produced by cars.

Government also controls science and technology by deciding which projects will get public funding. Federal funds are often used for research in agriculture, space, cancer, and military weapons. Other decisions that involve the government include the conservation of natural resources and the preservation of wildlife. Handling toxic wastes and garbage, and protecting the ozone layer are some other examples.

The products of technology solve many of society's problems. However, science and technology *alone* cannot solve every problem. The attitudes and habits of people are just as important in making changes.

▬ Section 1-4 Review ▬

Write the definitions for the following terms in your own words.

1. **technology**
2. **global effect**
3. **benefits and burdens**
4. **trade-off**

Answer these questions.

5. Explain the difference between science and technology.
6. Give two specific examples of how science and technology help advance one another.
7. List several ways in which society controls technology and research.
8. What things should be considered in evaluating any new technology?

SCIENCE, TECHNOLOGY, & SOCIETY

Where Do Scientists Work?

Two hundred years ago, scientific research was a hobby. That is, most scientists had other jobs. Today, however, most research is done by full-time, professional scientists.

Professional scientists have college degrees in some area of science or mathematics. They usually work for companies, colleges, hospitals, and government agencies. Figure 1-19 shows scientists at work.

Large companies employ many scientists. Most company research is aimed at improving products made by the company.

The government employs scientists to test products for quality and safety. Governments also support international agencies that sponsor research on big projects, such as world food production.

Some people think that scientists just want to discover and invent things. This is only partly true. In addition to discovering and inventing, scientists work to provide for their families and for recognition and promotion. By publishing their research in journals, scientists receive recognition from other scientists. Sometimes scientists can patent an invention, which means they own the invention for a certain number of years. If they so choose, scientists can sell the rights to use their inventions to companies.

Science is big business, which means it uses and makes considerable amounts of money. In addition to providing work for scientists and technicians, science provides jobs for many other people. For example, office staff, equipment designers, laboratory builders, and electrical experts are are just a few of the positions needed to make research possible.

Follow-up Activity

Look for articles or TV programs on science research or scientists today. What scientific, economic, and political issues combine to decide what kind of research is being encouraged?

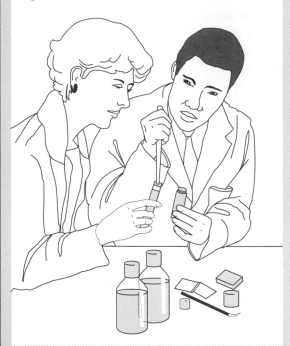

Figure 1-19

 **KEEPING TRACK**

Section 1-1 The Nature of Physical Science
1. Science is a collection of facts and ideas about how the world works. Science involves special ways of gathering facts.
2. Physical science studies materials and forces in the world.

Section 1-2 The Method of Science
1. The scientific method can be used to state and solve problems.
2. A controlled experiment changes one variable at a time, keeping all others constant.

Section 1-3 Science Process Skills
1. Science requires such skills as observing, measuring, estimating, calculating, organizing, analyzing, making models, predicting, and evaluating.

Section 1-4 Science, Technology, and Society
1. Science seeks to gain understanding of the natural world. Technology applies science to meet the needs of individuals and society.
2. Science and technology help each other advance.
3. Everyone interacts with the products of technology in some way.
4. Every technological process or device affects the environment in some way.
5. Technology affects society, the standard of living, businesses, and jobs.
6. Science and technology alone cannot solve society's problems.
7. Technological devices can have a global effect.

8. Every technological device should be judged by its benefits and burdens, its advantages and disadvantages, and its short-term and long-term consequences.
9. Making decisions and choices about technological issues usually requires trade-offs, or compromises.

BUILDING VOCABULARY

Write the term from the list that best matches each statement.

analyze, controlled experiment, data, global effects, model, physical science, science, trade-off, vacuum, variable

1. its methods are as important as its facts
2. studies materials and forces of our world
3. something that can influence the results of an experiment
4. one thing changes, but everything else does not
5. observations and measurements
6. absence of air or other material
7. look for patterns in the results
8. helps understand and explain results
9. one aspect of technology
10. a compromise

Explain the difference between the terms in each pair.

11. chemistry, physics
12. hypothesis, conclusion
13. theory, law
14. qualitative observation, quantitative observation

15. measure, estimate
16. science, technology
17. benefits, burdens
18. short-term consequences, long-term consequences

SUMMARIZING

Write the missing term for each sentence.

1. An experiment is said to be controlled if one variable is changed while all other variables are _____.
2. _____ is the science skill of using previous results to suggest what will happen under new conditions.
3. Microscopes are technological devices that _____ the study of science .
4. Each technological device has an affect on people and on the _____ in which we live.
5. Balancing the _____ of different choices is called making trade-offs.

INTERPRETING INFORMATION

Figure 1-20 shows some data collected by students as they interviewed many classmates.

*These are averages of a number of students in each group. The students interviewed were all doing about the same in the course so far, from good to very good. The grades were all taken from the same exam. All students had the same teacher.

1. What appears to be the main question this research is trying to answer?

Figure 1-20

Hours of Study for Exam*	Grade on Exam*
0	60
$\frac{1}{2}$	75
1	85
$1\frac{1}{2}$	95
2	95
$2\frac{1}{2}$	90

2. Write a possible hypothesis for this research.
3. Is this a controlled experiment? What was changed? What was kept the same?
4. What conclusion or conclusions can be made from these data?

THINK AND DISCUSS

Use the section number in parentheses to help you find each answer. Write your answers in complete sentences.

1. Two physical scientists are heating a mixture of water and sugar in a pot. What are two questions the chemist could ask? What other two questions could the physicist ask? (1-1)
2. What questions can be asked after an experiment is over and the conclusions are made? (1-3)

3. List four qualitative observations and four quantitative observations you can make about today's weather. (1-3)

GOING FURTHER

1. Choose a technological device like a refrigerator, video camera, or lie detector. Then investigate how it works and sketch how it is put together. List any special science discoveries or theories that make the device work. Prepare a report on your findings.

2. Put together a collection of articles about the topics of science, technology, and society. You may also take notes from television shows. You might find examples of how society is being affected or changed by a new invention. You might find examples of how society is concerned about and tries to control research or the application of results in technology.

COMPETENCY REVIEW

1. Science consists of
   a. just facts.
   b. just theories.
   c. just doing experiments.
   d. all of the above.
2. Light and motion are topics studied in
   a. physics.   b. biology.
   c. chemistry.   d. earth science.
3. The first step in approaching a problem is to
   a. make conclusions. b. form hypotheses.
   c. record data.   d. find the key issue.
4. An experiment is controlled if
   a. nothing is allowed to change.
   b. only one variable is allowed to change.

c. exact measurements are made.
d. all the equipment is new.

5. Which scientific statements are most certain?
   a. hypotheses
   b. conclusions
   c. theories
   d. laws
6. "This sand box has about a million billion grains of sand in it." This statement is an example of
   a. a qualitative measurement.
   b. a quantitative measurement.
   c. an estimate.
   d. a model.
7. Data from an experiment show that a student's test results decrease as study time decreases. This statement is an example of
   a. observing.
   b. stating a problem.
   c. a conclusion.
   d. calculating.
8. You want to find out whether weight affects falling speed. You should use objects
   a. of equal mass.
   b. of different shapes.
   c. of the same shape but different weights.
   d. of the same weight but different shapes.
9. Which statement about science and technology is most correct?
   a. Either can solve any problem.
   b. Neither is involved with society's problems.
   c. Both are part of the problem, not the solution.
   d. They cannot solve every problem.

CHAPTER 2

CHEMICAL SAFETY

Look around and list ten things you see that are made of chemicals. Did you list yourself? You are made of carbon, oxygen, iron, water, and many other chemicals as shown in Figure 2-1. Rocks, air, water, plants, and animals are also made of chemicals. A **chemical** is any of the substances or materials that make up all things. This chapter is about two types of chemicals, those found in households and those commonly found in laboratories. These chemicals might be thought of as friends. They can make your life easier and help unlock the secrets of nature. Handled incorrectly, however, many chemicals can become dangerous enemies.

Do you know the *Arabian Nights* tale about the genie in the bottle? A genie is a fictional spirit that is supposed to grant the wishes of the person who finds it. Sometimes, though, the genie also does something unexpected that catches the person off guard with an undesired result.

As you handle chemicals at home or in the lab at school, remember the genie in the bottle, and be careful to guard against unexpected, harmful results. You will find that being careful does not take the fun out of discovering the wonders of chemicals and working with them. It only helps make discovering safer.

2-1 Safety with Household Chemicals

■ *Objectives*

☐ *Give examples of household chemicals.*

☐ *List dangers of and safety rules for using household chemicals.*

☐ *Find first-aid information on labels.*

Most people use a variety of chemicals in and around their homes. These **household chemicals** include medicines as well as

Figure 2-1

People are made of chemicals.

products for cleaning and other jobs. You probably have chemicals such as floor wax, detergents, and snow salt at your house.

Chemicals are helpful, but some can be dangerous too. Chemicals that are possibly dangerous usually carry a warning. Read some labels on products in your home. Warnings often begin with the word *warning* or *caution* in bold or colored print. The message below is from the label of a bottle of cleaning liquid.

> **CAUTION.** KEEP OUT OF REACH OF CHILDREN. In case of eye contact, flush thoroughly with water. If swallowed, drink a glassful of water. Contains ammonia. To avoid fumes, do not mix with products containing chlorine bleach.

When handling household chemicals, the most important rule is, *always follow the label or package directions carefully.* You should also know how to handle the most common kinds of dangerous chemicals.

Toxic and Corrosive Chemicals

Most household chemicals are **toxic**, meaning that they are poisonous. Toxic chemicals can cause illness or death. Many household chemicals are **corrosive**, which means they can eat away or damage skin, clothing, and other surfaces. An example is lye, which is used to clean drain pipes. Lye is both toxic and corrosive. The message below is from a lye label.

> **POISON.** Contains lye. Corrosive: Causes severe eye and skin damage; may cause blindness. Keep away from eyes, skin, and clothing. Harmful or fatal if swallowed.

Some materials may give off **fumes** or **vapors** that are also toxic. Fumes and vapors are the gaseous forms of substances. For example, ammonia cleansers give off ammonia gas, and bleaching liquids give off chlorine gas.

Flammable Chemicals

Other household chemicals are **flammable**. *Flammable* means able to be easily set afire and to burn quickly. Many brands of floor wax are flammable. A number of flammable chemicals, such as gasoline, oil-based paints, paint remover, and lighter fluids, give off fumes that are also flammable. For example, a fire could start if paint fumes collect near the pilot light on a gas oven or stove. Flammable products should be used only away from any open flame, and in **well-ventilated** places. In other words, they should only be used where fresh air can flow in and polluted air can flow out steadily through open windows, open doors, or by means of ventilation fans.

The warning below appears on a can of kerosene, which is both a cleaner and a lamp fuel. Do you think that kerosene is flammable?

> **DANGER!** Keep away from heat, sparks, or open flame. Keep container tightly closed. Use ventilation. Avoid breathing of vapor and repeated contact with skin.

Avoid Mixing Chemicals

Mixing chemicals may make them more dangerous. Mixing certain chemicals sometimes creates toxic fumes. For example, products containing ammonia should not be mixed with products that contain chlorine, such as bleach. Read the warnings on the labels in Figure 2-2. These labels are from household chemicals, one from bleach and one from a cleaner containing ammonia. People who may not know about chemicals sometimes mix two different cleaners, hoping to make a stronger product. Mixing chlorine-type bleaches and products containing ammonia produces poisonous fumes.

When some chemicals are mixed, they can boil and spatter. This occurs when lye is mixed with vinegar, an acid. To be safe, never mix any household chemicals.

Medicines are chemicals that can be dangerous when mixed. Some people do not realize that taking medicines together, or within several hours, is mixing, and can be dangerous. The following message appears on the label of a cold medicine.

DRUG INTERACTION PRECAUTION. Do not take if you are taking a drug for high blood pressure or depression. Consult your doctor. Avoid alcoholic beverages when using this product.

Even mixing pain-relief medicines that can be bought without a prescription can be

Figure 2-2

CHLORINE BLEACH

DANGER Do not mix with other household chemicals such as products containing vinegar, acid, or amonia.

AMMONIA WATER

SUPER CLEAN

Do not mix with chlorine type bleaches or other household chemicals.

Why is it dangerous to mix these two household chemicals?

dangerous. Always read the labels on medicines before taking them. Call your doctor if you need more information.

Pressurized Containers

Andy left a can of spray paint in his car. The car was parked in the sun on a very hot day. Why was this a very dangerous thing to do?

Spray paint comes in an aerosol container. Aerosol containers are those in which a liquid or solid material is stored under very high pressure. These containers are also called **pressurized containers.** Insect sprays, air fresheners, and underarm deodorants often come in aerosol, or pressurized, containers. You should never **puncture**, or pierce, containers that are pressurized. Even a tiny hole could release the high pressure inside. As the aerosol rushes out the hole, the force of the pressure could cause the can to burst. Always keep pressurized containers away from heat. Heat increases the pressure inside, which could make the can explode. Below is a warning from an aerosol container.

CAUTION. Do not set on stove or radiator or keep where temperature will exceed 120° F, as container may burst. Do not puncture or throw into fire or furnace.

Prevention and First Aid

Have you heard the old saying, "An ounce of prevention is worth a pound of cure"? It means you should try to prevent accidents so that you will not have to repair the damage they cause. Here are some additional safety rules to follow when you use household chemicals.

- Read all directions, warnings, and first aid notices on products you buy.
- Handle household chemicals carefully.
- Store chemicals and medicines in closets out of reach from children and away from pets. Store chemicals and medicines securely so that they cannot pop open or fall down.
- Keep all containers tightly closed to avoid spills and evaporation.
- Throw away household chemicals by the suggested and legal way. It is dangerous and illegal to pour some chemicals, such as paint removers, down the drain.

Even when you take precautions, accidents sometimes occur. It is important to know what first aid is necessary if a household chemical gets into your eyes, on your skin, or into your body.

If a household chemical splashes onto you, wash the area immediately with plenty of water. If you swallow a chemical by accident, usually you should drink plenty of water. Water removes or weakens toxic chemicals and reduces the immediate danger to your health. Sometimes when you swallow a chemical, however, the advice is to drink milk, which coats your stomach.

First aid suggestions are printed on the label of dangerous chemicals, usually next to the warning. In many cases, a label warns you not to try to make yourself vomit, or throw up. Chemicals with such labels can burn your throat. They burn on the way down to your stomach. If you throw up, the chemical will burn you again as it moves back up your throat. The warning below is from the label of an oven cleaner in a pressurized can.

> **CAUTION!** If swallowed, drink a large quantity of water or milk and call a doctor or hospital at once. Do not induce vomiting.

Make a habit of reading all warnings and first aid suggestions on household chemicals before you use them. You should also keep a book on first aid immediately available at home along with emergency phone numbers.

How Can You Use Household Chemicals Safely?

Process Skills observing; recording information

Materials paper and pencil

Procedures
1. Make a table like the one in Figure 2-3.
2. Find at least five different household chemicals and five different medicines at your house.
3. On your table, record the name of each product and where it is stored in your home.

4. Look at the label on each container and use the information on each label to complete your table.

Conclusions
5. Ask an adult in your home to discuss your table with you. Do you think these products are stored in safe places in your home? Why or why not? If not, decide together on the best place to keep each chemical. Correct your table.
6. List the safety rules you should follow when using each of the products on your table.

Figure 2-3

Substance	Where stored in your home	Information Where to store	Label Warnings

▬▬ Section 2-1 Review ▬▬

Write the definitions for the following terms in your own words.

1. **corrosive**
2. **pressurized container**
3. **fumes, or vapors**
4. **puncture**
5. **ventilated**

Answer these questions.

6. Briefly state the safety rules for using
   a. toxic and corrosive chemicals
   b. flammable chemicals
   c. chemicals in pressurized containers
7. Tell why you should avoid mixing
   a. household chemicals, or
   b. household medicines.
8. What are some safety tips for storing chemical products at home?

2-2 Safety with Laboratory Chemicals

■ *Objective*
☐ *State safe laboratory rules and procedures.*

Scientists and science students use laboratory chemicals in many experiments. **Laboratory chemicals** are substances used by scientists in their work.

The chemicals used in a laboratory or classroom are often both strong and dangerous. As with household chemicals, the most important rule for using laboratory chemicals is to follow directions carefully.

In the laboratory, you will usually find the directions in your textbook or on an instruction sheet. Your teacher will give you additional information.

General Laboratory Rules

When you work in the laboratory, you should keep the following rules in mind at all times. Following these rules will keep you and your classmates safe. Carrying out experiments can be fun if you stay safe.

- Follow all instructions. If something is unclear, ask! Never go ahead until you know what you are doing and why.
- Work in a clean area away from unnecessary materials. Keep movement and talking to a minimum. No horseplay or running in the lab!
- Work only when the teacher is present.
- Do not perform extra experiments, such as mixing or heating chemicals, on your own. If you have an idea for an experiment, explain it to your teacher and ask for permission to go ahead.
- Know the location of the safety equipment in the lab. Safety equipment may include sinks, eye-wash stations, a safety shower, fire blankets, fire extinguishers, sand, an exhaust fan, and other items.
- Dispose of waste materials as instructed.
- Report all accidents or hazards immediately.

A **hazard** is any situation or substance that might cause harm to people or equipment.

Preventing Injuries

No one wants to get hurt in the lab. Laboratory chemicals, however, can create hazards that can cause accidents. Remember that laboratory chemicals are pure and strong. Some experiments will require you to make a **concentrated solution**. Concentrated solutions are strong, but not so strong as the pure laboratory chemical. You can make a concentrated solution by mixing a large amount of the pure chemical with a small amount of water. You can also dilute such a solution. **Diluting** makes a concentrated solution weaker by mixing it with more water.

In making and diluting concentrated solutions, you will be working with hazardous chemicals. An accident could result in serious injury. You can follow additional rules to avoid accidents in the lab.

- Do not eat or drink in the laboratory. By eating or drinking, you risk swallowing toxic chemicals.
- Protect your clothing with a rubber apron. Make sure your hair, jewelry, sleeves, ties, and other articles of clothing are tucked out of the way.
- Protect your eyes with goggles.
- Wash your desk, equipment, and hands before and after the required work. Wash away spills on your skin, clothing, or desk.
- Use only clean tools, test tubes, and glassware.
- Study Figure 2-4. Practice removing and replacing bottle stoppers as shown in the figure. While pouring liquids, hold the bottle's stopper or cork out of the way. This will keep liquid off the desk and the stopper clean.
- When diluting a concentrated solution, pour it into the water and stir. For example, pour concentrated acid into water, not water into concentrated acid. When water and acid are mixed, heat is

produced. Pouring acid into the water will prevent spattering and fumes. It is not safe to pour water into a concentrated solution.

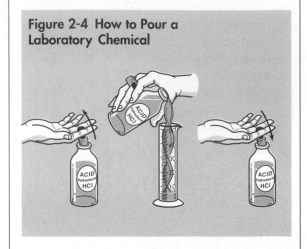

Figure 2-4 How to Pour a Laboratory Chemical

- When smelling something, use your hand to fan or wave the odor gently toward your nose. Do not inhale odors directly because the fumes may irritate your nose and lungs. Strong odors can even knock you out!
- Never taste or touch any laboratory chemical!
- Use materials only from labeled containers.

Hazard Symbols

Figure 2-5 shows hazard symbols used in all laboratories. They are like the warning messages on containers of household chemicals. You probably know how to read familiar symbols such as stop signs and no-smoking signs. Study the symbols in Figure 2-5 so that you will recognize them in the lab. Be familiar with what they mean and with related safety rules.

Figure 2-5 Laboratory Hazard Symbols

 Radioactive chemicals. Follow directions exactly.

 Poisonous (toxic) chemicals. Avoid skin contact. Do not breath vapors

 Lab coat must be worn.

 Corrosive chemicals. Keep away from eyes, skin, clothing. Do not breathe vapors.

 Fragile. Be especially careful.

 Follow animal safety rules.

 Safety goggles must be worn.

 Electrical shock. Follow instructions carefully. Unplug equipment when you are done.

 Keep hair, clothing, and papers away from flame.

 Explosive chemicals. Follow directions exactly.

 Toxic vapors. Do not breathe vapors.

ACTIVITY

Why Are Laboratory Chemicals Dangerous?

Process Skills *observing; stating conclusions*

Materials knife; glasses and beakers; water; water-soluble ink; paper; two slices of bread; two jars with lids; onions; perfume or food flavoring

Procedures

1. Spread some drops of water-soluble ink over a piece of paper. When it has dried, place the paper in a glass or beaker of water. Observe the ink streams flowing from the paper into the water.
2. Place slices of bread into two jars. Each jar should contain one slice of bread. Into one jar also add a drop of perfume, or food flavoring like vanilla, or a bit of onion. Be sure the bread does not directly touch the liquid or onion. Cover both jars for ten minutes and then compare the odors of the bread slices.
3. Ask your teacher to pour an equal amount of distilled water and sodium hydroxide into two beakers. Can you tell any difference by sight or by smell? Ask your teacher to pour a few drops of phenolphthalein or litmus into each beaker. What do you observe?

Conclusions

4. What facts about chemicals and chemical dangers do the results of these activities show?
5. What safety rules are related to these activities?

▬ Section 2-2 Review ▬

Write the definitions for the following term in your own words.

1. **laboratory chemicals**
2. **hazard**
3. **concentrated solution**
4. **diluting**

Answer these questions.

5. List any three of the general laboratory rules in your own words. Explain why each is important for laboratory safety.
6. Give a related safety suggestion for each of the five senses.
   a. sight   b. smell
   c. hearing   d. taste
   e. touch
7. What is the safest way to dilute a concentrated solution with water?
8. You find a bottle that is not labeled but contains the symbols in Figure 2-6.

Figure 2-6

What hazards does the substance inside present? What is the safest thing for you to do about the bottle?

2-3 Proper Use of Laboratory Tools

▬ Objectives

☐ *Use a Bunsen burner and other laboratory equipment properly.*

☐ *Recognize and name key laboratory equipment.*

Science laboratories contain many special tools and pieces of equipment. You will use burners, glassware, and ceramicware in the laboratory. These are safe if handled properly. Laboratories also contain protective equipment you can use to improve safety. This section explains safe ways to use basic laboratory equipment.

Burners

You will use a burner to heat materials in the laboratory. The standard burner is the **Bunsen burner,** which consists of a metal barrel and base and uses natural gas for fuel. Bunsen burners give out a high amount of heat. Figure 2-7 shows a Bunsen burner.

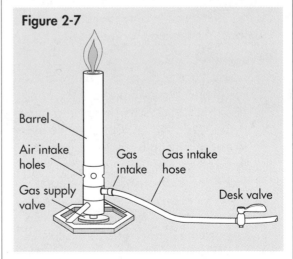

Figure 2-7

Barrel
Air intake holes
Gas intake
Gas intake hose
Gas supply valve
Desk valve

Notice the air intake holes and the valves. Other sources of heat can be used if you require less heat. In decreasing order, from most heat producing to least heat producing, there are Bunsen burners, electric-coil heaters, and candles.

Proper Flame in a Bunsen Burner A proper mixture of air and gas in a Bunsen burner results in a flame that looks like two cones, one inside the other. The inner cone is a strong blue and is called the **reducing flame** because it can remove oxygen from many materials that it burns. The hottest part of the flame is at the tip of the inner cone. The outer cone is pale blue and is called the **oxidizing flame** because it is rich in oxygen and can give oxygen to materials. The outer cone is not as hot as the inner.

Examine Figure 2-8. When the Bunsen burner is set properly, enough air enters through the air holes to completely burn the gas. If too much air enters the air holes, the flame is unstable and noisy, and you will see only the outer, pale blue cone. If too little air enters the air holes, the flame flutters and is not very hot. It burns with orange-yellow colors and produces black smoke.

Safety Rules for Using the Bunsen Burner The Bunsen burner can cause fires, so using it safely is important. Get to know the parts of the burner and how they are adjusted. Be sure to ask your teacher any questions you have about using the burner. The following are general guidelines.

1. **Do not use a Bunsen burner unless you are supervised by a teacher.**

Figure 2-8 The Bunsen Burner's Flame

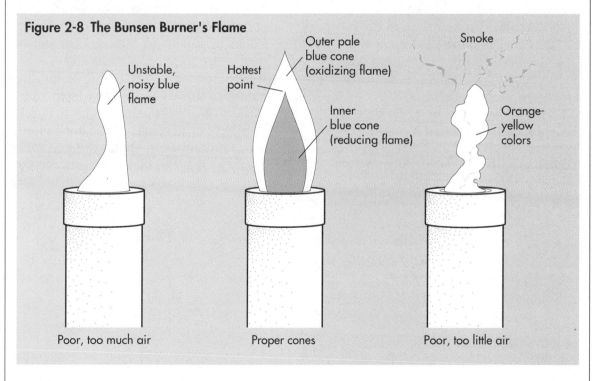

Unstable, noisy blue flame

Hottest point

Outer pale blue cone (oxidizing flame)

Inner blue cone (reducing flame)

Smoke

Orange-yellow colors

Poor, too much air

Proper cones

Poor, too little air

2. First light a match, and then turn on the gas. This avoids a possible explosion from gas that may collect in or around the burner.
3. When lighting the burner, be careful to keep your hands and face to one side so that you will not be burned by the flame.
4. Bring the match *up* along the side of the barrel to start the gas burning at the top. If you bring the match downward, you may be burned by the flame.
5. Look again at Figure 2-7 and locate the valves. To adjust the height of the flame, turn the gas supply valve underneath. To adjust the quality of the flame, that is, to make the flame have the proper shape and color, turn the air intake to allow more or less air through the air holes.
6. Immediately shut off the desk valve attached to the gas intake hose if gas has accumulated somewhere in the room, if the flame is blown out, or if the flame goes out by itself.
7. When you are finished using the Bunsen burner, turn it off.

Safety Rules for Heating Materials When you heat chemicals in the lab, you cause reactions in the chemicals. These reactions can affect not only the chemicals, but also the containers. For this reason, you will need to use glass and ceramic holders in the lab. This section provides safety rules for using holders. First, study these general rules about heating.

1. **Do not heat anything in the laboratory unless you are being supervised by a teacher.**
2. Study Figure 2-9. When heating materials, use a test tube holder, a clamp, or tongs to hold glassware. Never heat an object while holding it with your bare hands.
3. Apply heat evenly and gradually to the container. You can do this by slowly moving the container into and out of the flame and by gently shaking the material. Rapid heating at one spot can crack glass containers such as test tubes. For this reason, use a wire gauze or ceramic pad under any glassware being heated on a tripod or ringstand. Figure 2-10 shows how to use a tripod or ringstand.
4. When heating a test tube, point the mouth of the test tube away from yourself and others. As liquid begins to boil, it sometimes can spurt out and burn someone. Remember to shake the test tube gently so that the material heats evenly.

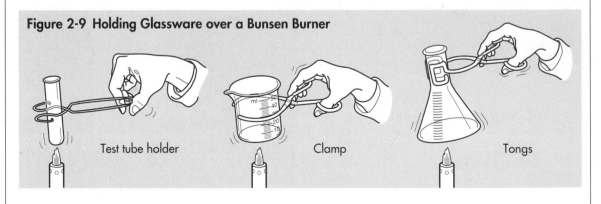

Figure 2-9 Holding Glassware over a Bunsen Burner

Test tube holder   Clamp   Tongs

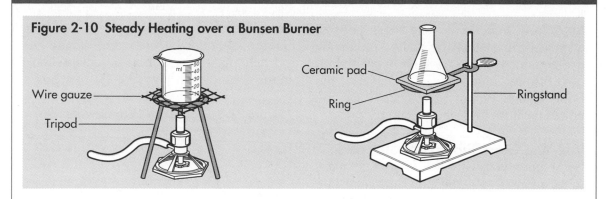

Figure 2-10 Steady Heating over a Bunsen Burner

Wire gauze

Tripod

Ceramic pad

Ring

Ringstand

5. Follow directions about the amount of heat required. Heating chemicals requires caution because most chemical changes occur at a faster rate as the temperature increases. With too much heat, the chemical changes may get out of control and cause an accident.

6. Never leave a burner, an electric hot plate, or any heated object unattended.

Safe Use of Glassware

Most glassware used in the lab is made of **heat-resistant glass**. Heat-resistant glass expands very little when heated and usually does not crack. Do not use chipped or cracked glassware in any experiment. In some experiments, however, your teacher may instruct you to bend or break some glass

tubing. Figure 2-11 illustrates the safe way to bend and break glass. Refer to Figure 11 while reading the instructions below.

Caution Do not bend or break glass unless you are being supervised by a teacher.

First, attach a wing tip to the burner to create a longer flame. With a wing tip, you can heat more glass tubing at one time. Then, to soften and bend glass tubing, heat the glass tubing over a Bunsen burner flame. Do not touch or wet the hot glass; it might crack. When the tube has softened, bend it on ceramic tile.

Cut glass tubes only when your teacher is present. Use a small file to make a notch at the cutting point. Wearing gloves and goggles, break the glass in the direction

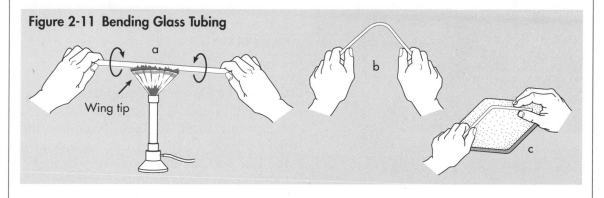

Figure 2-11 Bending Glass Tubing

a

Wing tip

b

c

away from your body. The sharp ends of all glass tubes must be made smooth by fire polishing. Place each end near the tip of the inner blue cone of the flame, the hottest part. Remove the glass once the edges have melted and are round.

Inserting Glass Tubing into Stoppers

Sometimes you will have to insert a glass tube, a funnel tube, or a thermometer into a hole in a rubber stopper or cork. Remember to be careful, because tubing breaks easily. Look at Figure 2-12 and follow these instructions:

1. Make sure the hole is large enough.
2. Moisten the tubing with soapy water or glycerine. Protect your hands by wrapping the tubing and stopper in paper or cloth towels, or by using cloth gloves.
3. Push the tube into the hole gently with a slight twisting action. Keep your hands close together to prevent the tubing from snapping.
4. Ask for help if the tubing does not go in easily.

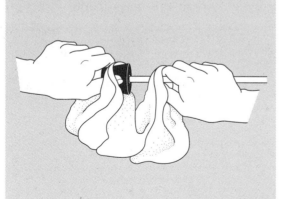

Figure 2-12 How to Insert Glass Tubing into a Stopper

Safe Use of Ceramicware

Some laboratory tools are made of ceramic materials such as porcelain. Look at the examples in Figure 2-13.

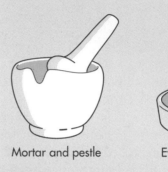

Figure 2-13 Laboratory Ceramicware

Mortar and pestle   Evaporating dish

You can use ceramic evaporating dishes to heat a solution and remove all the water, leaving behind the solid materials. Instead of heating an evaporating dish with a direct flame, place it on a wire gauze or ceramic tile.

A mortar and pestle are a ceramic bowl and grinding stick used to crush lumps of solid materials into powder. Be careful, because some solids explode when they are ground. Never mix solid chemicals by grinding. They may react violently because of the pressure you apply.

Catalog of Standard Equipment

Figure 2-14 shows some of the standard equipment used in labs. Examine the figure. How would you use each item in the lab? You can improve safety by knowing the name and purpose of each tool.

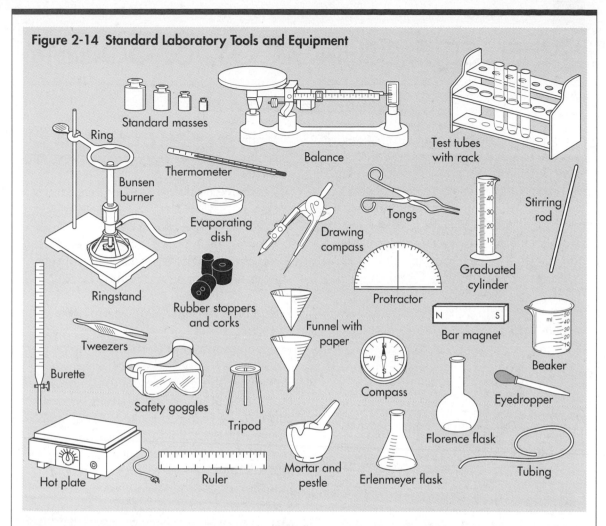

Figure 2-14 Standard Laboratory Tools and Equipment

Standard masses

Ring

Bunsen burner

Thermometer

Balance

Test tubes with rack

Stirring rod

Evaporating dish

Drawing compass

Tongs

Graduated cylinder

Ringstand

Rubber stoppers and corks

Protractor

Bar magnet

Beaker

Tweezers

Burette

Funnel with paper

Compass

Safety goggles

Tripod

Mortar and pestle

Florence flask

Eyedropper

Hot plate

Ruler

Erlenmeyer flask

Tubing

Section 2-3 Review

Write the definitions for the following terms in your own words.

1. **Bunsen burner**
2. **reducing flame**
3. **oxidizing flame**
4. **heat-resistant glass**

Answer these questions.

5. How does a Bunsen burner operate?
6. Describe a proper Bunsen burner flame.
7. What are some safety rules for lighting and using a Bunsen burner?
8. What are some safety rules for heating a material in a test tube?
9. What are some safety rules for bending, breaking, or inserting glass tubing?
10. Sketch and describe three pieces of laboratory equipment.

Science, Technology, & Society

Chemical Safety on the Job

Many jobs require people to use chemicals. In order to assure the safety of workers, several government agencies make rules for using hazardous chemicals. These agencies include the Occupational Health and Safety Administration (OSHA); the National Institute for Occupational Safety and Health (NIOSH), and the United States Consumer Product Safety Commission (CPSC).

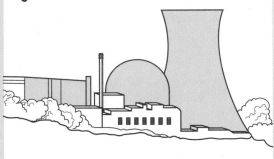

Figure 2-15

People who work at nuclear power plants handle hazardous materials.

Most agencies list chemicals under certain classifications. The United States Congress has defined nine classes of hazardous materials:
1. Explosive materials cause unstable, rapid energy release.
2. Carcinogenic materials cause cancer.
3. Mutagenic materials can cause birth defects.
4. Highly toxic materials, even in small amounts, cause death, severe illness, or disability.
5. Corrosive materials corrode steel and skin.
6. Irritant materials cause skin sores.
7. Flammable materials catch fire easily.
8. Reactive materials produce violent reactions when mixed improperly or exposed to air and water.
9. Radioactive materials give out harmful radiation.

Companies must follow rules for the proper use, storage, and disposal of hazardous substances. Safety notices must be posted. Protective equipment for handling chemicals may be required by law. Also, companies must hold training sessions for all workers to review safety rules. Companies that fail to obey safety practices may be punished under the law.

The laws regarding chemical safety are examples of how society affects science and technology. By creating and enforcing laws, society is saying that safety is Job One!

Follow-up Activities
1. Interview an adult who works with industrial chemicals. Find out how rules involving chemicals are followed at work.
2. Look for newspaper or magazine articles that refer to dangers and safety regarding chemicals in the workplace. Gather them together for a report or a display.

KEEPING TRACK

Section 2-1 Safety with Household Chemicals

1. The natural world is made of chemicals.
2. Many household chemicals are hazardous, and you should follow safe procedures in using them.
3. Many chemicals give off vapors that are toxic or flammable. These should be used only in the absence of a flame and in a well-ventilated place.
4. Unless directed to do so, do not mix chemicals. Mixing chemicals may produce a hazardous material.
5. Aerosols in pressurized containers may explode when punctured or heated. Use them according to directions.
6. Store household chemicals in a closed area out of the reach of small children and pets.
7. You can find first aid advice on labels, in instructions, and in first aid books. If you accidentally spill or swallow a household chemical, you will usually have to wash with water or drink plenty of water.
8. Follow rules for proper and legal disposal of chemicals.

Section 2-2 Safety with Laboratory Chemicals

1. Many laboratory chemicals are hazardous, and you should follow safe procedures in using them.
2. Follow laboratory rules and directions.
3. Report accidents and hazards immediately.
4. Use safe procedures when pouring, diluting, and smelling chemicals.
5. Be familiar with the common laboratory hazard symbols.

Section 2-3 Proper Use of Laboratory Tools

1. Follow safety procedures in handling laboratory tools and equipment.
2. Know the safe ways to use burners, heat materials, and work with glass.
3. Know the names and safe uses of laboratory equipment.

BUILDING VOCABULARY

Write the term from the list that best completes each sentence.

diluting, ventilated, Bunsen, corrosive, heat-resistant glass, flammable, toxic, aerosols, concentrated, pressurized containers

Most household chemicals are ___1___, that is, poisonous. Battery acid and lye can cause skin sores or blindness because they are ___2___. Use ___3___ materials away from any flame and in areas that are ___4___. Hair spray and Mace are examples of aerosols found in ___5___. When ___6___ with water, pour the ___7___ solution into the water and stir. In the ___8___ burner, the hottest point is at the tip of the inner blue cone of the flame. This inner part is the reducing flame and the outer part is the ___9___. Most laboratory glassware is made of ___10___.

SUMMARIZING

Write *true* if the statement is true. If the statement is false, change the *italicized* term to make the statement true.

1. *Paint remover* fumes are flammable.
2. Many brands of floor wax *are not* flammable.
3. Chemicals containing ammonia and chlorine *may* be safely mixed.
4. It is *unwise* to mix chemicals or medicines.
5. Pressurized containers are dangerous if they get too *hot*.
6. Usually, if you swallow a corrosive chemical, you *should not* try to throw up.

INTERPRETING INFORMATION

Look at Figure 2-16 below. Identify the things and actions that are not safe in the laboratory. Try to find at least eight.

THINK AND DISCUSS

Use the section number in parentheses to help you find each answer. Write your answers in complete sentences.

1. Describe the difference between household and laboratory chemicals. Tell their uses and give an example of each type of chemical. (2-1, 2-2)
2. Explain the dangers of mixing chemicals. (2-1, 2-2)
3. To clean an old sink, Mischa wants to scrub it with ammonia detergent and a scouring powder containing chlorine. What advice would you give Mischa? (2-1)
4. Pressurized containers should not be placed into garbage compactors or garbage incinerators at home. Why? (2-1)
5. What hazard symbols appear on laboratory chemicals? (2-2)

GOING FURTHER

Following is a warning printed on a can of a popular liquid furniture wax. Read the warning. Then answer questions 1 through 3.

Figure 2-16

Danger Contains petroleum distillates. Harmful or fatal if swallowed. Do not induce vomiting. Give 4 glasses of water or other fluid except milk. Call physician immediately. Avoid prolonged skin contact & inhaling. Vapor harmful. Use with adequate ventilation. Wash hands after using. Keep out of reach of children.

Caution Flammable. Keep away from heat or open flame. Keep container closed when not in use. Oily cloths should be soaked in water after use to avoid spontaneous combustion.

1. Name the hazardous characteristics of this chemical using the terms of this chapter.
2. Use both a dictionary and the glossary to find the meanings of the following terms:
   petroleum
   distillate
   spontaneous combustion
3. Write a short report that relates the three terms to each other and explains why the presence of petroleum distillates is hazardous. Conclude by explaining what might happen to oily cloths if they are thrown into a pile in a closet, and why the label might be better if it said, "Wash hands *with soap* after using."

COMPETENCY REVIEW

Circle the letter of the correct answer to each question.

1. All things in the natural world are made out of
   a. energy.   b. perceptions.
   c. chemicals.   d. intelligence.

2. Safe procedures must be followed in using
   a. household chemicals.
   b. laboratory chemicals.
   c. laboratory equipment.
   d. all the above.

3. Poisonous materials are also called
   a. toxic.
   b. corrosive.
   c. flammable.
   d. pressurized.

4. **Caution** Keep away from heat, sparks, and open flame.
   Such a warning indicates a material is
   a. flammable.
   b. poisonous.
   c. strongly acidic.
   d. electric.

5. The statement below is printed on a common household item.
   Caution Do not use full strength. Mix with four parts water. Keep windows open for ventilation while drying.
   The contents of this container probably
   a. are radioactive.
   b. give off flammable vapors.
   c. are highly pressurized.
   d. give off toxic fumes.

6. **Caution** Do not set on stove or radiator or keep where temperatures exceed 100° F, as container may burst. Do not puncture or throw in fire. Keep out of reach of children.
   This warning indicates the material is
   a. very corrosive.
   b. radioactive.
   c. flammable.
   d. pressurized.

CHAPTER 3

MEASUREMENT IN SCIENCE

Do you like to talk in general terms or in precise terms? Would you say, "I caught a big fish," or "I caught a five-pound fish that was 13 inches long"? Measurement allows you to give accurate descriptions of an object or an event. Science is built on measurement.

Consider the question, Is an elephant big? You can only answer this question if you make a comparison. Compared to you, an elephant is big, but compared to the earth it is tiny. Because terms like *big, small, fast,* and *slow* are vague, scientists use measurements. A scientist might weigh the elephant and measure its length, height, and speed. With accurate measurements of one elephant, it is easier to make comparisons with other elephants, other animals, and other objects. Look at Figure 3-1.

Scientists often use tables and graphs to communicate information so that people can compare two or more things easily. Becoming skilled at reading and making tables and graphs will help you be successful in this course and in more-advanced studies.

3-1 The Metric System

■ *Objectives*

☐ *Use a ruler, a graduated cylinder, and a triple-beam balance.*

☐ *Calculate areas and volumes.*

☐ *List metric units for length, area, and volume.*

☐ *Distinguish between weight and mass.*

You are familiar with measuring in inches, feet, gallons, and pounds. These are the units commonly used in the United States. Most countries, however, use the metric

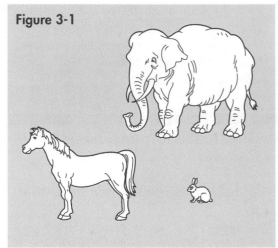

Figure 3-1

Is a horse big?

system of measurements, with measurements made in meters, liters, and grams. In all countries scientists use the metric system.

In all measurement systems, every measurement has a number and a unit. A **unit** is the standard amount or quantity on which measurement is based. Suppose it takes Kim 32 seconds to swim the length of a pool. In this measurement of time, the number is 32, and the unit is seconds. Thus, it took 32 times the standard quantity of 1 second. There are different units to measure length, area, volume, temperature, weight, and mass.

Measuring Length

The basic unit for measuring length in the metric system is the meter. Other units of length are defined in terms of this basic unit. For example, one kilometer is 1000 meters. Look at Figure 3-2. Which is larger, one centimeter or one millimeter?

Figure 3-3 shows a typical ruler with centimeters. The longer lines show centimeter lengths. Each centimeter is divided into ten parts, so the distance between the shortest lines is one tenth of a centimeter, or one millimeter.

Figure 3-2 Metric Units of Length and their Abbreviations

1 meter (1 m)	= basic unit
1 kilometer (1 km)	= 1000 m
1 centimeter (1 cm)	= 1/100 m
1 millimeter (1 mm)	= 1/1000 m
1 micrometer (1 µm)	= 1/1,000,000 m

In the ruler in Figure 3-3, the number below each arrow shows the distance from the line above the number 0 to the spot where the arrow points. You can change a fraction measurement into its decimal form before using it in calculations. For example, $6^6/_{10}$ centimeters can be written as 6.6 cm. Notice that the reading for the 2-cm-plus-4-mm spot can be written as 2.4 cm. Because there are ten millimeters in one centimeter, the reading can also be written as 24 mm.

Look again at Figure 3-3. How long are the two needles? How would you write their lengths in centimeters and millimeters? If an object falls between two millimeter marks, round your measurement to the nearest millimeter. The short needle is 6.6 cm, or 66 mm. The long needle is 7.4 cm, or 74 mm.

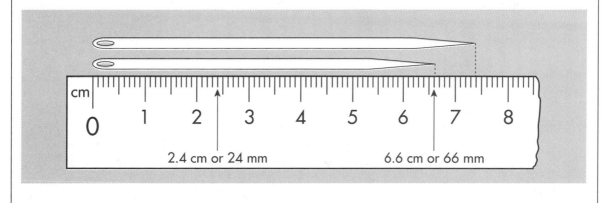

2.4 cm or 24 mm   6.6 cm or 66 mm

Measuring Area

Measure the square in Figure 3-4. It is exactly one centimeter on each side. The area

Figure 3-4 One Square Centimeter

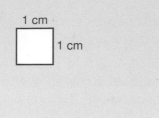

of the square is one square centimeter. This is abbreviated 1 cm². What is the area of the rectangle in Figure 3-5? The area of the rectangle is the number of one-square-centimeter squares that will fit inside the rectangle. Look at the dashed lines inside the rectangle. They make six one-square-centimeter squares. The area of the rectangle, therefore, is six square centimeters. You can also measure the area of something in other units. For example, you can use square meters (abbreviated m²) and square millimeters (mm²).

You can find the area of a square or rectangle by using the formula

area = length × width

For example, the rectangle in Figure 3-5 has an area of

3 cm × 2 cm = 6 cm²

This is the same answer as the one you got by counting the squares that fit inside.

Figure 3-5 Finding the Area of a Rectangle

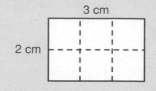

For practice, find the area of the large rectangle which surrounds all the words on this page. The length of the rectangle is closer to 19 centimeters than it is to 20 centimeters. So, to the nearest centimeter, the length is 19 centimeters. The width is 15 centimeters to the nearest centimeter. So the area of the rectangle is about 19 cm × 15 cm, which is 285 square centimeters.

Use a meter stick to measure the length, width, and height of a room. Record each measurement to the nearest meter. What is the total area of all the walls and the ceiling? Can you think of a reason you might want to know that total area?

Measuring Volume

Look at the cube in Figure 3-6. It is exactly one centimeter on each side. The **volume** of the cube is 1 cubic centimeter. Imagine that you are filling a shoebox with these cubic centimeter cubes. The volume of the box is the number of one-cubic-centimeter cubes that will fit inside it.

Look at the box in Figure 3-7. You can see that 12 cubic centimeters will fit in the bottom layer, and 12 cubic centimeters will fit on the top layer. The volume of the box, therefore, is 24 cubic centimeters.

Figure 3-6 One Cubic Centimeter

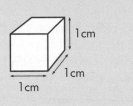

1cm
1cm
1cm

You can also find the volume of a cube or a box by using the formula

volume = length × width × height

For example, a cube that is 2 cm long on each side has a volume of

2 cm × 2 cm × 2 cm = 8 cubic centimeters, or 8 cm³

Figure 3-7 Finding the Volume of a Box

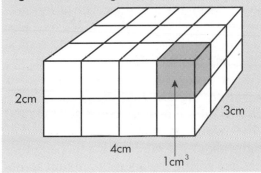

2cm
4cm
3cm
1cm³

The volume of the box in Figure 3-7 is

4 cm × 3 cm × 2 cm = 24 cm³

You may measure the volume of objects in other units. For example, you can use cubic meters (m³) and cubic millimeters (mm³).

Liters and Milliliters   Scientists usually measure the volume of liquids and gases in liters (L) and milliliters (mL). One **liter (L)** is the volume of a cube that is 10 cm on each side. Look at the cube in Figure 3-8. The volume is

10 cm × 10 cm × 10 cm = 1000 cm³

Figure 3-8 The Liter and the Milliliter

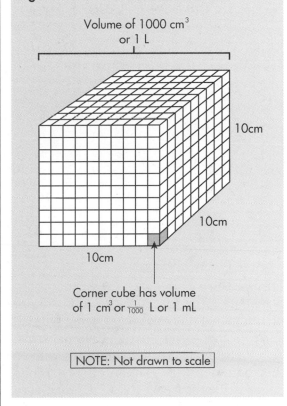

Volume of 1000 cm³ or 1 L

10cm
10cm
10cm

Corner cube has volume of 1 cm³ or $\frac{1}{1000}$ L or 1 mL

NOTE: Not drawn to scale

One liter, therefore, is equal to 1000 cubic centimeters, and one milliliter is equal to one cubic centimeter.

The units cm³ and mL can replace one another. Doctors and nurses, in addition, often use the abbreviation *cc* for cubic centimeters. Thus, volumes of 25 cm³, 25 cc, and 25 mL are identical.

You can measure volumes of liquids in **graduated cylinders**. These are cylinders with volume lines, or gradations, printed on them. Look at Figure 3-9. Scientists use graduated cylinders of various sizes, with various scales. Note that the markings on graduated cylinders are in milliliters (mL).

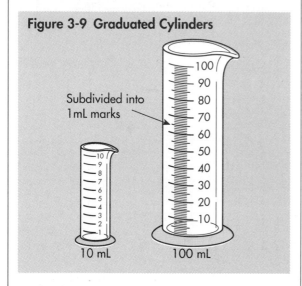

Figure 3-9 Graduated Cylinders

Subdivided into 1mL marks

10 mL   100 mL

Temperature

Scientists often measure the temperature of things. In the metric system, temperature is measured on the **Celsius** scale. Scientists measure temperature with a thermometer that is marked in Celsius degrees. Water freezes at 0° C, read zero degrees Celsius. Water boils at 100° C. Therefore, there are exactly 100 degrees between the temperature at which water freezes and the temperature at which it boils. The normal body temperature of most people is 37° C. A comfortable room temperature is about 21° C.

Weight and Mass

All matter has both mass and weight. An object's **mass** is the amount of matter it contains. You have a mass. Your mass comes from all the water, food, bone, skin, muscle, and fat in your body. If you go up in a plane or even to the moon, your mass will remain the same because your body will have the same matter in it.

An object's weight depends not only on its mass, but also on the pull of gravity on the object. Gravity is a force that acts to pull one object to another. The pull of gravity high in the sky or on the moon is less than on the earth. In a plane or on the moon, you will weigh less than you weigh on earth.

You can see that although an object's mass does not change, its weight changes, depending on the pull of gravity. The pull of gravity is slightly less on the fiftieth floor of a building than on the first floor. An object on the fiftieth floor, therefore, weighs slightly less than on the first floor. The difference, however, is very tiny. Only sensitive scales can detect an object's change in weight at such small distances from the ground.

Measuring Weight and Mass Scientists measure weight in units called newtons (N) in the metric system. An ordinary apple weighs about 1 newton on Earth. Scientists often use spring balances to obtain an object's weight.

The basic unit for mass in the metric system is the **kilogram (kg)**. One liter of water at 20° C has a mass of one kilogram. One gram (g) is 1/1000 of a kilogram.

A triple-beam balance like the one shown in Figure 3-10 is commonly used in laboratories to measure an object's mass in grams.

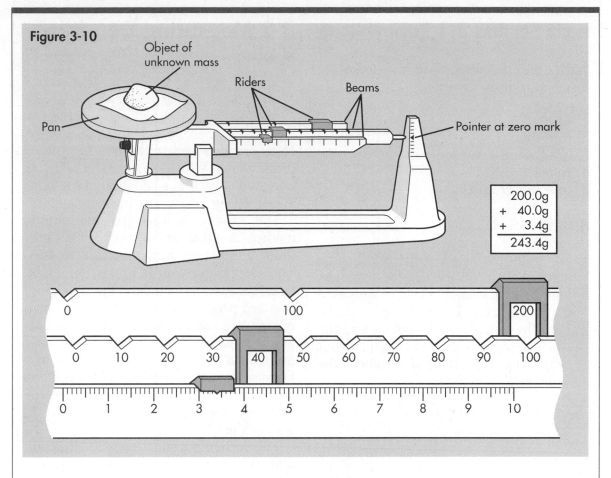

Figure 3-10

Object of unknown mass

Riders

Beams

Pan

Pointer at zero mark

$$
\begin{array}{r}
200.0g \\
+ 40.0g \\
+ 3.4g \\
\hline
243.4g
\end{array}
$$

Here are the steps for using a triple-beam balance:

1. Start with all the riders to the left.
2. Check that the pointer is on zero.
3. Place the object to be measured on the pan.
4. Move the largest rider to the right, one notch at a time, until the pointer drops down. Then move it to the left one notch.
5. Repeat step 4 for the rider on the middle scale.
6. Move the front rider until the pointer on the right side of the balance comes to the zero point.

7. Add the readings from the three beams to obtain the object's mass.

Did you notice that the balance in Figure 3-10 is similar to the scales in some doctors' offices? You can see that the balance breaks the mass of the object in Figure 3-10 into three units, 200 g, 40 g, and 3.4 g. If you add these amounts, you will find a total mass of

200 + 40 + 3.4 g

which is 243.4 g.

Section 3-1 Review

Write the definitions for the following terms in your own words.

1. **unit**
2. **one liter (L)**
3. **graduated cylinder**
4. **mass**
5. **one kilogram (kg)**

Answer these questions.

6. For each unit listed, write its full name and what it measures, for example, length, area, volume, mass, or weight. g, mL, cm³ , km, L, m³ , N, kg
7. Find the length, width, and area of the two objects in Figure 3-11, using your centimeter ruler. Read your ruler to the nearest tenth of a cm.

Figure 3-11

a.

b.

8. A box measures 12.0 cm long, 6.0 cm wide, and 1.5 cm thick. Determine its volume.
9. After a satellite has been launched into space, how does its mass and weight compare with what they were before launch? Explain your answer.

3-2 Measuring Density

■ *Objectives*
☐ *Use the density formula.*
☐ *Find the density of solids and liquids in the laboratory.*

Here are two trick questions for you to think about. Which has greater mass, a kilogram of rock or a kilogram of feathers? Which has greater mass, iron or paper?

The answer to the first question is neither. A kilogram of rock and a kilogram of feathers have identical masses. The second question is impossible to answer because, to compare the masses of iron and paper, you must know how much of each to measure. For example, a dictionary is made of paper, yet it has greater mass than a hairpin that is made of iron because the dictionary is much larger. You therefore need to know the size, or volume, of the iron and paper in order to answer the question.

The second question leads to an important idea. It is sometimes helpful to use **density** measurements. The density of a material equals its mass divided by its volume.

density = mass/volume

Recall that the measurement for mass is given in grams (g) or kilograms (kg). Volume is given in cubic centimeters (cm³) or cubic meters (m³).

Example: A cube of solid brass measures 3.0 cm on each side and has a mass of 216 grams. Find the density of this brass cube.

Volume = length × width × height
Volume = 3 cm × 3 cm × 3 cm
Volume = 27 cm^3 , or 27 cubic cm
Density = mass/volume
Density = 216 g/27 cm^3
Density = 8 g/cm^3, or 8 grams per cubic centimeter

This brass cube has a density of 8 grams of mass for each volume of 1 cubic centimeter that you measure.

Example: The volume of a tank is 2 cubic meters. The mass of water needed to fill it is 2000 kilograms. Find the density of the water.

Density = 2000 kg/2 m^3
Density = 1000 kg/m^3, or 1000 kg per cubic meter

Density of Regular Solids

Boxes and cubes are examples of **regular solids**. Regular solids have familiar shapes. You can find the density of a box by first finding its volume.

Example: A solid box of aluminum measures 5 cm by 4 cm by 3 cm. Its mass is 162 grams. Find the density of aluminum.

Volume = length × width × height
Volume = 5 cm × 4 cm × 3 cm
Volume = 60 cm^3

Density = mass/volume
Density = 162 g/60 cm^3
Density = 2.7 g/cm^3, or 2.7 grams per cubic centimeter

Density of Irregular Solids

Solids that have an unusual shape are called **irregular solids.** Examples of irregular solids include apples, shoes, and rocks. There is no easy formula for finding the volume of an irregular solid. You can, however, find the volume of irregular solids by putting them under water in a graduated cylinder. First note the level of water in the cylinder. Then add the object. The change in the water level will be the volume of the object. This method of finding volume is called the **water-displacement method.**

Example: A rock with a mass of 41.6 grams is placed into a graduated cylinder containing 15 mL of water. After the rock is placed in the water, the water level rises to the 31 mL mark. What is the density of the rock?

Volume of rock = level after − level before
Volume of rock = 31 mL − 15 mL
Volume of rock = 16 mL = 16 cm^3
Density = Mass/ Volume
Density = 41.6 g/ 16 cm^3
Density = 2.6 g/ cm^3, or 2.6 grams per cubic centimeter

Density of Liquids

You can measure the density of a liquid with a graduated cylinder. First weigh the empty cylinder on a balance and record the weight. Then pour the liquid into the same cylinder. Determine the volume of the liquid from the markings on the cylinder. Then measure the liquid's mass by putting the cylinder on the same balance. Subtract the mass of the empty cylinder to find the mass of the liquid.

Making Predictions

Everyone knows that iron sinks and cork floats. Yvette, Ron, and Elvio decided to make some scientific observations about floating. They used not only a variety of objects, but also liquids that do not mix with water. The table in Figure 3-12 is what the students observed. They looked up the densities in a reference book and added the information to their table.

1. What pattern do you see in these results? Remember that the density of water is 1 g/cm³.
2. Write a hypothesis that explains when a material will float in water.
3. Material X has a density of 1.3 g/cm³. Predict whether it will sink or float in water.
4. Using Appendix 3, find 3 materials that will float in water.

Figure 3-12

Material	Floats in water?	Density g/cm³
iron	no	8.0
cork	yes	0.2
graphite (carbon)	no	2.2
mercury liquid	no	13.6
kerosene liquid	yes	0.8
ebony wood	no	1.1
cherry wood	yes	0.9

Example:
Given: volume of liquid = 28 mL
mass of liquid and cylinder = 125.7 g
mass of empty cylinder = 92.1 g

Mass of liquid = 125.7 g - 92.1 g
Mass of liquid = 33.6 g
Density = mass/ volume
Density = 33.6 g/ 28 mL = 1.2 g/mL

You could also give the answer as 1.2 g/cm³, but the preferred unit for a liquid is g/mL.

Using the Density Formula

In most situations, you would not determine the density of an object. You would look it up in a book instead. You then would use the density to calculate an object's mass or volume. You can do this by rearranging the formula for density.

formula #1: **Density = mass/volume**
formula #2: **Mass = density × volume**
formula #3: **Volume = mass/density**

Example: A typical milk carton contains 946 mL of milk with a density of 1.05 g/mL. What is the mass of the milk in the carton?

Mass = density × volume
Mass = 1.05 g/mL × 946 mL
Mass = 993 g

Example: A silver bracelet has a mass of 252 grams. What volume of silver is in the bracelet? The density of silver is 10.5 g/cm³.

Volume = mass/density
Volume = 252 g/ 10.5 g/cm³
Volume = 24 cm³

■■■ Section 3-2 Review ■■■

Write the definitions for the following terms in your own words.

1. **density**   2. **irregular solid**
3. **water-displacement method**

Answer these questions.

4. Imagine cutting a slice from a loaf of bread. How does the slice compare with the loaf in mass, weight, volume, and density? Explain your answers.
5. A 342-gram solid object is put into a graduated cylinder containing water. The water level was at 25 mL without the object, and is at 55 mL after the object is added. Determine its volume and density. Then use Appendix 3 to determine the object's possible composition and tell what it might be.

3-3 Tables and Graphs

■ *Objectives*
☐ *List some purposes of tables and graphs.*
☐ *Use data tables.*
☐ *Analyze mass-volume graphs of liquids.*
☐ *Prepare line and bar graphs.*
☐ *Interpret circle graphs.*

You encounter tables and graphs often. Tables and graphs are used to present facts and measurements. Facts and measurements are called **data**. The calculations you have carried out in the earlier sections of this chapter could be easily presented in the form of a table or graph. Newspapers, books, and magazines use tables and graphs to present data on money, politics, health, and even sports. Some of the main purposes of tables and graphs are to help people collect data in an organized way, present data clearly, and analyze data for their meaning.

Data is a plural word; *datum* is the singular form. If you refer to only one piece of information it is a datum. If you refer to two or more pieces of information, they are data. Because it is a plural word, it is correct to say "the data *are*..." and "the data *show* that."

Tables

You can use a table to organize data into categories and to present data neatly. A scientist's actual laboratory notebook may seem quite disorganized to other people. The scientist must often reorganize his or her notes into tables so that others can better understand the observations and results.

Tables come in a wide variety. You read some from left to right and others from top to bottom. A proper table includes headings and measurement units for all the categories of data it contains.

Imagine that two seventh-grade students, Tony and Gina are taking measurements of

Figure 3-13 Raw Data in a Lab Notebook

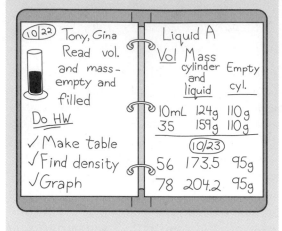

Figure 3-14	
Volume of Liquid A (mL)	***Mass of Liquid A (g)**
10	14
35	49
56	78.4
78	109.2

*The mass is obtained by subtracting the mass of the empty cylinder from the mass of the cylinder with the liquid in it.

a liquid labeled Liquid A. Their laboratory notebook might look like the one shown in Figure 3-13. The notebook shows the data the students measured directly, without any adjustments or calculations. These kinds of data are called **raw data**. Now these students must present the important data in a table like the one shown in Figure 3-14.

The table in Figure 3-14 contains headings, units, the important data, and a footnote for explanation. Notice that raw data about the mass of the empty cylinder and the mass of the cylinder with liquid have been left out. They are not important, but the mass of the liquid by itself is important.

Graphs

A graph is a very meaningful way to present data. Graphs can help you see trends in measurements and make comparisons. Graphs come in a wide variety, but three types appear most commonly. They are line graphs, bar graphs, and circle graphs.

Features of a Line Graph

Figure 3-15 shows key features of a good line graph. A line graph has two axes, one horizontal and one vertical. As you can see in Figure 3-15, the horizontal axis runs from side to side and the vertical axis runs up and down. Scientists use each axis for a specific reason, and they label each axis with numbers to indicate the **scale** they are using.

Once each axis is labeled with a scale, you can plot data points on the graph. For example, in Figure 3-15 the following five data pairs have been plotted. The top dot shows a value of 103 on the vertical axis and 95 on the horizontal axis.

time (in seconds)	10	30	59	74	95
distance (in meters)	4	39	61	90	103

Suppose you wanted to investigate the heating ability of your stove at home. You placed a pot of water over a flame and

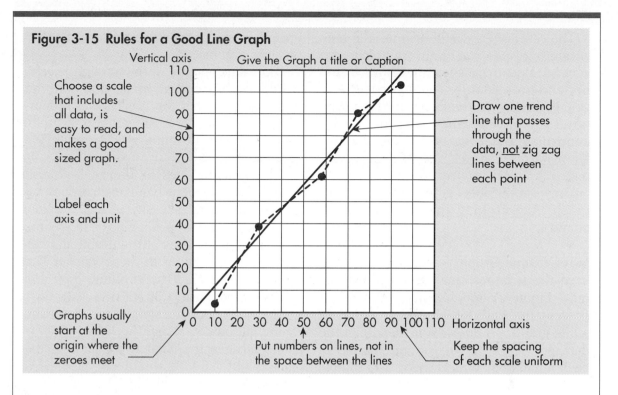

Figure 3-15 Rules for a Good Line Graph

Vertical axis

Give the Graph a title or Caption

Choose a scale that includes all data, is easy to read, and makes a good sized graph.

Draw one trend line that passes through the data, <u>not</u> zig zag lines between each point

Label each axis and unit

Graphs usually start at the origin where the zeroes meet

Horizontal axis

Put numbers on lines, not in the space between the lines

Keep the spacing of each scale uniform

measured the water temperature every two minutes with a thermometer. You obtained the results shown in Figure 3-16.

For practice, put these results in a graph. Read the next three paragraphs before you begin. You will need to get some graph paper and decide on the size of graph you want. You must also decide how many lines will be available for the graph, upward and across, before you can decide how to number the lines. The scale you choose must make the graph size you desire, show all the data points, and be easy to read. Line numberings that are easiest to read increase by numbers like 0.1, 1, 2, 5, 10, or 20.

In making this graph, a researcher would put time, in minutes, along the horizontal axis, because time is the **independent variable**. The independent variable is the part of the experiment you can control. *You* decided to read the water's temperature every two minutes.

Figure 3-16					
Time of heating, in minutes:	0	2	4	6	8
Water temperature, °C:	30	43	56	69	82

The researcher would next put the temperature of the water, in °C, along the vertical axis because temperature is the **dependent variable**. The dependent variable is the part of the experiment you cannot control. In this experiment, you controlled the *time of heating, but the thermometer told you the water temperature. You had no control over that.*

Draw and label both axes for your graph. After setting the scale, plot your points.

Mass-Volume Graphs   Suppose you measured the volumes and masses of two liquids, Liquid A and Liquid B. You might present the data for Liquids A and B in a line graph. To do so, you would plot the data points for the mass and volume of each liquid and connect them. The sample graph in Figure 3-17 is called a mass-volume graph.

The data points form straight lines, which indicate that mass and volume increase together by the same proportion. For instance, if the volume is multiplied by two, the mass is also multiplied by two.

The straight lines in Figure 3-17 also show that the density of each liquid remains constant. Take any point on a line, such as point P. Read its mass and volume values from the axes of the graph and calculate the density from these values. The answers for point P are: volume = 60 mL; mass = 36 g; density = 36 /60 mL = 0.6 g/mL.

You will obtain the same value of density from all points along this line. This is true because graphing the data resulted in straight lines. It would not be true if graphing the data had produced curved lines.

Look at the mass-volume table for water in Figure 3-18.

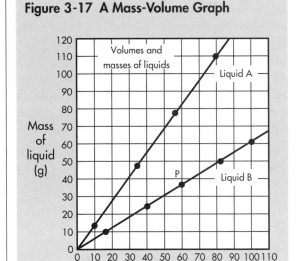

Figure 3-17 A Mass-Volume Graph

Volumes and masses of liquids

Liquid A

Liquid B

P

Mass of liquid (g)

Volume of liquid (mL)

Figure 3-18				
Water				
Volume, mL	10	29.6	55	90
Mass, g	10	29.6	55	90
Density, g/mL	1	1	1	1

On your own, choose a scale and plot the points in the figure on a graph. Do the data produce a straight line or a curved line?

What Do Mass-Volume Graphs Look Like?

Process Skills *measuring; making a graph*

Materials various liquids such as tap water, sea water, milk, alcohol, cooking oil, and concentrated solutions such as $MgSO_4$ and $NaNO_3$; 100-mL graduated cylinders; triple-beam balance.

Procedures

1. Find the mass of an empty graduated cylinder and record it.
2. Choose one of the liquids. Pour 20 mL of the liquid into the cylinder.
3. Place the cylinder with its liquid on a triple-beam balance to determine the combined mass. Take care not to get the balance pan wet.
4. Repeat steps 2 and 3 for four additional volumes spread between 20mL and 100mL.
5. Repeat steps 2, 3, and 4 with other liquids.
6. Complete a separate table for each liquid you test. Use the following headings. Volume taken; Mass of liquid and cylinder; Mass of empty cylinder; Mass of liquid alone; Density. Be sure to label each table with the name of the liquid being tested.
7. Make a mass-volume graph of all your data.

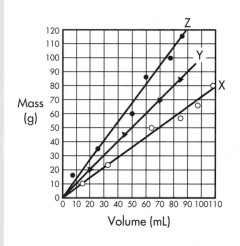

Figure 3-19

Use your own graph paper and the scale size shown in Figure 3-19.

To avoid confusion, plot the data for each liquid with different types of data points. These could be dots, circles, Xs, triangles, or different colors. The line you draw should not be zig-zagged. It should pass as closely as possible through the points, with some to one side of the line and some to the other side.

Conclusions

8. How close together are the density values for each liquid?
9. What small problems with the equipment or procedures might account for their not being exactly equal?

Evaluating Line Graphs The graphs in Figure 3-20 show ways to organize the data in Figure 3-17. Look for typical graphing errors discussed in the comment made about each graph.

Figure 3-20 Evaluating Graphs

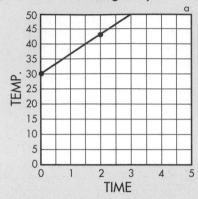

Graph A
Comment Off the chart! The scales on both axes are too small. All the data cannot fit into the graph. Only two data points are within the desired area of the graph. The line would have to extend beyond the graph to include all the data points.

Graph B
Comment A postage stamp! The scales on both axes are too large. The data are squeezed into a small part of the area available for the graph. Points on the line will be read with less accuracy because the scales are so large. A slight shift changes the reading a lot.

Graph C
Comment Awful! The scale along the temperature axis is totally unacceptable. It begins with 30, but it should begin with zero. Worse, the scale is not kept constant. It goes by 2s, 5s, 10s, and even by 20 in the last space. Data that should form a straight line are now distorted into a curve by the inconsistent scale on the vertical axis.

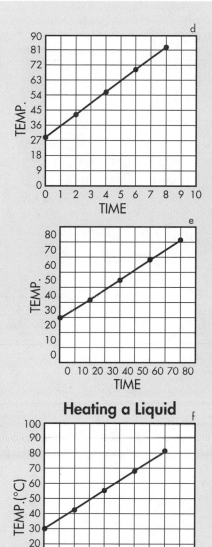

Graph D

Comment Looking better! The data nicely fill the desired area. However, the temperature scale goes by 9s, and it is a difficult scale to read. Where, for example, is the value of 78 C located? You can figure it out, but only with difficulty. A scale should be easy to read. The scale on this vertical axis is not easy to read.

Graph E

Comment Almost there! Unfortunately the numbers on both scales were positioned in the spaces between lines. Numbers should be positioned so the line runs through them. Numbers in the spaces are confusing. Do they apply to the line on the left or on the right? The graph becomes harder to read.

Graph F

Comment A model graph! The scales go by 1s and by 10s, making them very easy to read. The data points adequately fill the graph area. "Time of heating" says more about the horizontal axis than just "Time" alone. Both axes give the units of each measurement, in minutes and C. Someone has finally titled the graph.

Figure 3-21 Bar Graphs

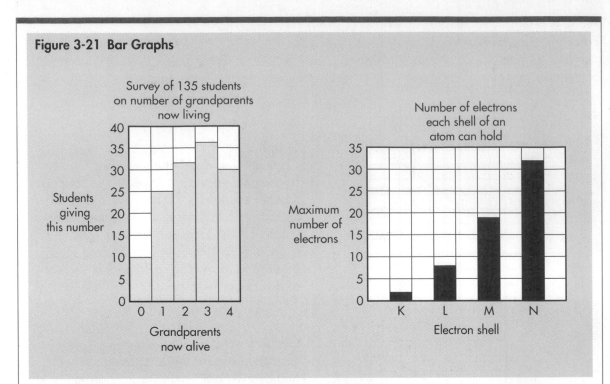

Figure 3-22 Circle Graphs (Kinds of Elements Found on Earth)

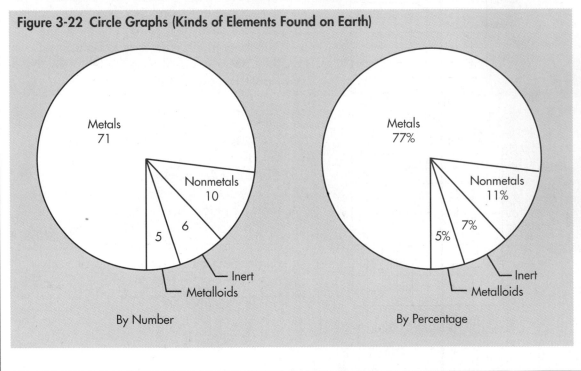

Features of a Bar Graph   Most of the rules for making a line graph apply to making a bar graph too. There are some differences. In a bar graph, you write the numbers along the horizontal axis in the spaces between lines. This is because the number is for the space, not the line. In addition, you can separate bars by skipping spaces to give the graph a less-crowded appearance.

Use bar graphs to organize data that have very certain values, such as whole numbers. For example, 34 students, eight electrons, five calendar years, or seven atoms in a molecule. For continuous data, line graphs are best. Continuous data are data that can take on any values, including fractions, such as 2.3 minutes or 59.6° C. Study the two examples of bar graphs in Figure 3-21.

Features of a Circle Graph   You can use circle graphs to compare amounts, percentages, and how different events occur. To do this, divide a circle into wedge-shaped parts. A circle graph is also known as a pie graph.

The two circle graphs in Figure 3-22 compare the types of chemical elements that occur on the earth. One gives amounts, and the other gives percentages. The size of each section compared with the whole circle is roughly the same as the percentage of the thing each represents.

■■■ Section 3-3 Review ■■■

Write the definitions for the following terms in your own words.

1. **data**

2. **raw data**
3. **scale**
4. **independent variable**
5. **dependent variable**

Answer these questions.

6. List three purposes of tables and graphs. For each, give an example.
7. Study the mass-volume graph in Figure 3-23. Then answer the following questions.
   a. Which liquid has the lowest density?
   b. Which might be water?
   c. What is the density of Liquid A? Take any point on the line and read its values.

Figure 3-23

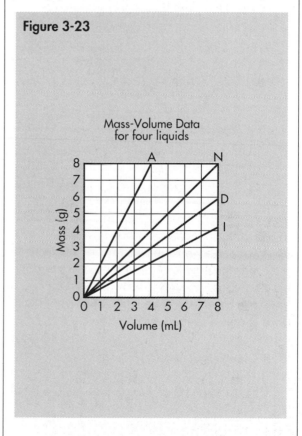

SCIENCE, TECHNOLOGY, & SOCIETY

The Story of Archimedes and Density

Science gives to society stories about discoveries, heroes of science, and amazing events. Science gives people exciting ideas and a deeper appreciation of the universe. In these ways, science enriches our culture. The story of Archimedes is an example.

Archimedes was a scientist who lived in Greece around 250 B.C. The king asked him to find out whether his crown was pure gold or not. The king suspected that the crown maker had kept some of the gold he was given and, instead of using pure gold mixed in some other metal to make the final weight the same. The king told Archimedes to uncover the facts, but in a way that would not scratch or harm the crown.

One day, while taking a bath, Archimedes thought about the water that overflowed as he stepped into his tub and about his seeming loss of weight in the water. In a flash he had an idea. Archimedes was so excited that he ran out naked into the street crying, "Eureka! eureka!" "I have found it! I have found it!"

Today his idea is called the "principle of Archimedes." His idea was that a submerged object loses an amount of weight that is equal to the weight of the water it has displaced. According to his idea, if you displace 40 pounds of water, you will weigh 40 pounds less in the water.

This discovery led Archimedes to the discovery of density. He tested equal weights of pure gold, copper, silver, lead, and various mixtures of these metals to find out their volumes. He did this by submerging them in a container of water and looking at the rise in the water level.

When Archimedes submerged the crown, the water level rose higher than it should have risen if the crown had been pure gold. The gold must have been mixed with another metal that was less dense. Eventually, Archimedes was able to tell the king that silver had been mixed with the original gold, and even how much gold was stolen.

Follow-up Activities

1. Make your own list of great ideas or stories that come from science or technology. You may find some ideas in library books about science history. Avoid listing things. Just list ideas and stories.
2. Recall that you can calculate the volume for a certain weight of material by using the formula volume = mass/density. Suppose that the king's crown had a mass of 1500 grams. Find what its volume would be if it were (a) pure gold, density = 19.3 g/cm^3; (b) pure silver, density = 10.5 g/cm^3; (c) 50 percent gold and 50 percent silver, density = 14.9 g/cm^3.
3. Suppose that the crown was found to have a volume of 88 cm^3. What might you conclude about its composition?

CHAPTER REVIEW

KEEPING TRACK

Section 3-1 The Metric System
1. Every measurement has a number and a unit.
2. The basic unit of length is the meter. From this are derived other units for length, area, and volume.
3. Mass is the amount of matter in an object.
4. Weight depends on mass and the pull of gravity.
5. The basic unit of mass is the kilogram.
6. You may use rulers, graduated cylinders, and balances to measure objects.

Section 3-2 Measuring Density
1. Density = mass/volume.
2. Density for regular solids, irregular solids, and liquids can be easily measured.
3. You can find the quantity of mass, volume, or density by measuring the other two quantities.

Section 3-3 Tables and Graphs
1. People use tables and graphs to collect, present, and analyze data.
2. Line, bar, and circle graphs are common.
3. A graph must include all the data, be easy to read, and be of a good size.

BUILDING VOCABULARY
Write the term from the list that best matches each description.

data, water-displacement method, gram, liter, meter, raw data, scale

1. metric unit of length
2. metric unit of volume
3. metric unit of mass
4. submerging an object in water
5. facts and measurements
6. numbers recorded during an experiment
7. numbering of lines on axes of a graph

Explain the difference between the terms in each pair.

1. millimeter, kilometer
2. cm^3, mL
3. mass, weight
4. regular object, irregular object
5. data, raw data
6. independent variable, dependent variable

SUMMARIZING
Write the missing word for each sentence.

1. In "60 kg" the "kg" part is the ___ of the measurement.
2. There are ___ mm in 1 m.
3. There are ___ cm in 1 km.
4. A cubic centimeter is a unit of ___.
5. When selecting laboratory equipment, you should measure liquid volumes in a ___ and masses on a ___.
6. If an object is taken into space, its mass will ___ and its weight will ___ .
7. In a bar graph the numbers for the horizontal axis are placed ___ .
8. In a pie graph the ___ of each slice illustrates the percentage of each item.
9. The ___ variable is usually plotted along the horizontal axis.

INTERPRETING INFORMATION

Figure 3-24 shows a milk carton that has been cut and filled with liquid. Answer the questions below.

1. What is the length, width, and height of the carton?

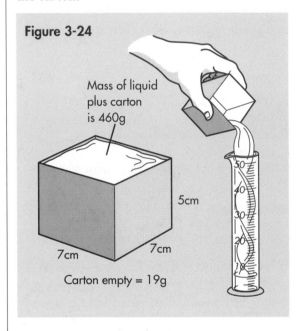

Figure 3-24

Mass of liquid plus carton is 460g

5cm

7cm   7cm

Carton empty = 19g

2. What is the volume of the carton in cm³?
3. Determine the mass of liquid in the carton.
4. If the liquid is poured into a graduated cylinder, what reading could you expect, in mL?
5. What is the density of the liquid in g/mL.

THINK AND DISCUSS

Use the section numbers in parentheses to help you find each answer. Write your answers in complete sentences.

1. By examining the metric units for lengths, figure out what these metric prefixes mean: *centi-*, *milli-*, *micro-* , and *kilo-*. (3-1)
2. Find the density of a solid metal cube of mass 750 g with sides 5 cm in length. (3-2)
3. A solid piece of plastic measures 40 cm by 40 cm by 3 cm. The plastic's density is 1.5 g/cm³. Find its volume, in cm³; its mass, in g; and its mass, in kg. (3-2)
4. Imagine putting a graduated cylinder containing 50 mL of a liquid on a balance. The reading is 100 g. Why is the density of the liquid impossible to determine from just this information? (3-2)

GOING FURTHER

Practice measuring things at home. Record the measurements of various objects, using a ruler with the metric scale. Studying your list will help you start to "think" metric.

COMPETENCY REVIEW

1. The unit of length in the metric system is the
   a. kilogram   b. second
   c. meter   d. cubic meter
2. The amount of matter in any object is its
   a. mass   b. weight
   c. volume   d. density
3. The weight of an object depends on
   a. only its mass
   b. only the local gravity
   c. its mass and its speed
   d. its mass and the local gravity
4. What is the volume of a box that is 2 cm long, 2 cm wide, and 2 cm deep?
   a. 2 cm   b. 2 cm³
   c. 6 cm³   d. 8 cm³

5. What is the volume of liquid in the cylinder shown in Figure 3-25?
   a. 2.7 mL   b. 20 mL
   c. 20.7 mL   d. 27 mL

6. What is the density of a liquid if 10 mL have a mass of 20 g ?
   a. 0.5 g/mL   b. 2 g/mL
   c. 30 g/mL   d. 200 g/mL

Figure 3-25

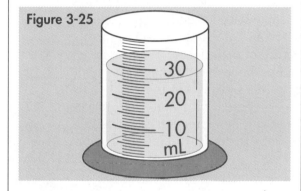

7. Figure 3-26 shows a graph received with an electric bill. How much more energy was used by this household in October compared with the average use?

Figure 3-26

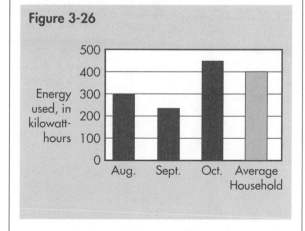

a. 50 kilowatt hours
b. 300 kilowatt hours
c. 400 kilowatt hours
d. 450 kilowatt hours

8. Which of the graphs in Figure 3-27 shows 50 percent A, 40 percent B, and 10 percent C?

9. What is the mass of 15 mL of liquid X from the graph in Figure 3-28?
   a. 10 g   b. 14 g
   c. 15 g   d. 18 g

Figure 3-27

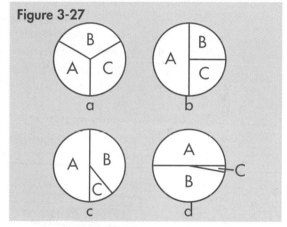

Figure 3-28

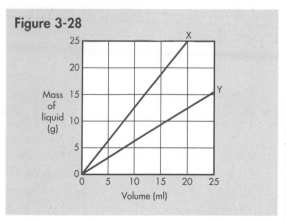

CHAPTER 4

INTRODUCTION TO CHEMISTRY

You have probably watched wood burn. The fire seemed to consume the wood, changing it into smoke, ashes, and black soot. In some mysterious way, the smoke, ashes, and soot were in the wood before it burned. How does one thing change into another? People have asked this question for thousands of years. Attempts to answer this question led to chemistry, the study of the composition and structure of materials.

Aristotle, an ancient Greek philosopher, believed all things are made from just four substances: earth, air, water, and fire. In the 2300 years since Aristotle lived, people have experimented with materials to see what they are made from and how they can be changed. For example, people have learned how to get iron from rocks and make glass from sand.

The science of chemistry has come a long way during the last 200 years. Today we know that all materials are made up of one or more of 108 substances.

4-1 Types of Matter

■ *Objectives*
□ *Define matter.*
□ *Distinguish among elements, compounds, and mixtures.*
□ *Name some common elements and compounds.*

In science, **matter** is anything that occupies space and has mass. It may be easy to see that rocks and water are matter, but air is matter too. Like rocks and water, air also

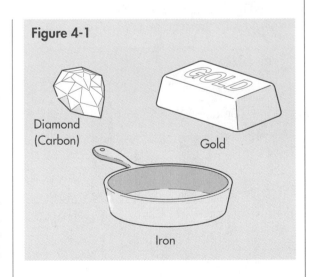

Figure 4-1

Diamond (Carbon)

Gold

Iron

takes up space and has mass. Electricity, sound, and light are not matter. They neither take up space nor have mass. All matter can be divided into three types: elements, compounds, and mixtures, as described below.

Elements

The most basic form of matter is an **element.** An element contains only one kind of matter and cannot be broken down into simpler materials by ordinary means. Elements are the building blocks for other types of matter. Scientists know of 108 elements. Fewer than half of these, however, are commonly found on Earth. Figure 4-1 shows some common items made of just one element.

Scientists use special symbols to stand for the names of the elements. Below are some common elements and symbols:

Element	Symbol
Aluminum	Al
Nickel	Ni
Carbon	C
Oxygen	O
Chlorine	Cl
Silicon	Si
Gold	Au
Silver	Ag

Compounds

Many common materials are **compounds**. A compound is made from two or more elements that are chemically combined. Water is a compound. It is made up of two elements, hydrogen and oxygen. When hydrogen and oxygen are chemically combined, they form a new substance.

Sugar is a compound that has three elements: carbon, hydrogen, and oxygen.

Elements lose their individual characteristics when they join to make a compound. For example, two harmful elements may become harmless when combined. Think of table salt. It is made from a dangerous silver-colored metal (sodium) joined with a poisonous green gas (chlorine). Yet usual amounts of table salt are not harmful.

Mixtures

On Earth, many substances are **mixtures.** The elements or compounds in a mixture are not chemically joined. They keep their individual characteristics, unlike those in a compound. Also, unlike a compound the amount of each ingredient can differ from mixture to mixture. You can recognize the individual elements or compounds in a mixture because they still have their original characteristics. The ingredients in mixtures can be rather easily separated.

Think about a box containing apples and oranges. You can mix them together in any percentages you want, yet you can still spot the individual apples and oranges. You can separate them easily. The apples and oranges are mixed, not combined.

The same is true of mixtures in chemistry. For example, air is a mixture of oxygen, nitrogen, carbon dioxide, and water. In three different air samples, a scientist could find one sample with 18 percent oxygen, one with 20 percent oxygen, and one with 22 percent oxygen.

The characteristics of each of the elements that make up air do not change. For example, pure oxygen helps things burn. Oxygen in the air helps things burn.

Interpreting Data on Graphs

Scientific data are often shown in graphs. Figure 4-2 contains three circle graphs that show the approximate percentage by weight of elements in living things, in the earth's crust, and in the universe. Study the graphs and then answer the following questions.

1. What is the most common element, by mass, in living things? In the earth's crust? In the known universe?
2. What element makes up 5% of living things but is not shown on the graphs for the earth's crust and the universe?
3. About what percent of elements in the earth's crust are metals?

Figure 4-2 Percentage of Elements by Mass

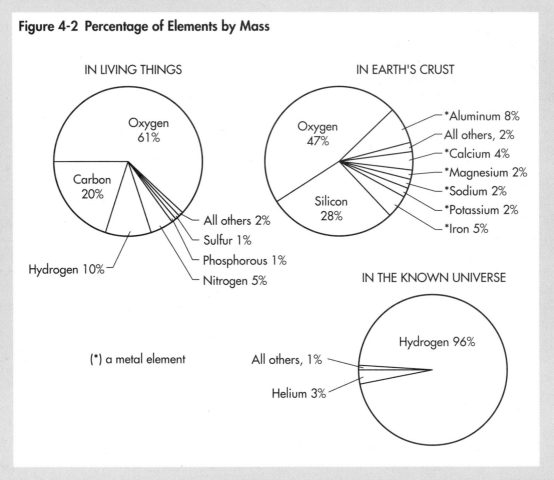

IN LIVING THINGS

Oxygen 61%
Carbon 20%
Hydrogen 10%
All others 2%
Sulfur 1%
Phosphorous 1%
Nitrogen 5%

IN EARTH'S CRUST

Oxygen 47%
Silicon 28%
*Aluminum 8%
All others, 2%
*Calcium 4%
*Magnesium 2%
*Sodium 2%
*Potassium 2%
*Iron 5%

IN THE KNOWN UNIVERSE

Hydrogen 96%
All others, 1%
Helium 3%

(*) a metal element

Sand is also a mixture. If you look closely at sand, you can see light and dark grains mixed together. Other examples of mixtures include a glass of orange juice, salad dressing, and a cup of milk.

■■ Section 4-1 Review ■■

Write the definitions for the following terms in your own words.

1. **matter**   2. **compound**
3. **element**   4. **mixture**

Answer these questions.

5. What happens to the individual characteristics of elements in a compound? In a mixture?
6. Name the elements the following symbols represent: Ag, Si, Al, Au, C, Cl, Ni, O.

4-2 Physical Properties

■ *Objectives*
☐ *Describe the three phases of matter.*
☐ *Relate the phases of matter to the arrangement of particles within the matter.*
☐ *Describe what occurs at the melting point and boiling point of a substance.*
☐ *List the standard physical properties of matter.*

Since birth you have had daily contact with water. Water is a liquid, so it feels wet. It is clear and colorless. It has no odor and no special taste. Water freezes at 0° C, and boils at 100° C. These are some of the important **physical properties** of water. A physical property can be observed and measured. Can you name three physical properties of sugar?

Particles in Matter

One way scientists study the physical properties of a substance is to look for the smallest pieces of the substance. You can try this at home. Take a piece of paper. Cut it into two pieces. Take one of the pieces and cut it into two more pieces. Keep doing this until your scissors or fingers cannot handle the small pieces. Could you continue to cut the paper forever if you had a microscope and special tools?

Ancient thinkers argued about this question. Some believed that matter could be cut into smaller and smaller pieces forever and still be the same material. Others believed that eventually they would reach a final tiny piece that could not be cut or broken further. Scientists have discovered that this second idea is true.

All elements are made of very tiny pieces of matter. Each of these pieces, which are known as particles, have all the properties of the element. Look at Figure 4-3.

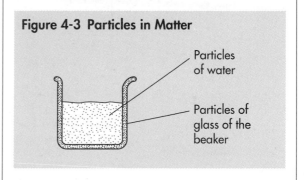

Figure 4-3 Particles in Matter

Particles of water

Particles of glass of the beaker

The particles in the glass beaker are different sizes and different distances apart than those in the water. Particles help to explain many of the physical properties of matter.

The Phases of Matter

Matter exists in one of three **phases,** or states. A material's phase depends on the distance between particles and on how fast the particles move. Look at Figure 4-4 and compare the particles in solids, liquids, and gases.

A material with a definite volume and a definite shape is a **solid**. For example, a mountain and a pebble are solids because the volume and shape of both, for the most part, do not change. The particles in a solid are close together. They are locked in position, so they move very little.

A **liquid** is a material that has a definite volume but no definite shape. Liquids take the shape of their containers. Pour a cup of water into a glass, then into a pot, and then into a jar. Notice that the volume of the water remains the same, but the water fills each container to a different level. Liquids take the shape of their containers because their particles are freer to move about and move more rapidly than particles in a solid.

A material with no definite volume or shape is a **gas.** If you put gas into a container, it spreads throughout the volume and shape of the container. Odors are gases. If you spill perfume, its odor will spread throughout the room. The volume and shape of the room is the volume and shape of the gas. This is because the particles in a gas are far apart, have complete freedom to move, and move with much greater speed than particles in a liquid or solid.

Melting and Freezing   Most solids have a **melting point**, the temperature at which they turn into a liquid. Most liquids have a **freezing point**, the temperature at which

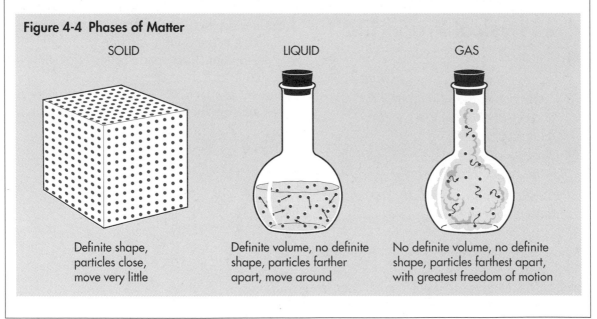

Figure 4-4 Phases of Matter

SOLID

Definite shape, particles close, move very little

LIQUID

Definite volume, no definite shape, particles farther apart, move around

GAS

No definite volume, no definite shape, particles farthest apart, with greatest freedom of motion

they turn into a solid. The melting point and freezing point of a substance are identical. For example, the melting point and freezing point of water are both 0° C. At this temperature, ice, which is water in its solid form, will melt, or become liquid. At this same temperature, liquid water will freeze, or become solid.

A solid melts when heat is applied, as when you leave an ice cube in the sunshine. Heat gives the particles in the solid more energy, so they can move about with greater speed. When this happens, the solid becomes a liquid.

A liquid freezes when heat is removed, as when you put an ice tray in a freezer. Removing heat reduces the energy available to the particles in the liquid. The particles come together and move with less speed. When this happens, the liquid becomes a solid.

You can see how particles move in melting and freezing by looking at Figure 4-5. Notice the role heat plays. Heat is a form of energy. Whenever heat enters or leaves, it affects all of the particles in a substance.

Boiling and Condensing Every liquid has a **boiling point**, the temperature at which a liquid turns into a gas. Every gas has a **condensation point**, the temperature at which a gas turns into a liquid. The boiling point and condensation point of a substance are the same. The boiling point and condensation point of water are both 100° C. At this temperature, water will boil and become a gas. Gaseous water, or steam, will condense and become a liquid at 100° C.

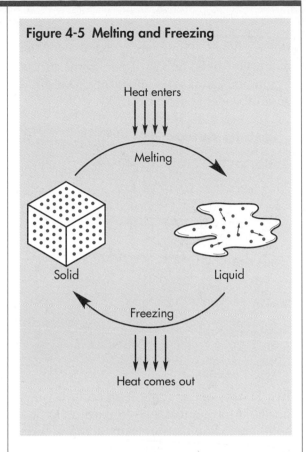

Figure 4-5 Melting and Freezing

Heat enters

Melting

Solid

Liquid

Freezing

Heat comes out

A liquid boils and changes to gas when heat is added. For example, when you put a pot of soup over a burner, heat gives the liquid particles in the soup more energy. The particles move farther apart and with greater speed. The soup steams, meaning some of the liquid becomes a gas.

A gas condenses when enough heat is removed. For example, steam from water boiling on a stove forms beads of liquid on cold kitchen windows. Removing heat leaves the particles in the steam with less energy, so they come closer together and move with less speed. The steam turns into liquid water, forming droplets.

Study Figure 4-6. Notice how heat affects the particles. Think about what happens when you boil soup in a pot. Some of the liquid boils and turns into steam. Does that steam condense later? Why?

Figure 4-6 Boiling and Condensing

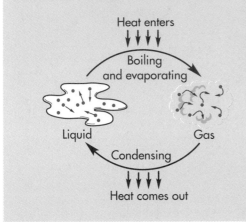

Two Phases at the Same Temperature It is possible for a substance to exist as both a solid and a liquid at the same temperature, at the melting/freezing point. This occurs in a glass of water that contains ice cubes. See Figure 4-7. If you measured the temperatures of both the ice cubes and the water, both will measure 0° C.

A substance can exist as both a liquid and a gas at the same temperature, at the boiling/condensation point. An example occurs inside a covered kettle with boiling water. See Figure 4-7 again. The temperature of both the water and the steam will measure 100° C.

Standard Physical Properties

The standard physical properties of a material include its phase, or state, its

Figure 4-7 Two Phases at One Temperature

melting point, its boiling point, and its density. Scientists use standard physical properties to describe materials. Color, odor, taste, hardness, and texture are also standard descriptions.

Scientists describe complicated physical properties as well. One example is a material's viscosity, which is how fast a liquid can flow. Another example is malleability, which is a material's ability to be hammered into shape. Metals such as copper and gold have great malleability.

■■■ Section 4-2 Review ■■■

Write the definitions for the following terms in your own words.

1. **solid**   2. **boiling point**
3. **liquid**   4. **condensation point**
5. **gas**

Answer these questions.

6. List the *obvious* physical properties of any two materials (not mixtures).

Consider salt, sugar, sand, olive oil, ammonia water, or others.

7. Describe the shape and volume of a solid, of a liquid, and of a gas.
8. What happens to the particles in matter during melting? During boiling?

4-3 Physical Changes

■ *Objectives*

☐ *Distinguish between a chemical change and a physical change.*

☐ *List various types of physical changes.*

☐ *Contrast a solution and a suspension.*

☐ *Describe the energy changes that occur during changes of phase.*

If you help with cooking at home, you may realize that in preparing a meal, you cause changes in food. For example, heating food causes it to change. Later, as you eat and digest your meal, the food changes even more. One amazing change occurs when the food eventually becomes part of you. Food, like all matter, can undergo two types of changes.

Two Types of Changes

Whenever a substance undergoes a change so that one or more new substances with different properties are formed, a **chemical change** has taken place. For example, when you bake a cake, a chemical change takes place. Flour, sugar, milk, and other ingredients combine in the heat of the oven to make a new substance, cake.

In a **physical change,** the form of a substance changes, but the materials that make up the substance do not change. No new materials are created.

Preparing a cup of tea illustrates several physical changes. First, you heat water until it boils. This is a physical change because even though the water has changed to steam, no new materials have been created. Next, you pour the hot water over the tea bag to remove the tea-flavoring chemicals from the tea leaves. This also is a physical change because the tea-flavoring chemicals were not created, they were only moved from the tea leaves to the water. Finally, you might mix lemon juice with the tea. The juice mixes with the water, but no new materials are created.

Making Changes

Most of the changes you make to substances are physical. There are many ways to make physical changes.

Tearing and Breaking Tearing and breaking are physical changes. For instance, tearing a sheet of paper into strips does not change the paper's composition. The strips have the same properties as the original paper. Making powder out of rock is also a physical change. Breaking the rock merely changes its appearance. Figure 4-8

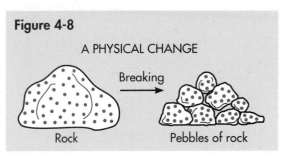

Figure 4-8

A PHYSICAL CHANGE

Breaking

Rock   Pebbles of rock

Breaking does not change the composition of the material.

shows particles in a rock and particles in pebbles broken from the same rock.

Mixing and Stirring When you mix and stir two materials together, you cause a physical change. You make a mixture, yet the original materials still exist and have their own properties. You can still recognize these individual properties. Mixing simply means bringing two or more materials together in very close contact. Figure 4-9 shows particles of two materials being mixed. They have been brought into close contact, but they have not changed.

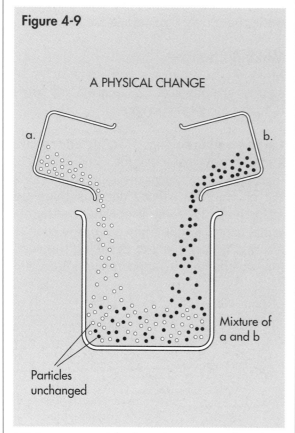

Figure 4-9

A PHYSICAL CHANGE

a.

b.

Mixture of a and b

Particles unchanged

Mixing puts particles into close contact but does not change their composition.

You can separate the materials in a mixture by physical means. Sometimes this may be easy to do, but sometimes it can be very difficult. For example, look at Figure 4-10. You can easily separate a mixture of sulfur powder and iron filings by using a magnet. The iron is metal, which is attracted to the magnet. The iron filings cling to the magnet, leaving the sulfur powder behind.

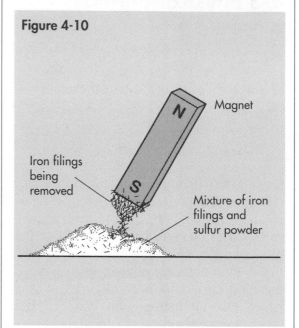

Figure 4-10

N Magnet

Iron filings being removed

S

Mixture of iron filings and sulfur powder

A mixture can be separated by physical means.

On the other hand, you would have difficulty separating a mixture of sugar and salt.

Dissolving When you **dissolve** a material, you mix it with water or some other liquid. The material seems to disappear, but it is still there. Particles of the material have simply been spread out between the particles of the water, as you can see in Figure 4-11.

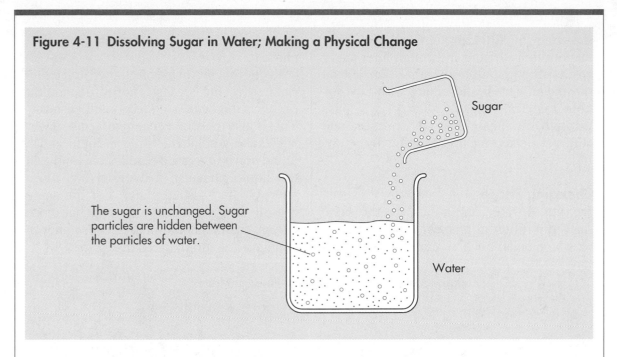

Figure 4-11 Dissolving Sugar in Water; Making a Physical Change

Sugar

The sugar is unchanged. Sugar particles are hidden between the particles of water.

Water

The mixture that is made when a material dissolves in water is called a **solution**. A cup of tea is a solution made of tea flavorings, water, and anything else you may add, like sugar or lemon.

Dissolving is a physical change. A dissolved material keeps its composition. Even when a material becomes invisible in a solution, you can separate it from the solution. One way to do this is to let the liquid evaporate. A liquid evaporates when it is left for a long time. During that time it gradually changes to a gas. To try this out, mix a little salt in a half glass of warm water. Pour some of the solution onto a clean plate. After several hours, the water will evaporate. You will see the salt left behind on the plate.

Materials that dissolve in water are called **soluble**. Materials that do not dissolve in water are **insoluble**. Sugar is soluble, and sand is insoluble. You can recover both soluble and insoluble solids from a solution by evaporating the liquid.

Filtering and Decanting Even though sand is insoluble, you could stir it vigorously in water to form a mixture. The sand does not dissolve and disappear into the water, but swirls throughout. This type of mixture is called a **suspension**, which means the insoluble material is suspended in, or held up by, the water. Usually, a suspension looks cloudy, but a solution looks clear. In a suspension, the insoluble materials often settle at the bottom of the container. In a solution, the soluble materials never settle.

Once the insoluble materials in a suspension have settled, you can try to pour out the liquid without disturbing the solids at the bottom. This process is called **decanting**. You can also remove insoluble materials by **filtering.** One way to filter something is to pour it through filter paper so that the insoluble materials are left on the paper. The liquid, along with any soluble materials dissolved in it, will pass through the paper. You can compare and contrast decanting and filtering by studying Figure 4-12.

Changing Phase

Ice, liquid water, and steam are the same matter in different phases. When ice melts or water boils, no new material results. A change of phase, therefore, is a physical change.

Recall that in a physical change, the material may receive or lose energy, depending on whether heat is added or taken away. This is also true for a change of phase. Heat will make a solid melt into a liquid and a liquid boil into a gas. Because heat is added, the liquid phase of a material has more energy than its solid phase. For the same reason, the gas phase of a material has more energy than its liquid phase. These energy changes are summarized in Figure 4-13.

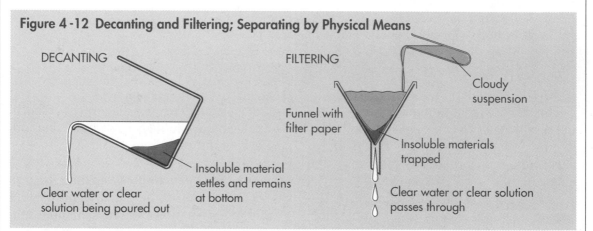

Figure 4-12 Decanting and Filtering; Separating by Physical Means

DECANTING

FILTERING

Funnel with filter paper

Cloudy suspension

Insoluble materials trapped

Clear water or clear solution being poured out

Insoluble material settles and remains at bottom

Clear water or clear solution passes through

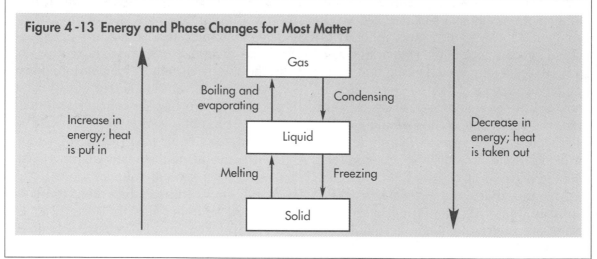

Figure 4-13 Energy and Phase Changes for Most Matter

Gas

Boiling and evaporating

Condensing

Increase in energy; heat is put in

Liquid

Decrease in energy; heat is taken out

Melting

Freezing

Solid

How Is a Suspension Different from a Solution?

Process Skills observing; using laboratory equipment.

Materials sugar, starch, filter paper, beakers, funnels, an evaporating dish or a watch glass; test tubes and a holder, stirring rod.

Procedures

1. Fold filter paper as shown in Figure 4-14 and insert it into a funnel.
2. Mix about a teaspoon of sugar in a beaker half full of water. Is the mixture clear? Has the sugar dissolved? Set the mixture aside.
3. Mix about a teaspoon of starch in a beaker half full of water. Is the mixture clear? Has the starch dissolved? Discard the mixture.
4. Add about one heaping teaspoon of starch to the sugar solution from step 2, and stir for several minutes. Pour the mixture through the filter paper. Collect the solution that comes through the funnel in a clean beaker.
5. Examine the filter paper. What do you find?
6. Place a small amount of the solution that came through the filter into a watch glass. Set the watch glass on top of a radiator or in the sunlight so that the water can evaporate slowly. What is left in the glass?

Conclusions

7. Which material was on the filter paper? How do you know?
8. Can you use filter paper to separate all mixtures? Why or why not?

Figure 4-14

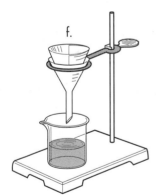

Section 4-3 Review

Write the definitions for the following terms in your own words.

1. **chemical change**
2. **physical change**
3. **solution**
4. **soluble**
5. **suspension**

Answer these questions.

6. What is the key difference between a chemical change and a physical change? For example, shredding a piece of paper is a physical change, but burning it is a chemical change. Why?
7. Copy and complete the table in Figure 4-15 on phase changes. Fill in the blanks, based on the example.

Figure 4-15

Phase change	What occurs?	Energy change
Melting	Solid becomes liquid.	Energy added
Freezing		
Boiling		
Condensing		

8. Copy and complete the table in Figure 4-16 to compare a solution with a suspension.

Figure 4-16

Characteristic	A solution	A suspension
Is it a mixture?		
Clear or cloudy?		
Will it settle?		
Can filtering remove material?		
An example		

4-4 Chemical Properties and Changes

■ *Objectives*
☐ *Give examples of chemical properties.*
☐ *Describe what occurs during chemical change.*
☐ *State the law of definite composition.*
☐ *Describe the energy changes in chemical reactions.*

Do you wipe away water from objects made of iron and steel, such as your bicycle? If you do not, there is a good chance rust will form on your bicycle. Wet iron and steel combine with the oxygen in the air and create a new substance, rust. Iron and steel rust, but gold and silver do not. Gold and silver do not combine with the oxygen in the air.

The ability of iron to combine with oxygen is one of its chemical properties. The action of iron combining with oxygen is a chemical change. This section describes chemical properties and chemical changes.

Chemical Properties

The ability of a substance to go through chemical changes by combining with other materials is known as a **chemical property**.

In order to learn about the chemical properties of a material you need to ask questions. Does the material burn? How much heat is released if the material does burn? How can the material be broken down into its elements? What happens if the material is heated? How quickly does it react with other chemicals? When you answer such questions, you are describing a material's chemical properties.

Chemical Changes

Recall that a chemical change produces one or more new substances, each with its own properties. The original material or materials no longer exist. In the example of rusting iron, the rust spots are neither iron nor oxygen. Before the chemical change, the iron was a gray magnetic metal. The oxygen was a colorless gas and was not magnetic. After the chemical change, the rust is a red powder. It is not magnetic. Thus the new material has a different color, texture, and magnetic property than its ingredients.

Careful measurements show that when rust forms, the mass of iron that combines is always $2\frac{1}{3}$ times the mass of oxygen. For example, exactly 112 g of iron will combine with 48 g of oxygen; 112 g is $2\frac{1}{3}$ times 48 g. You could also say that the compound called rust is formed from 70 percent iron and 30 percent oxygen.

Chemicals react according to definite combining ratios. This is the **law of definite composition**, also known as the law of definite ratios. For example, if you put eight grams of hydrogen together with eight grams of oxygen, only one gram of hydrogen can combine with the oxygen to make water. The other seven grams of hydrogen are left over.

Chemical Reactions and Energy

Scientists call a chemical change a **chemical reaction**. During a chemical reaction, energy is either given out or taken in, usually in the form of heat. Sometimes, energy in the form of light or electricity can be given out or taken in as well.

Many common chemical reactions, in nature and in the laboratory, release energy. A reaction that releases energy is called **exothermic**. For instance, heat energy and light energy are released when wood burns. Light energy is released by the chemical reactions in a firefly. Electrical energy is released as the chemicals inside a battery react.

Other chemical reactions absorb energy and are called **endothermic.** For example, heat energy is absorbed in the production of plastics. Plants absorb light energy during photosynthesis. Electrical energy is absorbed when water is broken apart into its elements.

Many chemical reactions require a certain amount of **activation energy** before the reaction can begin. Coal, wood, and oil release a large amount of heat once they start to burn. Before a chemical reaction can begin though, coal, wood, and oil must be heated to the proper temperature. You must, therefore, light them with a flame, perhaps from a match, a pilot light, or some burning paper and sticks. This flame adds activation energy that starts the reaction, as shown in Figure 4-17.

The Rate of Reaction A small piece of iron may take years to completely combine with oxygen and completely change to rust. A similar amount of magnesium metal may require just a week to combine with the oxygen in the air, and the same amount of sodium metal may require only one minute. One chemical property of a metal is found by determining how active the metal is when combining with oxygen. The most active materials react quickly. Less-active materials react more slowly. Inactive materials may not react at all.

When scientists are making a material by means of a chemical reaction, they measure the **rate of reaction**. The rate of reaction is the amount of material produced in a given amount of time. For example, a rate of reaction might be one kilogram of ammonia formed per minute.

Changing the Rate of Reaction In the laboratory, you might want to increase the rate of reaction in order to speed up an experiment. You can increase the rate of a reaction by using several methods.

You could try increasing the temperature of the reaction. An approximate rule

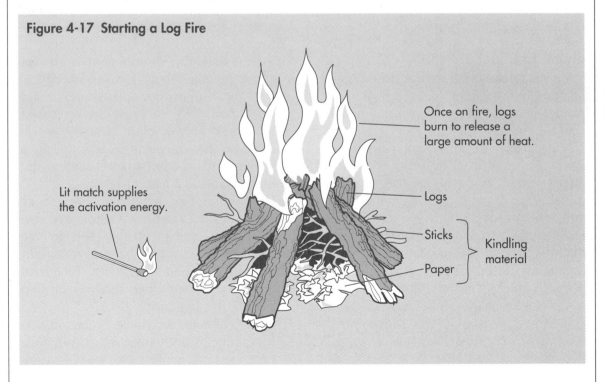

Figure 4-17 Starting a Log Fire

Lit match supplies the activation energy.

Once on fire, logs burn to release a large amount of heat.

Logs

Sticks

Paper

Kindling material

of thumb is that the speed of a reaction doubles for every 10° C rise in temperature. The water inside a pressure cooker is not boiling at 100° C but at around 110° C. This increases the speed of chemical reactions in the food so that the cooking time is decreased.

You could try breaking the material into smaller, finer pieces. A chunk of metal may take several minutes to react with an acid. However, if you grind it into a powder, it may react in only a few seconds. When you break something into smaller pieces, you do not change its mass. You do, however, increase the surface area that comes into contact with other chemicals. The rate of reaction increases because more area can react at one time. How does Figure 4-18 demonstrate this?

Each year, news reports from farming areas tell of explosions in grain storage tanks caused by grain dust and sparks. You can burn dry kernels of corn, but only slowly. However, if you grind the corn into powder and toss the powder in the air near a flame, it burns so rapidly it explodes.

Another way to increase the rate of reaction is to increase the concentration of the materials. Concentrated solutions have more chemicals dissolved in a given volume of water than do diluted solutions, so they react faster. For example, you can use a diluted solution of ammonia in water to remove grease and dirt around your home. In industry, though, people use a more concentrated, and therefore more powerful, solution to remove heavy grease and ink.

You can increase the pressure of gaseous

Figure 4-18

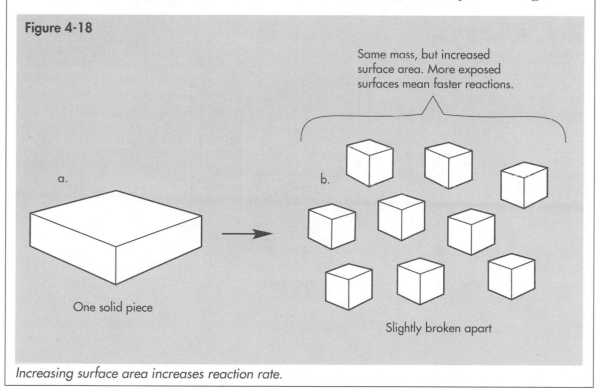

Same mass, but increased surface area. More exposed surfaces mean faster reactions.

a.

One solid piece

b.

Slightly broken apart

Increasing surface area increases reaction rate.

materials. Gases under high pressure react at a faster rate than gases under low pressure. This is because at high pressure, the particles of the gas squeeze closer together and can meet and combine in less time.

This rule is at work inside the cylinder and piston of a car engine. Study Figure 4-19. A spark ignites the gasoline vapor and air exactly when a piston has squeezed them together as much as possible. At that moment the pressure is the greatest, so a very fast reaction occurs in the cylinder. If the spark ignites before these gases are squeezed to the smallest volume, then the reaction is not as fast and the engine does not run as well. A car mechanic would say the engine's timing is poor.

You can also use a **catalyst** to speed up a reaction. A catalyst is a substance that can increase the rate of reaction without any change to its own composition. In other words, a catalyst helps a reaction to occur, but the catalyst itself is not changed by the reaction.

You have seen bubbles of carbon dioxide gas coming out of soda or seltzer water. You can produce more bubbles per second if you add some salt to the soda. The salt speeds up the chemical reaction, but it remains as salt, acting as a catalyst.

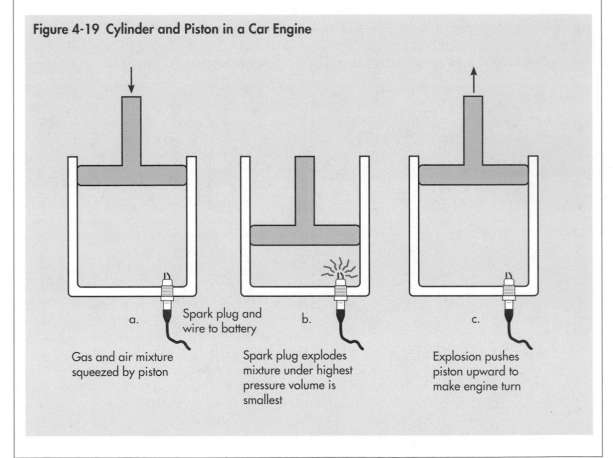

Figure 4-19 Cylinder and Piston in a Car Engine

a. Gas and air mixture squeezed by piston

Spark plug and wire to battery

b. Spark plug explodes mixture under highest pressure volume is smallest

c. Explosion pushes piston upward to make engine turn

How Does Baking Powder Work?

Process Skills *designing experiments; observing*

Materials baking powder (*not* baking soda) and vinegar; thermometer; three drinking glasses and teaspoons

Procedures

1. Fill three glasses about ¼ full with water. Use the same types of glasses and the same amounts of water. The first glass should contain cold water; the second should contain water at room temperature; and the third should contain hot water. Measure and record the temperature of each.

2. You and some helpers should then place one teaspoon of baking powder into each glass. Keep stirring the solutions with your spoons. Note which stops bubbling first, second, and last. If you perform this experiment alone, stir baking powder into the first glass and wait for the bubbling to stop. Record the time this took and then move on to the next glass.

3. Repeat the above experiment, but add four tablespoons of vinegar to each glass before putting in the one teaspoon of baking powder. Note the reaction.

Conclusions

4. What evidence indicates that you observed chemical reactions?

5. What happens to the rate of reaction when you increase the temperature or when you add vinegar, which is an acid, to the water?

6. Recipes for cake and bread call for baking powder. Without it, the dough will not rise, that is, it will not get bigger. Why does baking powder make the dough get bigger during baking?

▪▪▪ Section 4-4 Review ▪▪▪

Write the definitions for the following terms in your own words.

1. **law of definite composition**
2. **chemical reaction**
3. **endothermic**
4. **activation energy**
5. **rate of reaction**

Answer these questions.

6. In what ways is energy involved with chemical reactions? Try to explain at least three ways.
7. How can the rate of a chemical reaction be increased?

SCIENCE, TECHNOLOGY, & SOCIETY

Chemicals in Our Environment

Chemical technology provides new materials and faster production of industrial chemicals. Materials like plastic, nylon, and paints are products of chemical technology.

Chemical products have allowed people to live healthier and longer lives, to be more productive, and to enjoy more comforts than ever before.

Yet there is a dark side to all the benefits. Many chemicals have bad long-term consequences. One example is DDT, which for years was sprayed on plants to kill bugs. Using DDT improved the amount and quality of food produced by the plants. Eventually, researchers learned that DDT can cause cancer and other illnesses. It was also determined that DDT damages plants and animals. For these reasons, the United States Department of Agriculture banned DDT. Careful watch is needed over hundreds of other such chemicals still in use.

Another example is automobiles. Not only is the automobile beneficial to society, it has become a necessity for many people. Just like DDT, however, automobiles cause problems. Cars produce gases that go into the air. Eventually they dissolve into rain, which washes them from the air. The solution that results is known as acid rain. Acid rain harms lakes, forests, and farmland, and therefore it harms the fish, animals, and food you eat.

Figure 4-20

Governments and international organizations are considering ways to control the kinds and amounts of waste gases that are put into the air. This could reduce the amount of acid rain.

People now know that every technology has certain unintended effects on the environment. We must consider these potential hazards and constantly watch when a new technology is introduced.

Follow-up Activity

Prepare for a discussion or a debate on whether life would be better with or without modern chemicals. Or, consider the choices people and government have for improving the handling of toxic chemicals today.

Section 4-1 Types of Matter

1. Matter is anything that occupies space and has mass.
2. All objects are composed of one or more elements.
3. Fewer than half of the elements are common on Earth.
4. Compounds are made from elements united in definite percentages. The properties of compounds differ from the properties of the elements in them.

Section 4-2 Physical Properties

1. All substances have their own physical properties.
2. All matter consists of particles.
3. The phase in which a material exists depends on how close together its particles are and how fast they move.
4. A solid has a definite volume and a definite shape. In a solid, the particles are close together and resist changing their positions.
5. A liquid has a definite volume but no definite shape. Liquids take the shape of their containers. The particles in a liquid are farther apart and change position more readily than in a solid.
6. A gas has no definite volume and no definite shape. The particles in a gas are farther apart and change position more readily than in a solid or a liquid.
7. Most solids turn into liquids at a temperature called the melting point. Most liquids turn into solids at a temperature called the freezing point. The melting and freezing points of a material are equal.
8. Liquids turn into gases at a temperature called the boiling point. Gases turn into liquids at a temperature called the condensation point. The boiling and condensation points of a material are equal.

Section 4-3 Physical Changes

1. Matter can change chemically and physically.
2. Chemical change results in the formation of new substances with their own properties. The properties of the original material can no longer be detected.
3. Physical change alone does not produce any new substances. Breaking, mixing, dissolving, filtering, heating, and changing phase are physical changes.
4. Heat is absorbed when a solid becomes a liquid and when a liquid becomes a gas. Heat is released when a liquid becomes a solid and when a gas becomes a liquid.

Section 4-4 Chemical Properties and Changes

1. Chemical properties describe how a substance reacts with other substances.
2. Some chemical changes must be started by an amount of energy called activation energy. Other chemical changes can start without adding energy.
3. Exothermic changes release energy as heat, light, or electricity; endothermic changes absorb energy.
4. The rate of a chemical reaction usually increases as the temperature increases and when the material is broken into smaller pieces.

Write the term that best completes each sentence.

mixture, elements, freezing point, exothermic, compound, dissolve

___1___ are the building blocks from which all other matter is made. Pure water is a ___2___, but salty water is a ___3___. The temperature of 0° C is the melting point and the ___4___ of water. Sugar can ___5___ in water. A chemical reaction that releases heat is ___6___.

Write *true* if the statement is true. If the statement is false, change the *italicized* term to make the statement true.

1. *Energy* has mass and takes up space.
2. The materials in a *mixture* keep their own properties.
3. A *gas* will occupy the whole volume of the container in which it is put.
4. The *melting point* of water is 100° C.
5. Melting is a *physical* change.
6. Dissolving is a *chemical* change.

Look at Figure 4-21.

Caution Carbon disulfide is a flammable toxic substance. Do not attempt to do any experiments with carbon disulfide.

If iron shavings are mixed with sulfur powder, they can be separated by using a magnet or dissolving the sulfur in carbon disulfide liquid. When the mixture is heated, it begins to glow on its own for several minutes, even with the flame removed. The material now in the test tube is not magnetic and does not dissolve in carbon disulfide liquid.

1. Before the mixture of iron and sulfur is heated, what two ways could you separate the sulfur and iron? Describe what could be done using the magnet or the carbon disulfide liquid.

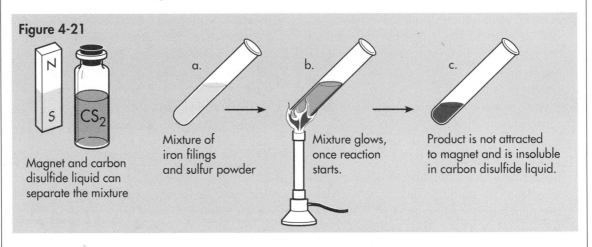

Figure 4-21

Magnet and carbon disulfide liquid can separate the mixture

a. Mixture of iron filings and sulfur powder

b. Mixture glows, once reaction starts.

c. Product is not attracted to magnet and is insoluble in carbon disulfide liquid.

2. What evidence is there that the heat produces a chemical change?

THINK AND DISCUSS

Use the section number in parentheses to help you find each answer. Write your answers in complete sentences.

1. Is heat matter? Explain. (4-1)
2. What is the difference between a physical change and a chemical change? Give an example of each. (4-2, 4-4)
3. Copper is the metal used for most electric wires and for many cooking pots. What physical properties must copper have to serve these purposes? (4-2)
4. Tell what takes place
   a. in endothermic reactions. (4-4)
   b. when saltwater evaporates. (4-3)

GOING FURTHER

Prepare a display of different kinds of matter. Include elements, compounds, and mixtures. For a mixture you might prepare some muddy water.

Caution Do not collect caustic substances or other substances that are dangerous.

COMPETENCY REVIEW

1. Every material must contain
   a. only one element.
   b. at least one element.
   c. at least two elements.
   d. only one compound.

2. How many elements occur commonly on Earth?
   a. millions   b. thousands
   c. 108   d. fewer than 54
3. Which is true of a liquid?
   a. Particles are fixed in place.
   b. Particles have more speed than in a gas.
   c. It has a definite volume.
   d. It has a definite shape.
4. Which is not a physical property?
   a. density   b. color
   c. hardness   d. the ability to burn
5. The main characteristic of chemical change is that
   a. heat is released.
   b. heat is absorbed.
   c. a new material is made.
   d. a new phase results.
6. Which is a chemical change?
   a. making a compound out of two elements
   b. boiling
   c. dissolving
   d. breaking into finer pieces
7. A material boils at 60° C. Which of the following is true about this material?
   a. It will melt at 30° C.
   b. It will freeze at 60° C.
   c. It will condense at 60° C.
   d. It will melt at 60° C.
8. A spark plug in a car engine makes the gasoline and oxygen burn because it
   a. provides energy to start the reaction.
   b. provides endothermic energy.
   c. provides exothermic energy.
   d. reacts with gasoline.

CHAPTER 5

INTRODUCTION TO ATOMIC STRUCTURE

Have you ever been bothered by static electricity? You walk across a carpet, reach for the door knob, and ZAP! You get an annoying spark of electricity at your fingertips. Or, you comb your hair, only to see, hear, and feel the static make your hair fluff out. Perhaps some of your clothes tend to pick up lint and cling to other clothing, especially after being in a drying machine. You may have noticed that static electricity is most bothersome during the winter months, when the air indoors is dry.

The sparks of static electricity reveal that particles of the carpet, your fingertips, your comb, and your hair contain electrical parts. This chapter is about the particles of matter. What are they like? How do they behave during chemical changes? When does old matter disappear and new matter appear? Scientists often ask such questions. Happily, they have found many interesting answers.

5-1 Atoms and Molecules

■ *Objectives*

- ☐ *Define an atom and a molecule.*
- ☐ *Compare atoms, using the atomic mass unit.*
- ☐ *State the facts that must be explained in a model of an atom.*

You have learned that everything is made from elements, which can join to make millions of compounds. You also know that all matter, whether elements or compounds, is

Figure 5-1

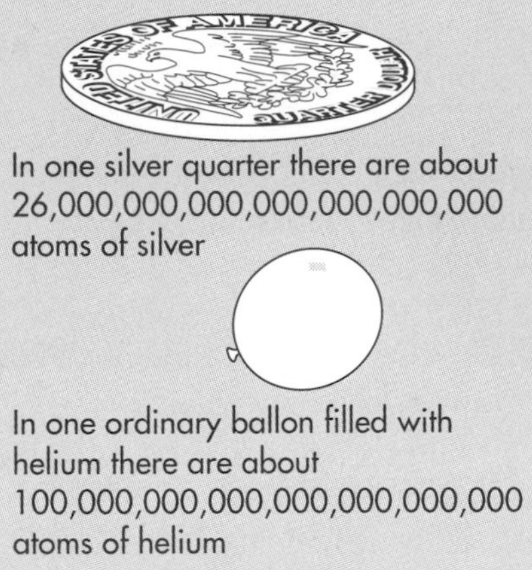

UNITED STATES OF AMERICA
QUARTER DOLLAR
In one silver quarter there are about
26,000,000,000,000,000,000,000
atoms of silver
In one ordinary ballon filled with
helium there are about
100,000,000,000,000,000,000,000
atoms of helium

Elements contain atoms.

built from tiny particles. Somehow, the particles in elements join together to make the particles in compounds.

The smallest possible particle of an element that still has the chemical properties of the element is an **atom.** Atoms are very tiny. There are, therefore, enormous numbers of atoms, as you can see in Figure 5-1.

Atomic Mass Units

Atoms are small in mass as well as in size. There is no laboratory balance delicate enough to show the mass of a single atom. In fact, a trillion atoms together have a mass of much less than one gram. It is inconvenient, therefore, to measure the masses of atoms in grams. That is why scientists express the masses of atoms in **atomic mass units (u).** One atomic mass unit is defined as $1/12$ the mass of a carbon atom. Carbon is used as the standard, and one atom of carbon has a mass of exactly 12 u. An atomic mass unit is 0.00000000000000000000001661 grams. This number in scientific notation is

$$1 \text{ u} = 1.661 \times 10^{-24} \text{ grams}$$

The table in Figure 5-2 shows the masses of various atoms in atomic mass units.

Figure 5-2 Mass of One Atom

Element	Mass of one atom in atomic mass units
Hydrogen	1 u
Helium	4 u
Carbon	12 u
Oxygen	16 u
Sulfur	32 u
Iodine	127 u
Uranium	238 u

Particles in Compounds

The smallest particle of a compound that still has the chemical properties of the compound is called a **molecule.** A molecule is made of atoms that are linked together, just as some toys have parts that can be joined together with sticks.

Figure 5-3 shows the compound called water. Water has two elements, oxygen and hydrogen. One cup of water contains about 8.4×10^{24} molecules of water. 8.4×10^{24} is the short way of writing

Figure 5-3

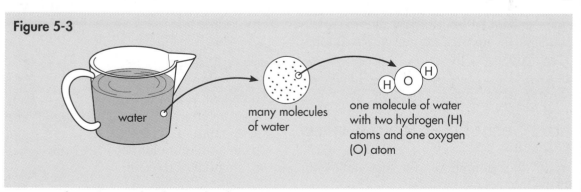

water · many molecules of water · one molecule of water with two hydrogen (H) atoms and one oxygen (O) atom

Compounds contain molecules made of atoms.

How Can You Show That Molecules Move?

Process Skills *observing; developing a model*

Materials cotton; alcohol or ammonia solution; potassium permanganate (KMnO₄); beakers; insoluble powder such as dust, pollen, lycopodium, or finely ground sulfur; microscope; microscope slides; dropper

Procedures

1. In a well-ventilated area, wet a piece of cotton with alcohol or ammonia solution. After five minutes, observe what happens. Has the odor spread? Write down your observations.

2. Fill a beaker with hot water. Drop a crystal of potassium permanganate into the beaker. Watch the beaker for three minutes and observe how a purple color spreads through the water. Write down your observations.

3. Stir a tiny pinch of your insoluble powder into a few spoonfuls of water. Recall that you are making a suspension. Using a dropper, place a drop of this suspension on a microscope slide and place the slide on the stage of a microscope. Following your teacher's instructions, look at some of the powder grains under the microscope. Do the powder grains make zig-zag motions? Do they seem to be bumped around continuously? Write down your observations. Wait for the swirling of the water to stop. Look at the slide again. Write down your observations.

Conclusions
Using diagrams of molecules, make a model that explains your observations.

8,400,000,000,000,000,000,000,000. Look again at Figure 5-3. Each water molecule contains one atom of oxygen and two atoms of hydrogen, so the cup of water contains about 8.4×10^{24} oxygen atoms. It contains twice as many hydrogen atoms!

■■■ Section 5-1 Review ■■■

Write the definitions for the following terms in your own words.

1. **atom**   2. **atomic mass unit (u)**
3. **molecule**

Answer these questions.

4. A copper atom has a mass of 64 u. How does a copper atom compare in mass to
   a. an atom of hydrogen?
   b. an atom of helium?
   c. an atom of oxygen?

5-2 Uncovering the Atom's Structure

■ *Objectives*

☐ *Distinguish among protons, neutrons, and electrons.*

☐ *Describe Bohr's model of the atom.*

☐ *Define atomic number, mass number, and isotopes.*

Before 1910, chemists knew that atoms had to exist, even though atoms could not be seen. The idea of atoms helped explain the physical and chemical properties of the elements. Then, between 1910 and 1930, scientists made great strides in understanding the atom. The model of an atom that they developed during those years has remained basically unchanged, although scientists continue to make improvements.

Protons and Electrons

During the 1800s scientists discovered that objects can be neutral or can have an electric charge. For example, the sparks you see when you rub blankets together show electric charges. They also discovered that the electric charge could be of two types, which they called positive and negative. Two objects that have the same charge repel one another. Objects with unlike charges attract one another. Objects that do not have a positive or negative charge are said to be neutral. Neutral objects do not attract or repel other objects.

The years 1909 and 1910 were important for the study of atoms. Before 1909, scientists noticed that a few elements gave off radiation. Radiation is a stream of invisible particles that seems to shoot out of the material. To find out more about atoms, the British physicist Ernest Rutherford performed a series of experiments in 1909 and 1910. In the experiments, he aimed radiation at very thin sheets, or foils, of gold. The radiation he used was made up of tiny, positively charged particles.

Rutherford used a detector to discover what happened to the particles when they hit the gold foil. Amazingly, he found that over 99 percent of the radiation passed straight through the gold. A few particles were deflected, or turned aside, and even fewer were reflected, or turned back.

From this experiment, Rutherford stated astounding conclusions about atoms.

First, atoms are mostly made up of empty space. Rutherford based this conclusion on the fact that most of the radiation particles passed right through the gold foils, as if the gold atoms contained almost nothing. Second, atoms have very small, very hard centers. Rutherford based this conclusion on the small number of particles that were reflected. These particles must have approached the hard center of a gold atom and been bounced backward, like a ball hitting a wall. Rutherford said this result was like firing a cannon at tissue paper and seeing the cannonball bounce back.

The hard center of the atom is now called the **nucleus**. The nucleus contains practically all of the atom's mass, although it is small compared to the whole atom.

Rutherford's third conclusion was that the atom's nucleus must contain positive particles. He based this conclusion on the particles that were deflected. Rutherford guessed that the nucleus of each gold atom contained positive particles. To Rutherford, the deflections resulted from positive radiation particles coming close to positive particles in the nucleus of the atom. The positive radiation particles were pushed away by the positive particles in the nucleus. These positive particles in the nucleus are now called **protons**.

Rutherford's fourth conclusion was that the nucleus is surrounded by negative particles having practically no mass. The number of these negative particles had to match the number of protons. This explained why the whole atom does not have a positive charge or a negative charge. These negative atomic particles are now called **electrons**. Because the nucleus contains most of an atom's mass, electrons must have very

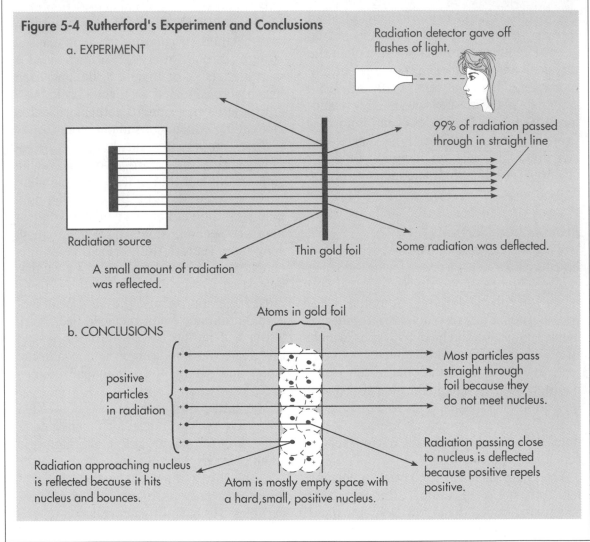

Figure 5-4 Rutherford's Experiment and Conclusions

a. EXPERIMENT

Radiation detector gave off flashes of light.

99% of radiation passed through in straight line

Radiation source

Thin gold foil

Some radiation was deflected.

A small amount of radiation was reflected.

Atoms in gold foil

b. CONCLUSIONS

positive particles in radiation

Most particles pass straight through foil because they do not meet nucleus.

Radiation approaching nucleus is reflected because it hits nucleus and bounces.

Atom is mostly empty space with a hard, small, positive nucleus.

Radiation passing close to nucleus is deflected because positive repels positive.

little mass. Look at Figure 5-4 for Rutherford's conclusions.

Many other scientists have reproduced Rutherford's results and confirmed his conclusions. Today, we know that the nucleus of an atom is only 1/10,000 of the total size of an atom. Suppose that an atom could be increased to the size of a football stadium. Its nucleus would be like a jelly bean in the middle of the football field. Still, the nucleus contains almost all the mass of an atom, because a proton's mass is about 1836 times an electron's mass.

The Neutron

After accepting Rutherford's conclusions, scientists discovered that the nucleus also contains neutral particles called **neutrons.** A neutron has almost exactly the same mass as a proton. As you can imagine, the discovery of the neutron required a lot of detective work, for neutral particles do not attract or repel other particles. Because the proton and neutron are both located in the nucleus of an atom, scientists call them nucleons. The table in Figure 5-5 summarizes the particles in an atom, which are called subatomic particles.

Spectra

Between 1913 and 1915, Niels Bohr improved the model of the atom. He investigated how the electrons in an atom are arranged. He did this by studying the **spectra** of elements, which are lines of color found in the light made by a heated element.

If you hold a penny in tongs and heat it, the flame turns green because of the copper in the penny. When you sprinkle table salt into a flame, a yellow color appears because of the sodium in the salt. These colors are caused by the spectra of each element in the material you are heating.

The Bohr Model of the Atom

Bohr stated a hypothesis based on the spectra he observed. He suggested that an atom looks like the solar system. The electrons go around the nucleus just as the planets orbit the sun. He also believed that electrons move in fixed **electron shells.** Each shell is a different distance from the nucleus and surrounds the nucleus like a ball. Bohr labeled these shells with the letters K, L, M, N, O, and P.

In addition, Bohr suggested that each electron shell can only hold a certain number of electrons. He also concluded that electrons in orbits near the nucleus have

Figure 5-5

Subatomic particle	Located	Mass	Charge
Proton	In nucleus	1 u	Positive
Neutron	In nucleus	1 u	Neutral
Electron	Around nucleus	1/1836 u	Negative

less energy than those in orbits far from the nucleus. For this reason, the modern term for electron shell is energy level. The K, L, and M shells are shown in Figure 5-6. Examine the table in Figure 5-7. You may want to refer back to it as you continue studying this section.

Figure 5-6 Bohr's Model of the Atom

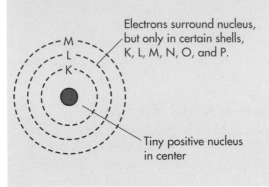

Electrons surround nucleus, but only in certain shells, K, L, M, N, O, and P.

Tiny positive nucleus in center

Electron Arrangement

An electron arrangement for an atom shows how the electrons are arranged in a Bohr model. Instead of drawing an atom diagram, you can write the number of electrons per shell, starting with the innermost shell. This means the arrangement follows the order of K-L-M-N-O-P. For example, an electron arrangement of 2-8-5 means that there are two electrons in the K shell, eight electrons in the L shell, and five electrons in the M shell. In this section, we will look only at elements containing the K, L, and M shells. Examine Figure 5-8. You can see that the diagrams show the number of protons and neutrons in the nucleus, as well as the electrons.

Notice that phosphorus has 15 protons and 15 electrons. The electron arrange-

ment is 2-8-5. To determine the electron arrangement for atoms with 1 to 18 electrons, use the following rules:
1. The first two electrons go into the K shell.
2. The next 8 electrons go into the L shell.
3. The next 8 electrons go into the M shell.

For example, an atom with 11 electrons has an electron arrangement of 2-8-1.

Atomic Number and Mass Number

As you have seen, an atom can be described by its electron arrangement. However, you can describe atoms in other terms. For example, scientists have assigned every element an **atomic number.** The atomic number of an element is the number of protons in the nucleus of each atom of the element. A hydrogen atom has 1 proton in its nucleus. The atomic number of hydrogen, therefore, is 1. A uranium atom has 92 protons in its nucleus, so its atomic number is 92.

The number of protons plus the number of neutrons equals an atom's **mass number.** Only protons and neutrons are added to get the mass of an atom, because the electrons have practically no mass. All this can be summarized by these two formulas.

Atomic number = numbers of protons
Mass number = number of protons + number of neutrons

Isotopes

Atoms of the same element have the same number of protons. Atoms of the same element, however, can have different numbers of neutrons. For example, all carbon atoms have six protons. However, some

Figure 5-7

Bohr's shell	Energy level	Energy	Largest number of electrons it holds
K	1	lowest	2
L	2	↑	8
M	3		18
N	4		32
O	5	↓	50
P	6	highest	72

Figure 5-8 Atom Diagrams

Element	Lithium	Oxygen	Phosphorous
Number of Electrons	3	8	15
K-L-M Electron Arrangement	2-1	2-6	2-8-5

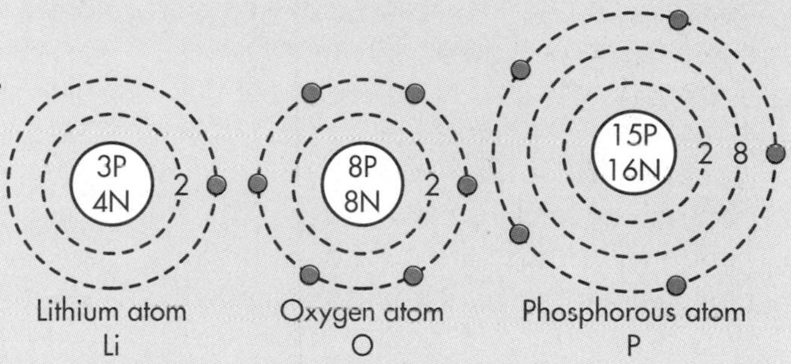

Lithium atom
Li

Oxygen atom
O

Phosphorous atom
P

carbon atoms have six neutrons, some have seven, and some have eight.

The different forms of carbon are called **isotopes**. An isotope of an element is a form of the element in which all the atoms of that isotope have the same number of neutrons. The isotopes of an element have the same atomic number and the same chemical properties.

More or fewer neutrons will change an atom's mass but will not change the element. Scientists have found many isotopes. To be clear about which isotope of an element you are referring to, you write the mass number after the element's name, as in carbon-12. Scientists have used the isotope carbon-14 to learn how old some fossils are.

The Simplest Atom

Hydrogen-1 is the simplest of all atoms. It consists of one proton, plus one electron in

Making a Model

An atom diagram shows the atom's nucleus and electron shells. It includes information about the number of protons and neutrons in the nucleus. The electrons are put in shells according to their electron arrangement.

To draw a diagram of an atom, you must know its atomic number and mass number. From these two numbers you can figure out the atom's structure using the Bohr model. Figure 5-9 shows the logical steps in drawing a diagram of an aluminum (Al) atom.

Step 1. The nucleus must contain 13 protons, since the atomic number is 13. Draw a small circle to represent the nucleus. Make it at least half an inch in diameter. Write 13 P in it.

Step 2. The nucleus must contain 14 neutrons, because the number of protons plus the number of neutrons must equal 27, the mass number. Write 14 N in the nucleus.

Step 3. There must be 13 electrons in this atom, because the number of electrons equals the number of protons. Using the rules on page 96, you can see that there are two electrons in the K shell, eight in the L shell, and three in the M shell. You can write this as 2-8-3.

Step 4. Draw three dashed circles around the nucleus to show the K, L, and M electron shells. Try to keep the circles an equal distance apart.

Step 5. On the inner shells, write the number of electrons they contain. On the outer shell, draw the electrons it contains as solid dots. Space your dots evenly around the outer circle.

Step 6. Place the name or symbol of the element below the completed atom diagram. Place the K-L-M electron arrangement to the side of, or underneath the diagram.

Figure 5-9 Steps in Drawing an Atom Diagram

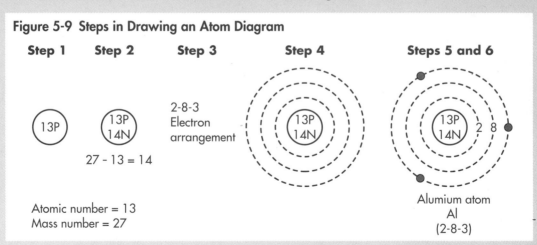

| Step 1 | Step 2 | Step 3 | Step 4 | Steps 5 and 6 |

2-8-3 Electron arrangement

27 – 13 = 14

Atomic number = 13
Mass number = 27

Alumium atom
Al
(2-8-3)

the K shell. Look at Figure 5-10. You can see that this hydrogen atom has no neutron. The second-most simple atom is the isotope hydrogen-2. It consists of one proton and one neutron in the nucleus, again with one electron in the K shell.

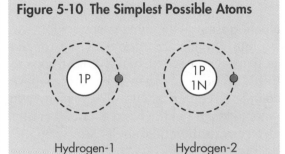

Figure 5-10 The Simplest Possible Atoms

Hydrogen-1
Atomic number = 1
Mass number = 1

Hydrogen-2
Atomic number = 1
Mass number = 2

▰▰ Section 5-2 Review ▰▰

Write the definitions for the following terms in your own words.

1. **neutron**
2. **spectra**
3. **electron shell**
4. **atomic number**
5. **isotope**

Answer these questions.

6. Explain Rutherford's experiment and his conclusions about atoms.
7. Compare the masses of protons, electrons, and neutrons and state what charges they have.

5-3 The Atom in Chemical Changes

■ *Objectives*

☐ *Explain why atoms give and take electrons.*

☐ *Contrast metal, nonmetal, metalloid, and noble gas elements.*

☐ *Define an ion.*

☐ *Distinguish between a chemical bond, an ionic bond, and a covalent bond.*

Have you been in stores where they keep a "penny dish" by the cash register? If you need a few cents, you take pennies from the dish. If you get pennies as change, you can put them into the dish for someone else to use. Your pennies belong to you, yet sometimes it is easier to give them away rather than keep them. Pennies can take up a lot of space, and they are worth very little in comparison to nickels, dimes, and quarters.

The chemical behavior of elements is actually based on a similar concept. Elements will either give away or take electrons. In doing so, elements release energy and join together to form compounds.

Completed Outer Shells

Most atoms "want" to have two or eight electrons in their outer shells. Suppose an atom has two electrons in the K shell and no electrons in other shells. Then the K shell is called a **completed outer shell**. Any other outer shell is a completed outer shell if it has eight electrons in it. Notice that shells M, N, O, and P are called completed with eight electrons even though they can hold more electrons. If an atom does not have a completed outer shell, it will give or take electrons until it does. This is the key to understanding the chemical behavior of elements. With a completed outer shell, an atom has less energy. With less energy, the atom becomes more stable, meaning that it resists chemical changes.

Elements with completed outer shells seldom react with other elements to form compounds. For this reason they are called the **inert elements**; the word *inert* means "not active." They are known as the noble gases. The word *noble* here means that these elements stay apart from other elements.

Classifying the Elements

Scientists classify elements based on the number of electrons in the outermost shell. They call the electrons in the outermost shell the **valence electrons.** Valence electrons are electrons that can move from one atom to another. They are the most important factors in the chemical behavior of an element.

Metals are elements that can conduct electricity. A material that conducts electricity allows electricity to pass through it easily. Metals are usually solid and have a shiny surface. Most metals are hard, but some can be hammered or pulled into shape. Mercury is the only metal that is liquid, instead of solid, at room temperature.

Many elements that are not metals are called nonmetals. Metals can react chemically with nonmetal atoms. Most elements are metals.

Nonmetals Nonmetals are elements that do not conduct electricity Nonmetals usually appear as a solid or a gas. The solid nonmetals are brittle and have dull surfaces. If you hammer them, they break. Nonmetal atoms have five, six, or seven valence electrons. Nonmetals will react with metal atoms. Bromine is the only liquid nonmetal at room temperature.

Noble gases The noble gases seldom enter into chemical reactions. They remain as gases even at extremely cold temperatures. This is because their atoms have completed outer shells. They have two outer electrons in the K shell, or eight outer electrons in the other shells. For example, helium has 2 electrons and they are in the K shell. Neon has 10 electrons, 2 in the K shell and 8 in the L shell.

Metalloids A few elements have the physical and chemical properties of both metals and nonmetals. These are called **metalloids**. Boron and silicon are examples of metalloids. Metalloids have three, four, or five electrons in their outermost shells.

Ions An atom is neutral because it has an equal number of electrons and protons. Losing electrons, however, produces a positively charged atom, and gaining electrons produces a negatively charged atom. An atom that has an electric charge, either positive or negative, is an **ion.**

Metal Atoms and Ions

Atoms with one, two, or three valence electrons are usually metals. Metal elements tend to give away all of their electrons to nonmetal elements. Look at the magnesium atom in Figure 5-11. It gives away the 2 electrons in its outer shell. Then the L shell becomes a completed outer shell. A completed outer shell makes the atom more stable.

Because the metal atom has given away some electrons, there are now more protons than electrons in the atom. This imbalance gives the atom a positive charge. The atom has become a positive ion. The table in Figure 5-12 describes some metal elements and ions. In each case, the electrons that are given away are the electrons in the outer shell.

Figure 5-11

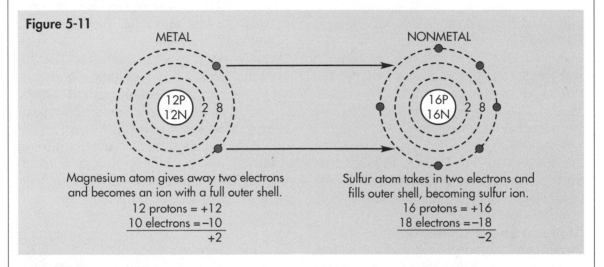

METAL

NONMETAL

Magnesium atom gives away two electrons and becomes an ion with a full outer shell.

12 protons = +12
10 electrons = −10
＿＿＿＿＿＿＿＿＿＿
+2

Sulfur atom takes in two electrons and fills outer shell, becoming sulfur ion.

16 protons = +16
18 electrons = −18
＿＿＿＿＿＿＿＿＿＿
−2

Figure 5-12

	Atomic number	Electron pattern	Gives away	Charge of ion
Hydrogen	1	1	1 electron	+1
Lithium	3	2—1	1 electron	+1
Beryllium	4	2—2	2 electrons	+2
Boron	5	2—3	3 electrons	+3
Carbon	6	2—4	4 electrons	+4
Sodium	11	2—8—1	1 electron	+1
Magnesium	12	2—8—2	2 electrons	+2
Aluminum	13	2—8—3	3 electrons	+3

Nonmetal Atoms and Ions

Recall that atoms with five, six, or seven valence electrons are usually nonmetals. Nonmetals tend to take electrons from metal elements. By doing so, the previously incomplete outer shell now becomes a completed outer shell. A completed outer shell of two or eight electrons makes the atom more stable. Look at the sulfur atom in Figure 5-11.

Because the nonmetal atom has taken in some electrons, there are now more electrons than protons in the atom. This unbalance gives the atom a negative charge. The atom has become a negative ion.

The table in Figure 5-13 describes some nonmetal elements and ions. In each case, the electrons that are taken in go into the outer shell to make a completed outer shell.

Atoms in Chemical Reactions

Recall that in chemical reactions, the properties of the materials change. In nature, the simplest type of chemical reaction occurs when a metal atom directly joins with a nonmetal atom. The metal gives away electrons and the nonmetal takes in electrons. In other words, the electrons go from the metal atom to the nonmetal atom.

After the electron has moved, the atoms become ions. The metal ion is positive and the nonmetal ion is negative. Because they have opposite charges, these two ions are attracted to each other. They stay together because of this attraction. Scientists call this attraction a bond.

Bonds occur in all compounds. The energy that joins and holds atoms together in molecules is called a **chemical bond**. Chemical bonds contain, or store, chemical energy. In an **ionic bond**, the energy holding the molecule together comes from the positively and negatively charged ions attracting each other. Compounds with ionic bonds can be formed with more than two atoms.

Sharing Electrons

Usually, atoms of different elements combine, but atoms or ions of the same element can combine. This happens when two atoms share electrons so that each

Figure 5-13

Nonmetal	Atomic number	Electron pattern	Takes in	Charge of ion
Nitrogen	7	2—5	3 electrons	−3
Oxygen	8	2—6	2 electrons	−2
Sulfur	16	2—8—6	2 electrons	−2
Chlorine	17	2—8—7	1 electron	−1

has a completed outer shell. Scientists call such a combination a **covalent bond**. A covalent bond is another type of chemical bond. Sometimes atoms of different atoms also combine with a covalent bond.

■■■ Section 5-3 Review ■■■

Write the definitions for the following terms in your own words.

1. **valence electrons**   2. **metal**
3. **nonmetal**   4. **ion**

Answer this question.

5. Why do some atoms give away their outer electrons while others take them in?

5-4 The Periodic Table of the Elements

■ *Objectives*
☐ *Define a family of elements.*
☐ *Describe how the periodic table is arranged.*
☐ *Obtain information about elements and their atoms from the periodic table.*
☐ *Distinguish between mass number and atomic weight.*

Have you ever read a bus, railroad, or airline schedule? You will recall that it gives you information such as the times you may leave and arrive, and any connections you must make.

In a similar way, you can use the **periodic table of the elements** to learn facts about the elements. The periodic table is a chart used to organize a large amount of information about the elements.

Families of Elements
By 1820, scientists knew that some elements had similar physical and chemical properties. For example, lithium, sodium, and potassium metals are very similar. All three are silver colored and soft enough to be cut with a knife. They are very active chemically. In air, they join with oxygen rapidly and even catch fire. In water they float, burn, and release hydrogen gas by breaking water apart.

Scientists say these three elements belong to a **family of elements** because they have similar chemical and physical properties. Scientists have found other families as well. In designing the first tables of elements, scientists put these families of elements together.

A number of chemists made element charts of various forms. In 1870, Dmitri Mendeleev made a chart that became widely accepted. The modern periodic table is basically the same as Mendeleev's chart, with some improvements.

Arrangement of the Periodic Table
Figure 5-14 on pages 104 and 105 gives a simplified periodic table. It shows the symbols for 108 elements. This includes 88 natural elements and 20 elements created in laboratories.

Each column is a family of elements, also called a **group.** Elements in the groups have the same number of electrons in their

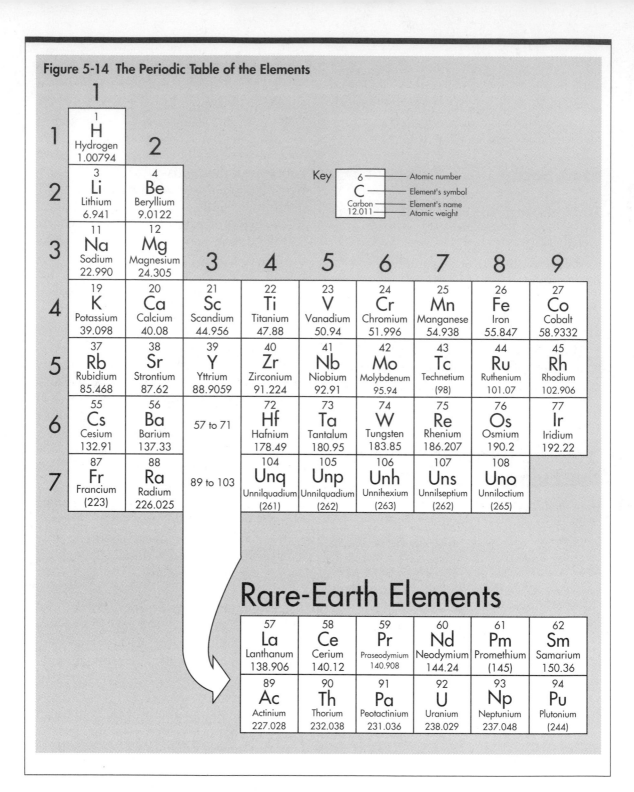

Figure 5-14 The Periodic Table of the Elements

10	11	12	13	14	15	16	17	18
								2 He Helium 4.003
			5 B Boron 10.81	6 C Carbon 12.011	7 N Nitrogen 14.007	8 O Oxygen 15.999	9 F Fluorine 18.998	10 Ne Neon 20.179
			13 Al Aluminum 26.98	14 Si Silicon 28.086	15 P Phosphorus 30.974	16 S Sulfur 32.06	17 Cl Chlorine 35.453	18 Ar Argon 39.948
28 Ni Nickel 58.69	29 Cu Copper 63.546	30 Zn Zinc 65.39	31 Ga Gallium 69.72	32 Ge Germanium 72.59	33 As Arsenic 74.922	34 Se Selenium 78.96	35 Br Bromine 79.904	36 Kr Krypton 83.80
46 Pd Palladium 106.42	47 Ag Silver 107.868	48 Cd Cadmium 112.41	49 In Indium 114.82	50 Sn Tin 118.71	51 Sb Antimony 121.75	52 Te Tellurium 127.60	53 I Iodine 126.905	54 Xe Xenon 131.29
78 Pt Platinum 195.08	79 Au Gold 196.967	80 Hg Mercury 200.59	81 Tl Thallium 204.383	82 Pb Lead 207.2	83 Bi Bismuth 208.98	84 Po Polonium (209)	85 At Astatine (210)	86 Rn Radom (222)

Numbers in parentheses are mass numbers
of the most stable or common isotope.

63 Eu Europium 151.96	64 Gd Gadolinium 157.25	65 Tb Terbium 158.925	66 Dy Dysprosium 162.50	67 Ho Holmium 164.93	68 Er Erbium 167.26	69 Tm Thulium 168.934	70 Yb Ytterbium 173.04	71 Lu Lutetium 174.967
95 Am Americium (243)	96 Cm Curium (247)	97 Bk Berkelium (247)	98 Cf Californium (251)	99 Es Einsteinium (252)	100 Fm Fermium (257)	101 Md Mendelevium (258)	102 No Nobelium (259)	103 Lr Lawrencium (260)

outermost shells. For example, the elements hydrogen, lithium, sodium, and so on in the first column have similar properties. The periodic table contains 18 groups. Each row of elements is called a **period**. The elements in each row across have the same outer electron shell, such as K, L, or M. The elements across a period are not chemically similar. They differ widely in their properties. The periodic table contains seven periods.

Arrangement by Atomic Number

The elements are arranged according to increasing atomic number. The elements are arranged so that their atomic numbers are in exact order when you read across from left to right, like reading a book.

Arrangement by Columns and Rows

The elements are arranged in columns and rows. The rows go across the table from left to right. The rows in the periodic table are called periods. There are seven periods.

The columns go down the table from top to bottom. The columns in the periodic table are called groups or families. There are 18 groups.

Arrangement by Chemical Families

Remember that groups contain elements with similar chemical properties and the same number of electrons in the outermost shell. As you read down a column of the periodic table, you are reading a group, or family, of elements. Look at the two families in Figure 5-15. For example, oxygen, sulfur, and selenium are in Group 16.

You know that the electrons in the outermost shell are called valence electrons. The

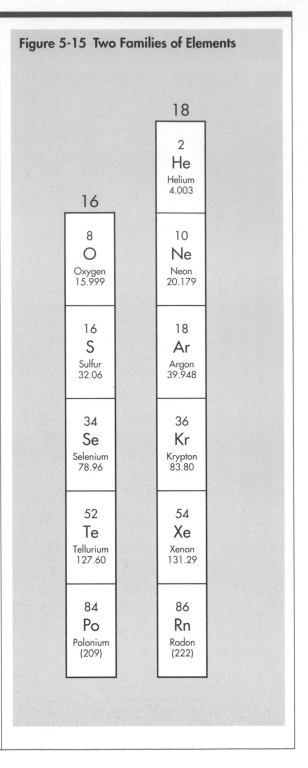

Figure 5-15 Two Families of Elements

number of valence electrons determines how the atom joins with other elements, and, therefore, the chemical properties of the atom. It is not surprising that groups of elements have the same number of valence electrons. This is what makes these elements similar in their chemical behaviors.

Scientists often call groups of elements by the name of the first element in the group. For example, Group 16 contains O, S, Se, Te, and Po. This family can also be called the oxygen group after its first element, oxygen. Other groups also have a name.

Group 1   the alkali metals
Group 2   the alkaline earth metals
Group 11   the coinage metals
Group 17   the halogens
Group 18   the noble gases

Arrangement by Electron Shells

Periods, as you will recall, contain elements that have the same outer electron shell, but the elements differ widely in chemical properties. Scientists often identify a shell by its energy level number instead of its shell letter. The period of an atom is the same number as the energy level of its outermost shell. Look at the table in Figure 5-16.

Elements in periods vary quite widely in properties. For example, sodium (Na); silicon (Si); and argon (Ar) are in period three, but they are totally different in chemical and physical properties. Their only similarity is that they have an outer M shell.

Elements within a period are about the same size. The size of an atom is given by its radius. The radius of an atom is the distance from the center of the nucleus to its outer shell. The radius of an atom is measured in ang-stroms. One angstrom is 0.00000001

Figure 5-16.

Period	Outer Shell	Energy level of outer shell
1	K	1
2	L	2
3	M	3
4	N	4
5	O	5
6	P	6

cm, or one one-hundred millionth of a centimeter. Each new shell adds to the radius of atoms.

Elements in period one are the smallest in size. Elements in period six are the largest in size.

Using the Periodic Table

You can learn a lot about an element from the periodic table. To begin with, you can find an element's atomic number. The atomic number determines what element it is, how it is structured, and what its properties are. Remember, the atomic number of an atom equals the number of protons in the atom's nucleus.

The periodic table does not give mass numbers. Instead, it gives **atomic weights.** The atomic weight of an element is the average of the mass numbers of the isotopes of the element as they occur in nature. For example, the element chlorine, on Earth, has two isotopes. About half are Cl-35 atoms and half are Cl-36 atoms. Therefore, the atomic weight of chlorine is close to the average of these mass numbers. The periodic table gives the atomic weight of chlorine as 35.453.

To obtain a mass number from the periodic table, round the atomic weight on the table to the nearest whole number. This is usually the mass number of the most common isotope of the element. For example, magnesium (Mg) has an atomic weight of 24.305. The mass number of its most common isotope is 24. The atomic weight 24.305 is the average based on many Mg-24 atoms, some Mg-25 atoms, and a few Mg-26 atoms.

You can learn the type of element you are considering by using the periodic table. Look at the heavy line that runs stepwise from boron (B) to astatine (At) in Figure 5-14. Now look at Figure 5-17.

All elements to the left of this line are metals. The elements in group 18 are noble gases. The elements in the shaded area are nonmetals. The elements to the right of the line and touching it are metalloids.

Trends in the Periodic Table

Each period begins with an active metal on the left, in group 1. An active element

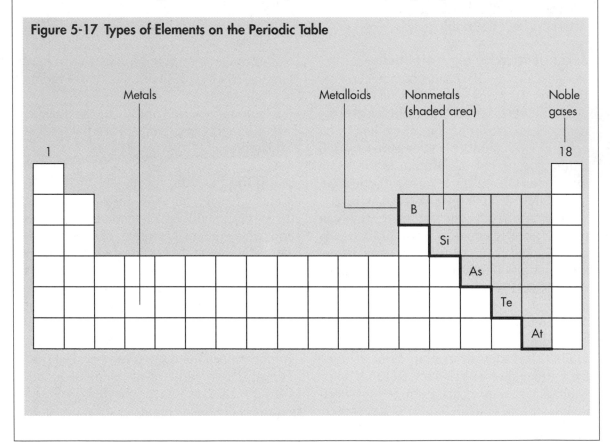

Figure 5-17 Types of Elements on the Periodic Table

can react chemically with other elements. The period ends with an inert noble gas element in group 18. Just to the left of the noble gas is an active nonmetal, in group 17. In going from left to right within a period, the elements change from being very metallic to being very nonmetallic. An element is more metallic than another element if it is a better conductor of electricity and it gives away electrons more readily. An element is more nonmetallic than another if it is a poorer conductor of electricity and it more readily takes in electrons.

In a group of metal elements, the metallic activity increases from top to bottom. For example, in group 2, Be is the least active metal and Ra is the most active metal. As you read down a group, you can see that the number of electron shells increases. Metals at the bottom of the table are the most active because their outer electrons are farthest from the nucleus and so are not held as tightly. The most active metals react most easily with nonmetals.

Nonmetals react by taking in electrons. When the electrons in the outer shell are close to the nucleus, the pull of the nucleus is strong and the atom can take in electrons. Therefore, nonmetals at the top of the table are the most active nonmetals because their outer shells are closest to the nucleus. They react most easily with metals. For example, in group 17, At is the least active nonmetal and F is the most active nonmetal.

These trends allow you to predict that the most active metal will be francium (Fr). You can also predict that the most active nonmetal will be fluorine (F).

▬ ■ Section 5-4 Review ■ ▬

Write the definitions for the following terms in your own words.

1. **periodic table of the elements**
2. **family of elements**
3. **group**
4. **period**
5. **atomic weight**

Answer the following questions.

6. Why are some of the elements described as being part of a family? Give an example of a family of elements.
7. How are elements arranged within the periodic table? Be specific about the arrangements of atomic numbers, groups and periods, and chemical families.
8. A silver (Ag) atom has a mass number of 108. However, on the periodic table its atomic weight is given as 107.868. What would explain the difference between these two numbers?
9. Using the periodic table, which is more metallic,
   a. barium (Ba) or calcium (Ca)?
   b. barium (Ba) or cesium (Cs)?
10. Use the periodic table in Figure 5-14 to find the following information about iron (Fe).
   a. its atomic number
   b. its atomic weight
   c. the mass number of its most abundant isotope
   d. its group number
   e. other members of the iron group

SCIENCE, TECHNOLOGY, & SOCIETY

Discovering Quarks

New technologies often help scientists do their basic research. You can see this in the efforts scientists are now making to study matter more deeply. Matter is made from elements, which are made from atoms, which are made from protons, neutrons, and electrons. What makes up protons and neutrons?

The answer seems to be quarks. Quarks are particles with part of an electric charge. Their charge is either $+2/3$ or $-1/3$. In this model, three quarks combine to form a proton or a neutron. See Figure 5-18.

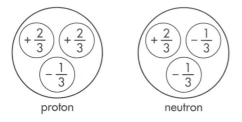

Figure 5-18

proton   neutron

Do the charges of the quarks add up to the correct charges for protons and neutrons?

Scientists have learned this by smashing protons, neutrons, and electrons together at extremely high speeds. This research requires very special, expensive equipment. It also involves large teams of scientists and technicians using advanced computers and electronics.

Somebody has to pay for the equipment, and for the salaries of the many scientists involved. Usually, the money comes from universities and industry. Governments, however, have also been giving financial help.

For example, the United States Congress has approved start-up funds for a huge set-up of equipment. It is called a superconducting supercollider, and it is to be built in Texas. This device will require a circular tunnel and 40 miles of pipe. The laboratory will be loaded with the finest technological equipment and computers. It will have hundreds of regular and visiting staff members. This entire project is being built to find out how atoms are put together. By paying for the research, society is helping scientists make their discoveries.

Follow-up Activities

1. When watching TV programs or reading about science research today, look for the use of high-technology equipment and computers. Also look for any mention of the money that supports the research. How much is required? Who pays? Were any trade-offs needed? What problems of funding were discussed? List your observations for later discussion.
2. How does this reading show that technology can advance science?

Section 5-1 Atoms and Molecules

1. An atom is the smallest particle of an element that still has the chemical properties of the element.
2. A molecule is the smallest particle of a compound that still has the chemical properties of the compound.

Section 5-2 Uncovering the Atom's Structure

1. Atoms are mostly empty space with an extremely small, hard nucleus.
2. Atoms contain subatomic particles called protons, electrons, and neutrons.
3. Electrons are in shells around the nucleus.

Section 5-3 The Atom in Chemical Changes

1. Atoms can give or take electrons to achieve a completed outer shell.
2. The properties of an element depend upon the number of electrons in its outermost shell, or its valence electrons.
3. Atoms of the same element or of different elements can combine. When combined, these atoms are joined together by chemical bonds that have chemical energy.

Section 5-4 The Periodic Table of the Elements

1. In the Periodic Table, atoms are arranged according to increasing atomic number.
2. Families of elements have similar chemical and physical properties.

On a sheet of paper write the term that best matches each description.

atomic mass unit, chemical bond, family of elements, inert elements, nucleus, valence electrons

1. unit of measurement for masses of atoms
2. center of an atom
3. electrons in the outer orbit of an atom
4. another name for the noble gases
5. energy holding a molecule together
6. elements with similar properties

Explain the difference between the words in each pair.

7. atom, molecule
8. proton, electron
9. atomic number, mass number
10. metal, nonmetal
11. group, period

Write the missing word for each sentence.

1. Protons have ___ charges and electrons have ___ charges.
2. In a nucleus with 12 protons and 13 neutrons, the atomic number is ___, and the atomic mass is ___.
3. An atom becomes an ion with a +3 charge by ___.
4. The spectra of heated elements convinced Bohr that electrons were arranged in ___.
5. The modern term for electron shell is ___.
6. The elements in a group have the same number of valence electrons, so the elements have almost identical ___.

CHAPTER REVIEW

▰▰▰ INTERPRETING INFORMATION ▰▰▰

Answer these questions by referring to the periodic table in Figure 5-14.

1. Write the symbol of the element in
   a. Group 4, Period 5
   b. Group 5, Period 4
2. List the symbols of all the elements
   a. in Period 2
   b. in Group 2
3. Write the name, symbol, and atomic weight of the elements with these atomic numbers:
   a. 4
   b. 15
   c. 23
   d. 53
4. How many protons and how many electrons are there in N, K, Zn, Sb, and W?
5. Write out the electron arrangements of C, Na, and Ar.
6. Which atom is largest in group 1? group 2? group 15?

▰▰▰ THINK AND DISCUSS ▰▰▰

Use the section number in parentheses to help you find each answer. Write your answers in complete sentences.

1. Figure 5-19 shows some molecules found in ammonia water. Use the figure to answer the following questions. (5-1)
   a. How many molecules are shown?
   b. What elements make up this compound?
   c. How many atoms of each element are shown?

2. Figure 5-20 shows a set of different nuclei. (5-2, 5-4)
   a. Write the atomic number and mass numbers for each.
   b. Which are isotopes of the same element?
   c. Write the element name and symbol for each.
3. Here are the electron arrangements of two atoms. (5-2, 5-3)
   Atom 1: 2-8-3
   Atom 2: 2-5
   a. What will each atom tend to do to get a completed outer shell?
   b. Diagram how atom 1 might join with atom 2, and write the ions that result from their transferring electrons.

Figure 5-19

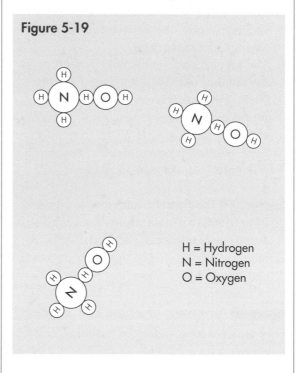

H = Hydrogen
N = Nitrogen
O = Oxygen

CHAPTER REVIEW

Figure 5-20

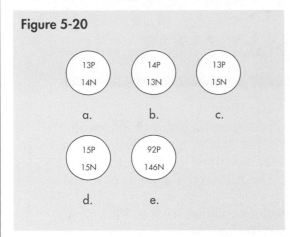

GOING FURTHER

Use your library to research a topic related to this chapter. Get details about its history and importance. Here are some additional possible topics related to this chapter.

1. John Dalton's atomic theory
2. William Crookes and cathode ray tubes
3. J. J. Thomson investigates electrons
4. James Chadwick's discovery of the neutron
5. Mendeleev and the periodic table
6. Rutherford's experiment
7. Bohr's experiments

COMPETENCY REVIEW

1. A molecule is the smallest particle of
   a. an atom
   b. an element
   c. a compound
   d. a mixture

2. Which particles are in the nucleus of an atom?
   a. only protons
   b. protons and neutrons
   c. only electrons
   d. protons and electrons

3. Which electron pattern has a completed outer shell?
   a. 2-2
   b. 2-7
   c. 2-8
   d. 2-8-1

4. Which electron pattern would indicate a nonmetal?
   a. 2-2
   b. 2-7
   c. 2-8
   d. 2-8-1

5. Where on the periodic table are elements with similar properties located?
   a. on the left side
   b. on the right side
   c. in a row
   d. in a column

6. Which particle has a negative electric charge?
   a. proton   b. neutron
   c. electron   d. atom

7. Where is the chemical energy stored in a molecule?
   a. in the nuclei
   b. in the atoms
   c. in the bonds between atoms
   d. in the movement of the molecule

CHAPTER 6

CHEMICAL FORMULAS AND EQUATIONS

IF U CN RD TH SENTNC, U CN GT A GD JB

Can you understand this? It is an advertisement for a shorthand writing school. Maybe you have seen this, or other similar advertisements, on buses or trains. It is an abbreviation of a sentence, and for people trained to write shorthand, its meaning is clear. The full sentence appears in the review at the end of this section.

Think of the large number of abbreviations you use, such as *TV* instead of *television* or N.Y. instead of New York. Take a few minutes to think of other abbreviations, such as USA, NBC, FBI, IRS, NFL, UN. What do they stand for?

Chemistry has its own shorthand. In chemistry shorthand, $5\ K_2CO_3$ means five molecules of the compound potassium carbonate, each containing two atoms of potassium (K), one atom of carbon (C), and 3 atoms of oxygen (O). By using chemical shorthand, you can describe a chemical and a chemical change very quickly and accurately.

6-1 Writing Chemical Formulas

■ *Objectives*

☐ *Identify element symbols.*

☐ *Explain a chemical formula.*

☐ *Distinguish a subscript from a coefficient.*

☐ *Obtain formulas for compounds, using the criss-cross method.*

You have already used symbols for the elements such as H for hydrogen and C for carbon. You will need to know the symbols for

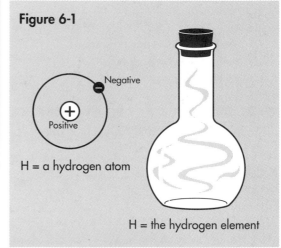

Figure 6-1

Negative

Positive

H = a hydrogen atom

H = the hydrogen element

H is the symbol for Hydrogen.

each element when you study compounds. Since many elements have names beginning with the same letter, some elements have two letter symbols. For instance, B is for boron, Be is for beryllium, Br is for bromine, and Ba is for barium. Many symbols are obvious. Others are not so obvious because they are abbreviations of foreign words, usually Latin words. For example, Na is the symbol for sodium. Na comes from *natrium,* the Latin word for sodium. Fe for iron comes from *ferrum,* and Pb for lead comes from *plumbum.*

The first letter of an element's symbol is always a capital. The second letter is always a small letter. When you write the symbol of an element, it could have two slightly different meanings. It could stand for the element, or it could stand for one atom of the element. Figure 6-1 illustrates the two slightly different meanings of symbols.

You will need to identify some common elements by their symbols. Important symbols to remember are Al, Ar, B, Be, Br, C, Ca, Cl, Cu, F, Fe, H, He, Hg, I, K, Li, Mg, N, Na, Ne, O, P, S, Sn, U, and Zn. When in doubt, check Appendix 4 or the periodic table in Chapter 5.

Formulas for Compounds

A **chemical formula** tells what elements are in a compound, as well as how many atoms of each element are in a molecule of the compound. For example, the chemical formula for baking soda is $NaHCO_3$. This formula tells you that baking soda contains the elements sodium, hydrogen, carbon, and oxygen. It also tells you that one molecule contains one atom of Na, one atom

of H, one atom of C, and three atoms of O. Figure 6-2 shows a model of one molecule of $NaHCO_3$.

Generally, the metal atom is written first in a formula. This is usually a positive ion. The nonmetal atom, which is usually a negative ion, is written next.

The small number written after and slightly below the symbol for an element is called the **subscript**. A subscript tells how many atoms of an element are in a molecule. If there is only one atom, no subscript is needed, as with the C in Na_2CO_3.

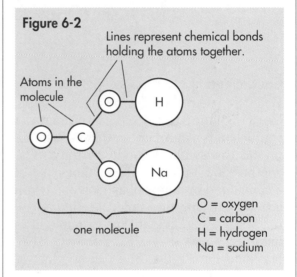

Figure 6-2

Lines represent chemical bonds holding the atoms together.

Atoms in the molecule

one molecule

O = oxygen
C = carbon
H = hydrogen
Na = sodium

Formulas with Parentheses

Many formulas contain a group of symbols inside parentheses. For example, one compound is $Al(NO_3)_3$. This means there is one atom of Al attached to three groups of NO_3. The total number of atoms in one molecule of $Al(NO_3)_3$, therefore, is 13. It contains one atom of aluminum (Al); 3×1 or three atoms of nitrogen (N); and 3×3 or nine atoms of oxygen (O). You can count the atoms in Figure 6-3, which is a molecule model of this compound.

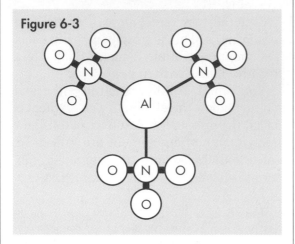

Figure 6-3

Coefficients

When you see a number before a formula, like 3 H_2O, the 3 is called the **coefficient**. A coefficient is a number by which you multiply all the symbols in the formula. The coefficient tells the number of molecules of a compound or atoms of an element. If you are studying only one molecule, then you do not write a coefficient. For example, it is incorrect to write 1 HCl; just write HCl. Examine Figure 6-4. You can see that in

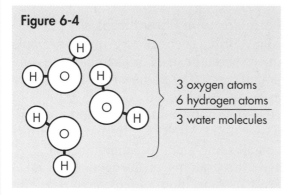

Figure 6-4

3 oxygen atoms
6 hydrogen atoms
───────────────
3 water molecules

three water molecules, you would find six hydrogen atoms and three oxygen atoms.

How many atoms of each element, and how many total atoms, are in 20 $(NH_4)_2S$? Check your answer with the table in Figure 6-5.

Finding Formulas

You can find the formula of a compound by drawing the atoms and showing how they transfer or share electrons. Recall what you learned in Chapter 5 about ions and sharing electrons. Then look at the example in Figure 6-6. One boron atom can join with three fluorine atoms to make BF_3. This figure shows how bonds are formed when electrons move from one atom to another. It also shows the model of a BF_3 molecule.

You might wonder if finding a formula is always this complicated. Usually, it is not. The **criss-cross method** is a faster way to find formulas for certain compounds. This method takes four steps.

Step 1 Find the usual ion charge of each atom or group of atoms in the compound.

Figure 6-5		
Atoms	**in 1 molecule**	**in 20 molecules**
N	2 x 1 = 2	40
H	2 x 4 = 8	160
S	1	20
Total	11	220

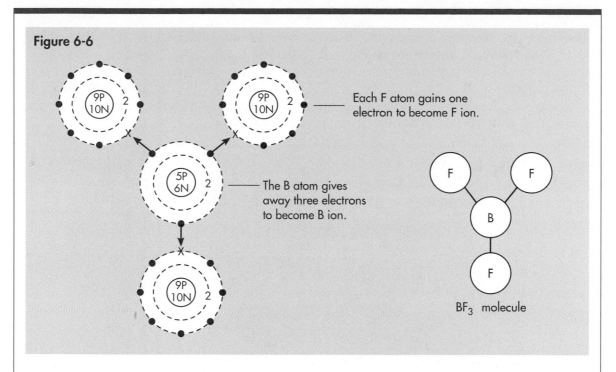

Figure 6-6

Each F atom gains one electron to become F ion.

The B atom gives away three electrons to become B ion.

BF_3 molecule

You can locate this in special tables that list ions. A short ion table appears in Figure 6-8. Such a table may be called an ion, oxidation-number, or valence table. For BF_3, you would find that B has a charge of $+3$ and F has a charge of -1.

Step 2 Write the charges above each ion. See Figure 6-7.

Step 3 Criss-cross the ion numbers, without the $+$ or $-$ signs, to make them into subscripts in a formula. Figure 6-7 shows ion numbers being criss-crossed.

Step 4 Write the formula with the simplest whole numbers. For B_1F_3, you would write BF_3. In some cases, you will have to reduce the subscripts. To do this, divide them by a common factor to get the simplest whole number. For example, with carbon and oxygen, C has a charge of $+4$, and O has a charge of -2. Then, using the criss-cross method in Step 3, write C_2O_4. Divide both subscripts by 2. C_2O_4 reduces to C_1O_2, or CO_2. This is the formula for carbon dioxide.

A List of Ion Charges

Remember that metals become positive ions, and nonmetals become negative ions when they combine to form a molecule.

Figure 6-7

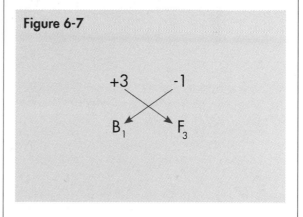

$$+3 \qquad -1$$
$$B_1 \qquad F_3$$

The table in Figure 6-8 gives the usual charge of ions and **many-atom ions**. A many-atom ion is a group of atoms, often found together, that behave as if they were a single ion. A many-atom ion can form molecules with metal or nonmetal ions.

Many-atom ions are usually written with parentheses.

Use the example in Figure 6-9 to see how the following question is answered. What formula results if iron joins with SO_4 ions?

Figure 6-8	
Ion Charge	**Some ions and many-atom ions**
+1	H, Li, Na, K, NH_4
+2	Be, Mg, Ca, Sr, Ba, Cu, Zn
+3	B, Fe, Al
+4	C, Si
−3	N, P, PO_4
−2	O, S, SO_4, CO_3
−1	F, Cl, Br, I, ClO_3, NO_3

Figure 6-9

Step 1: Fe   +3 SO_4 -2

Step 2: $\overset{+3}{Fe}$ $\overset{-2}{SO_4}$

Step 3: $\overset{+3}{Fe}_2$ $(SO_4)^{-2}_3$

Step 4: $\boxed{Fe_2(SO_4)_3}$

How Can You Make a Molecule Model?

Process Skill *designing models*

Materials A set of molecular models. If models are not available, construct them out of styrofoam balls with holes, spray paint, rigid and bendable sticks, and wire.

Procedures

1. Prepare a variety of metal atom models by inserting sticks or wire springs into the holes. Some balls will have just one stick. These represent metals that become +1 ions by giving away one electron. Others will have two sticks. These represent metals that become +2 ions by giving away two electrons, and so on.
2. Join the metal atoms with nonmetal atoms so that each stick of a metal atom is used to fill up a hole in the nonmetal atom. You are making a molecule model.
3. Use bendable sticks, if you need to, when atoms share two or three electrons. This will help the balls stay together easily.
4. Using balls with the correct number of holes and sticks, construct as many of the molecules given in this section as you can. You can also use the table in Figure 6-8 to make as many metal-nonmetal combinations as possible.
5. Label each molecule you make with a formula, as in Figure 6-10.

Conclusions

6. Compare your models to the Bohr models of atoms and molecules in Chapter 5. Which models do you think are better in showing how molecules are put together? Why?

Figure 6-10

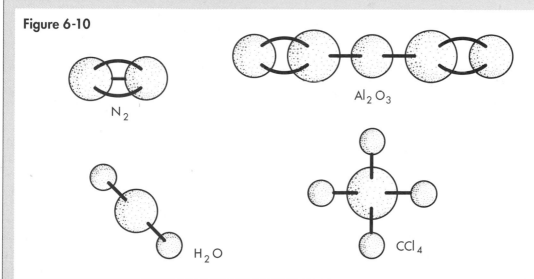

N_2

Al_2O_3

H_2O

CCl_4

Section 6-1 Review

Write the definitions for the following terms in your own words.

1. **chemical formula**
2. **subscript**
3. **coefficient**
4. **criss-cross method**
5. **many-atom ion**

Answer these questions.

6. The formula for milk of magnesia, a stomach medicine, is $Mg(OH)_2$.
   a. What elements are in this compound?
   b. How many atoms of each element are in one molecule?
   c. What is the total number of atoms in 300 $Mg(OH)_2$?
7. Write the following formulas on your paper. For each one, circle and identify the coefficients, the subscripts, and the many-atom ions if they are present.
   a. HF
   b. 4 Li_3N
   c. Fe_2O_3
   d. $(NH_4)_3PO_4$
8. Use the criss-cross method and the table in Figure 6-8 to find the formulas for these pairs.
   a. Ca joins with Cl
   b. K joins with Br
   c. C joins with S
   d. Mg joins with (PO_4)

The shorthand message at the start of this section is an abbreviation for, "If you can read this sentence, you can get a good job."

6-2 Writing Chemical Equations

■ *Objectives*
☐ *Interpret the information contained in a chemical equation.*
☐ *Balance equations.*

You know about equations from mathematics. Equations are a short way to describe how things are related. For example, read the sentence: *The volume of a box is equal to its length multiplied by its side and then multiplied by its height.* Simplified to an equation, the sentence becomes V = lwh. In the same way, equations can describe chemical changes more simply and directly than sentences can.

Word Equations

A word equation describes a chemical change by naming the materials that go into and come out of the reaction. For example, look at Figure 6-11. If you put sodium metal into water it floats and melts. The water fizzes, giving out hydrogen gas, which sometimes catches fire.
Caution Do not put sodium into water yourself. This should only be done by a teacher, wearing goggles, with a protective screen over the container.

This can be described using a word equation:

**sodium + water →
sodium hydroxide + hydrogen**

The **reactants** are the materials that go into a chemical reaction, in this case, the

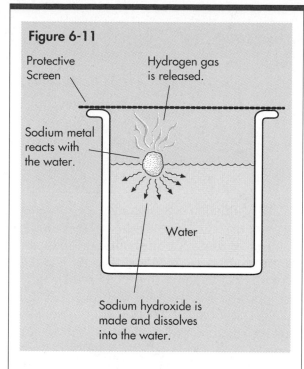

Figure 6-11

Protective Screen

Hydrogen gas is released.

Sodium metal reacts with the water.

Water

Sodium hydroxide is made and dissolves into the water.

reactants are sodium and water. They are used up to make the new materials, which are called the **products**. The products are sodium hydroxide and hydrogen. The arrow in an equation represents the chemical change. Reactants are put on the left side of the arrow and products are put on the right side. The arrow points to the right.

Inside your body, a sugar called glucose often joins with oxygen from your lungs. This reaction produces water, carbon dioxide gas, and energy. The chemical word equation makes this reaction clearer:

glucose + oxygen → carbon dioxide + water + energy

Chemical Equations

A **chemical equation** gives the formulas and conditions of the reactants and products in a chemical reaction. It also gives coefficients to show how many molecules of each material are needed or are made. Look at the following chemical equations.

$$2 \text{ Na (sodium)} + 2 \text{ H}_2\text{O (water)} \rightarrow 2 \text{ NaOH (sodium hydroxide)} + \text{H}_2 \text{ (hydrogen)}$$

$$\text{C}_6\text{H}_{12}\text{O}_6 \text{ (glucose)} + 6 \text{ O}_2 \text{ (oxygen)} \rightarrow 6 \text{ CO}_2 \text{ (carbon dioxide)} + 6 \text{ H}_2\text{O (water)} + \text{energy}$$

The sample chemical equations above have words to remind you of the material represented by each formula. Usually, chemical equations do not contain words. Instead, chemical equations just use symbols.

Balanced Equations

Sometimes a chemical equation is a direct abbreviation of a word equation. Look at the following equations.

word: **magnesium + sulfur → magnesium sulfide**
chemical: **Mg + S → MgS**

This is easy to picture, as you can see in Figure 6-12. The circles represent atoms.

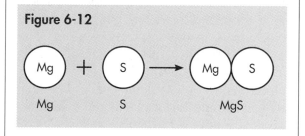

Figure 6-12

Mg + S → Mg S

Mg   S   MgS

Most chemical equations are not so simple to write because the formulas for the materials contain subscripts. Look at this pair of equations.

word: **lithium + sulfur →
lithium sulfide**
chemical: **Li + S → Li$_2$S**

An obvious problem appears in this example. The product must have the formula Li$_2$S. Where does the second Li come from? The equation above incorrectly shows one Li atom on the left and two Li atoms on the right. The equation is not balanced. In a **balanced equation**, the same number of atoms appear on both sides of the arrow. You can balance this equation by putting a 2 as the coefficient for Li on the left.

2 Li + S → Li$_2$S

To balance an equation you may change the coefficients in front of a formula. The coefficients tell how many molecules are needed, and that number can be changed. Subscripts tell how the molecule is made and subscripts cannot be changed without changing the molecule.

Diatomic Molecules

If you could examine the smallest particle in a sample of oxygen gas, you would discover it is not an oxygen atom. The smallest particle would be a pair of oxygen atoms joined together by sharing their electrons in a covalent bond. The two oxygen atoms form an oxygen molecule. Look at Figure 6-13.

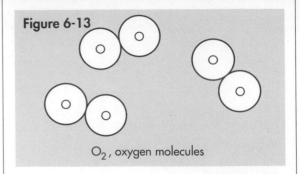

Figure 6-13

O_2, oxygen molecules

This situation exists for all gas elements, except the noble gases. Most gas elements consist of **diatomic molecules,** which are made of two identical atoms joined together. Elements that consist of single atoms are He, Ne, Ar, Kr, Xe, and Rn. These are Group 18 elements. The elements that consist of diatomic molecules are H$_2$, N$_2$, O$_2$, F$_2$, Cl$_2$, Br$_2$, and I$_2$. These diatomic molecules often appear in chemical reactions, and this affects balancing equations.

When silicon combines with oxygen, the chemical equation is balanced.

Si + O$_2$ → SiO$_2$

However, with other metals, balancing is required.

**4 Na + O$_2$ → 2 Na$_2$O
2 Mg + O$_2$ → 2 MgO
4 Al + 3 O$_2$ → 2 Al$_2$O$_3$
2 H$_2$ + O$_2$ → 2 H$_2$O**

Checking

You should check chemical equations to be sure they are balanced. Checking means counting all the atoms of each element on the left side and the right side

How Can You Balance Equations with Models?

Process Skills observing; designing a model; making predictions

Materials Use the same set of molecular models used for the first activity in this chapter.

Procedures

1. Choose a balanced equation given in this section or in a homework problem.
2. Make models of all the reactants. Make as many molecules as needed, based on the coefficients in the balanced equation.
3. Rearrange these atoms to form all the products. If the equation was balanced, the atoms of the reactants should rearrange perfectly to form all the molecules in the products. You should not need any extra balls to complete your models of the products. You should also have no unused balls left from the reactant side when you have made the product molecules. Look at Figure 6-14.
4. Repeat this procedure for several other balanced equations, as your teacher instructs.

Conclusions

5. What will occur if an equation is not balanced? Demonstrate this with the models.

Figure 6-14

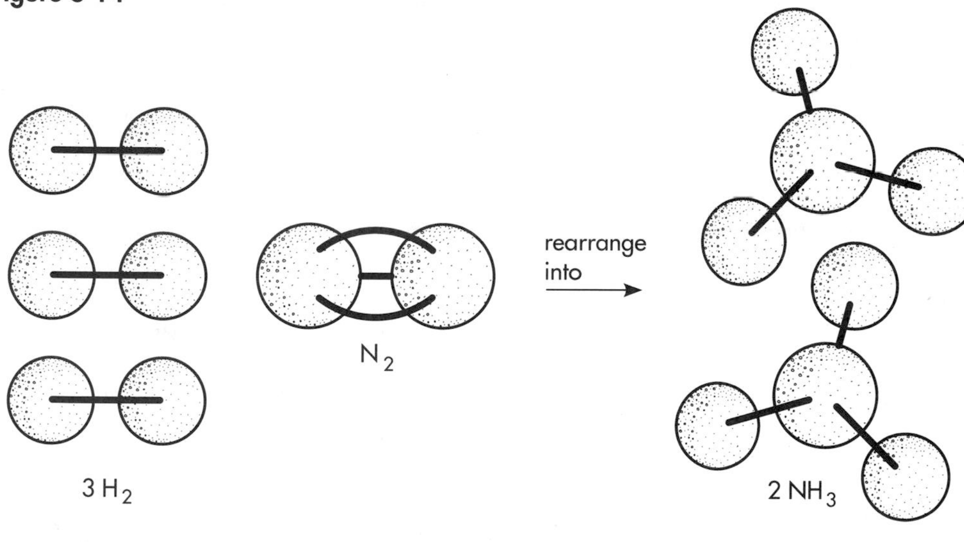

$3 H_2$ N_2 rearrange into $2 NH_3$

of an equation. The numbers must be equal. Figure 6-15 shows a check for the aluminum and oxygen reaction $4\ Al\ +\ 3\ O_2 \rightarrow 2\ Al_2O_3$.

Figure 6-15		
Check:	Left side	Right side
Al atoms	4	$2 \times 2 = 4$
O atoms	$3 \times 2 = 6$	$2 \times 3 = 6$

▬ Section 6-2 Review ▬

Write the definitions for the following terms in your own words.

1. **reactant**
2. **product**
3. **chemical equation**
4. **balanced equation**
5. **diatomic molecule**

Answer these questions.

6. Write these equations on your paper.

$$NH_4OH\ +\ HCl \rightarrow H_2O\ +\ NH_4Cl$$
$$CaCO_3\ +\ H_2O \rightarrow Ca(OH)_2\ +\ CO_2$$
$$2\ Fe\ +\ 3\ H_2O \rightarrow Fe_2O_3\ +\ 3\ H_2$$
$$16\ HCl\ +\ 2\ KMnO_4 \rightarrow 8\ H_2O\ +$$
$$2\ KCl\ +\ 2\ MnCl_2\ +\ 5\ Cl_2$$

List the reactants and the products for each reaction.

7. The equations given in Question 6 are all balanced. Show that this is true by checking each one.

8. Balance the equations below. Check each one to be sure it is balanced.

$$H_2\ +\ F_2 \rightarrow HF$$
$$H_2\ +\ S \rightarrow H_2S$$
$$H_2\ +\ N_2 \rightarrow NH_3$$
$$Li\ +\ N_2 \rightarrow Li_3N$$
$$Al\ +\ Br_2 \rightarrow AlBr_3$$

9. a. Why is hydrogen gas written H_2, not H?
   b. Nitrogen is an element, so why are the particles of nitrogen gas called molecules instead of atoms?
   c. What gas elements are made of atoms alone? Write the symbol for each.
   d. What gas elements exist as diatomic molecules? Write the formula for each.

6-3 Mass Relationships

■ *Objectives*
☐ *Calculate the formula mass of a compound.*
☐ *Define a gram-atomic mass (G.A.M.) and a gram-formula mass (G.F.M.).*
☐ *Explain the significance of Avogadro's number.*

Did you know that an airplane pilot figures out the weight of the plane just before taking off? To do this, the pilot adds the weight of the empty plane, the number of people times an average weight per person, the weight of the baggage, and the weight of the fuel. Likewise, in chemistry you will sometimes need to know the mass of a molecule. You can find it by adding together the masses of its atoms. You will use such mass values in the laboratory.

Gram-Atomic Mass (G.A.M.)

Recall that every atom has its own mass. The mass can be given with the unit u, the atomic mass unit.

Atom	Mass
H	1 u
C	12 u

You can find the mass of the most abundant isotope of each element in Appendix 4. For example, find Nickel (Ni). The mass of its most abundant isotope is 58 u.

Measuring samples in u is impossible because u is such a tiny mass. You must frequently use an amount of an element measured in grams. The **gram-atomic mass (G.A.M.)** of an element is an amount equal to its mass but written in grams, not in atomic mass units. For example, one atom of Ni has a mass of 58 u, or 58 atomic mass units. One G.A.M. of Ni is therefore 58 g. The table in Figure 6-16 shows the gram-atomic masses of some elements.

Formula Mass

As you know, you can find the mass of one molecule in a compound by adding the masses of its atoms. This is called the **formula mass** of the compound. It is also referred to as the molecular weight of the compound.

For example, to find the formula mass of water, add up the masses of the atoms in one molecule of water, H_2O. You can find the masses of the atoms in Appendix 4.

Notice that the formula mass is given in atomic mass units, not in grams. Look at Figure 6-17.

Figure 6-16

Atom:	Mass	G.A.M., Gram Atomic Mass
H	1 u	1 g
C	12 u	12 g
N	14 u	14 g
O	16 u	16 g
Al	27 u	27 g
S	32 u	32 g
U	238 u	238 g

Figure 6-17

Example: Find the formula mass of water. Add up the masses of the atoms in one molecule of water, H_2O.

Atom	Number in one molecule	x	Mass	=	Total Mass
H	2	x	1 u	=	2 u
O	1	x	16 u	=	16 u
Formula mass of H_2O				=	18 u

Figure 6-18

Example: What is the formula mass of H_2SO_4?

Atom	Number in one molecule	x	Mass	=	Total Mass
H	2	x	1 u	=	2 u
S	1	x	32 u	=	32 u
O	4	x	16 u	=	64 u
Formula mass of H_2SO_4				=	98 u

Example: Determine the formula mass of NH_4NO_3.

Atom	Number in one molecule	x	Mass	=	Total Mass
N	2	x	14 u	=	28 u
H	4	x	1 u	=	4 u
O	3	x	16 u	=	48 u
Formula mass of NH_4NO_3				=	80 u

Example: Find the formula mass of $Al_2(CO_3)_3$.

Atom	Number in one molecule	x	Mass	=	Total Mass
Al	2	x	27 u	=	54 u
C	3	x	12 u	=	36 u
O	9	x	16 u	=	144 u
Formula mass of $Al_2(CO_3)_3$				=	234 u

Now examine the examples in Figure 6-18 and practice finding the formula mass. Refer to Appendix 4 to make sure you can locate each of the correct masses. You may want to practice finding the formula mass for other, still larger, molecules.

Gram-Formula Mass (G.F.M.)

The **gram-formula mass** (G.F.M.) of a material is its formula mass taken in grams, not in u. For example, if you are told to take a gram-formula mass of $Al_2(CO_3)_3$, you would measure 234 grams of the material. This is because the formula mass of $Al_2(CO_3)_3$ is 234 u. Water has a formula mass of 18 u, which means it has a G.F.M. of 18 g.

The relationship between formula mass and gram-formula mass is very much like the relationship between atomic mass and gram atomic mass. You can see that atomic mass and formula mass are given in u, but the gram measures are given in grams. The gram measures are much easier to use in the laboratory because they can be detected by balances and scales.

Each single gram-formula mass contains 600,000,000,000,000,000,000,000 molecules. This is true for any compound because one gram contains this many atomic mass units. Thus, every G.F.M. contains the same number of molecules. Similarly, the G.A.M. of an element is important because every one gram-atomic mass contains this same number of atoms. This is true for any element.

This is a rather special number in science. It is known as **Avogadro's** (av-uh-GAD-roz) **number,** and it is usually written as 6×10^{23}. To see how gram-atomic mass, gram-formula mass, and Avogadro's number are related, look at Figure 6-19.

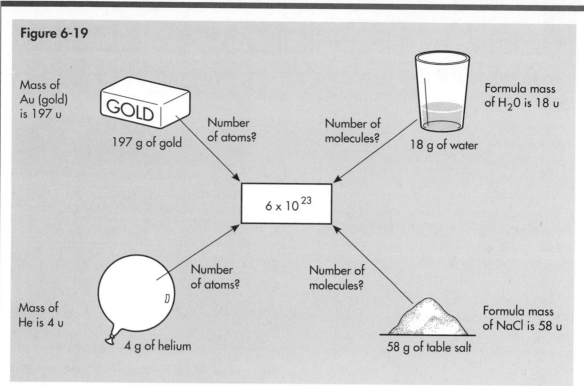

Figure 6-19

Mass of Au (gold) is 197 u

GOLD

197 g of gold

Number of atoms?

Number of molecules?

Formula mass of H_2O is 18 u

18 g of water

6×10^{23}

Number of atoms?

Number of molecules?

Mass of He is 4 u

4 g of helium

Formula mass of NaCl is 58 u

58 g of table salt

G.A.M., G.F.M., and Avogadro's number

■■■ Section 6-3 Review ■■■

Write the definitions for the following terms in your own words.

1. **G.A.M., gram-atomic mass**
2. **formula mass**
3. **G.F.M., gram-formula mass**
4. **Avogadro's number**

Answer these questions.

5. Determine the formula masses of the following compounds. Show your calculations.
   a. HF
   b. Al_2O_3
   c. $Mg(NO_3)_2$
   d. $(NH_4)_3PO_4$

6. You are told to use a balance to measure out one G.F.M. of NaOH and one G.F.M. of $Ca(OH)_2$. How much of each will you take?
7. Find either the G.A.M. or the G.F.M. of the following materials. Then list them in order of increasing mass.
   F_2, H_2S, K, Li_2O, P
8. a. Four grams of helium, 108 grams of silver, and 238 grams of uranium have something remarkable in common. What is that?
   b. Eighteen grams of water and 80 grams of NH_4NO_3 have something remarkable in common. What is that?

SCIENCE, TECHNOLOGY, & SOCIETY

Aluminum Cans and Trade-Offs

How often do you drink a can of soda? Did you know that people argue about the can itself? The issue is that most soda cans are made of aluminum.

One hundred years ago, aluminum was rarely used because it was hard to remove the element from its compounds. This changed around 1890, when a scientist, Charles Hall, discovered a process to obtain aluminum by using electricity.

As a result of Hall's technology, aluminum became widely available. Today, aluminum is used in airplanes, boats, household items, cans, and more. This widespread use of aluminum, especially the use of aluminum in cans, has led to arguments.

For example, large amounts of electricity are needed to make aluminum. Electricity is produced mainly by burning coal and oil; this increases air pollution. Electricity costs also make aluminum cans more expensive to produce than paper cartons, tin cans, and glass bottles.

Many people feel that the benefits of aluminum outweigh the problems. These people point out that aluminum cans are stronger than cartons, and so are easier to stack and ship. Aluminum is lighter than tin or glass, which greatly reduces shipping costs. Furthermore, aluminum metal always forms a tight layer of aluminum oxide, Al_2O_3, on its surface. This protects the metal against corrosion.

People must sometimes choose between two alternatives or make a compromise. This is called making a trade-off. Society makes decisions and trade-offs about technological issues through its purchases, attitudes, and laws. Aluminum cans are part of this continuing decision-making process.

Follow-up Activities

1. Use a science reference book or chemistry text to learn more about the Hall Process.
2. What makes you personally decide whether to buy juice in a bottle, in a can, or in a carton?

   Some people argue that a consumer is only interested in the *convenience* and the *cost* of an item, not whether it helps or hurts society. How do you respond? Cite some examples in giving your opinion.

Figure 6-20

KEEPING TRACK

Section 6-1 Writing Chemical Formulas

1. An element's symbol stands for the element or for one atom of the element.
2. A chemical formula tells how many atoms of each element are in a molecule.
3. In a chemical formula, a subscript shows how many atoms of an element are in a molecule. A coefficient shows the number of molecules of a compound or atoms of an element.
4. You can obtain formulas by using the criss-cross method.

Section 6-2 Writing Chemical Equations

1. A chemical equation gives the formulas and conditions of the reactants and products in a chemical change.
2. You can balance equations by adjusting the coefficients.
3. Except for the noble gases, all gas elements consist of diatomic molecules.

Section 6-3 Mass Relationships

1. You can find the formula mass of a molecule by adding the masses of its atoms.
2. The gram-atomic mass (G.A.M.) and gram-formula mass (G.F.M.) are taken in grams.
3. There are always 6×10^{23} atoms in one G.A.M. of an element and 6×10^{23} molecules in one G.F.M. of a compound.

BUILDING VOCABULARY

Write the term from the list that best completes each sentence.

subscript, balanced, diatomic, coefficient, products, chemical formula, formula mass, reactants, grams, gram-formula mass

In 3 H_2S, the term H_2S is called a ___1___, the *2* is called a ___2___, and the *3* is called a ___3___. The ___4___ in a reaction are written on the left side of the arrow, and the ___5___ are written on the right side. A ___6___ chemical equation has an equal number of atoms on the left and right sides. H_2, O_2, and Cl_2 are ___7___ gases. For the compound CH_4, the ___8___ is 16 u and the ___9___ is 16 g. Lithium has a mass of 7 u, so in 7 ___10___ of lithium there are 600,000,000,000,000,000,000,000 atoms.

SUMMARIZING

Write *true* if the statement is true. If the statement is false, change the *italicized* term to make the statement true.

1. S is the symbol for *sodium*.
2. *I* is the symbol for iron.
3. In 5 K_2CO_3, there are *three* elements.
4. In 5 K_2CO_3, there are *thirty* atoms.
5. NO_3 is an example of a *metal* ion.
6. To balance equations, you adjust the *subscripts*.
7. Helium gas consists of *single atoms*.
8. Chlorine gas consists of *diatomic molecules*.
9. The gram-formula mass of H_2 is *one gram*.
10. There are 6×10^{23} molecules in one *G.A.M.* of any compound.

INTERPRETING INFORMATION

Copy and complete Figure 6-21 on your own paper. It gives the charge of some positive and negative ions. Use these charges and the crisscross method to obtain the formulas for all the possible combinations.

Figure 6-21

Positive Ions and their charges	Negative Ions and their charges				
	Cl -1	ClO$_3$ -1	O -2	SO$_4$ -2	P -3
H $+1$					H$_3$P
NH$_4$ $+1$					
Ba $+2$					

Figure 6-22

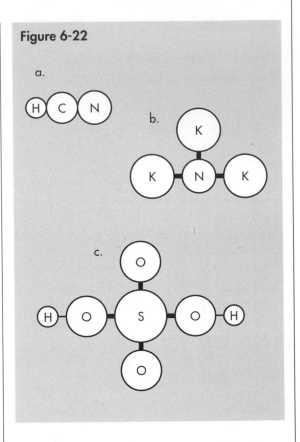

a.

b.

c.

THINK AND DISCUSS

Use the section number in parentheses to help you find each answer. Write your answers in complete sentences.

1. Write the formulas for the molecules in Figure 6-22. (6-1)
2. One mineral has the formula CaAl$_2$Si$_6$O$_{10}$
   a. What elements are in this mineral?
   b. How many atoms are in one molecule? (6-1)
3. How many atoms are in 100 Na$_2$SO$_4$ molecules? (6-1)

4. What formulas result if these are joined?
   a. Zn and I   b. K and O
   c. Si and P   d. Al and PO$_4$ (6-1)
5. The following reaction occurs in soda or seltzer.
   $$H_2CO_3 \rightarrow \quad H_2O \quad + \quad CO_2$$
   (soda water) (water) (carbon dioxide)
   a. Label the reactants and products.
   b. Is the equation balanced? (6-2)
6. Explain the difference between a nitrogen atom and a nitrogen molecule. (6-2)

7. Balance the following equations. Show your checking. (6-2)

$Cu + N_2 \rightarrow Cu_3N_2$

$B + Br_2 \rightarrow BBr_3$

$CH_4 + O_2 \rightarrow CO_2 + H_2O$

8. What is one gram-formula mass of these compounds? Begin by calculating their formula masses. (6-3)

a. HCl   b. Na_3N

c. NH_4ClO_3 d. $Mg(SCN)_2$

GOING FURTHER

1. Use reference books to find out why silver, tin, antimony, tungsten, gold, and mercury have unexpected symbols.

2. Find out more about Avogadro's work. How did he determine the number of atoms in one G.A.M. and the number of molecules in one G.F.M.?

COMPETENCY REVIEW

1. Look at Figure 6-23. The formula for the molecule shown in the figure is

a. C_2H_6 b. $2 CH_3$

c. $2 C_6H$ d. C_2H_6

Figure 6-23

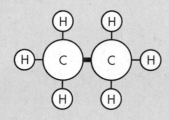

2. The formula for Epsom salts, a medicine, is $MgSO_4$. How many different elements are in Epsom salts?

a. 6   b. 2

c. 3   d. 4

3. The formula for sodium carbonate is Na_2CO_3. How many atoms are in one molecule of this material?

a. 6   b. 5

c. 3   d. 4

4. Water can be broken apart by electricity. The equation is

water $\xrightarrow{\text{electricity}}$ hydrogen gas + oxygen gas

Which is the reactant?

a. water   b. electricity

c. hydrogen gas   d. oxygen gas

5. Oxygen gas has the formula

a. O   b. 2O

c. O_2 d. OO

6. Which one of these items can you adjust to balance a chemical equation?

a. formulas   b. element symbols

c. subscripts   d. coefficients

7. Which equation is *not* balanced?

a. $B + P \rightarrow BP$

b. $B + N_2 \rightarrow BN$

c. $2 B + 3 S \rightarrow B_2S_3$

d. $2 B + 3 F_2 \rightarrow 2 BF_3$

8. The mass for H is 1 u, and for O it is 16 u. What is the formula mass of hydrogen peroxide, H_2O_2?

a. 17   b. 18

c. 34   d. 4

CHAPTER 7

COMMON CHEMICAL CHANGES

Do you like to watch or play baseball? If so, you probably know the names of players and the positions they play. You also know there are certain plays used during a game. Can you describe a hit-and-run, sacrifice flies or bunts, stealing a base, and walking a batter? Every sport or game has recognizable players and plays, actors and actions.

In chemistry there are millions of compounds and millions of reactions. All of these fall into certain categories and have specific names. With practice, you can recognize the type and name of a compound. With practice, you can also recognize the type and name of a reaction.

In this chapter, you will learn some principles involved in naming compounds. You will see, for example, that compounds formed from a metal and a nonmetal are named with one kind of ending in their name. Compounds having many-atom ions with oxygen have another kind of ending. You will also learn not only how compounds are formed but also how they may be broken down into simpler materials.

7-1 Naming Compounds

■ *Objectives*

☐ *Define a salt.*

☐ *Name compounds.*

☐ *Tell the meaning of* -ide, -ate, *and* -ite.

All compounds have a chemical name. Some compounds also have a common name. For example, H_2O has the chemical name hydrogen oxide, but everyone calls it water. A chemical name describes what is in a compound.

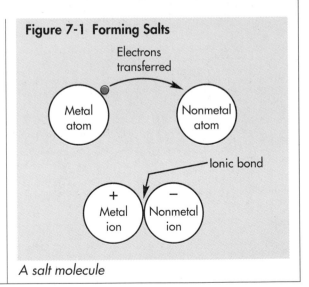

Figure 7-1 Forming Salts

Electrons transferred

Metal atom

Nonmetal atom

Ionic bond

+ Metal ion

− Nonmetal ion

A salt molecule

Compounds with Two Elements

A metal combines with a nonmetal to form a compound known as a **salt**. The metal atoms give electrons to the nonmetal atoms. In doing so, the metal ion becomes a positive ion and the nonmetal atom becomes a negative ion. They stay together because of their electric attraction. Look at Figure 7-1.

Ordinary table salt is one of thousands of salts. Table salt has the formula NaCl, which shows it to be a metal, sodium (Na), combined with a nonmetal, chlorine (Cl).

The chemical name of a salt has two parts. The first is the name of the metal. The second part is the name of the nonmetal. The names of most nonmetals, like chlorine and oxygen, end with *-ine* or *-en*. You change these endings to *-ide* in naming a salt made from two elements. The table in Figure 7-2 gives some examples.

Figure 7-3 shows electrons being transferred from aluminum to flourine to form the salt molecule aluminum fluor*ide* (AlF$_3$).

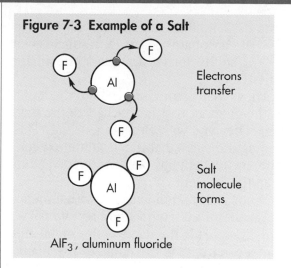

Figure 7-3 Example of a Salt

Electrons transfer

Salt molecule forms

AlF$_3$, aluminum fluoride

Compounds with Oxygen Groups

Many compounds contain three or more elements, one of which is oxygen. The oxygen is within a many-atom ion. As you learned in Chapter 6, a many-atom ion behaves like a nonmetal by joining with a metal. Some many-atom ions that contain oxygen include chlorate ions (ClO$_3$); chlorite

Figure 7-2			
Salt	**Metal**	**Nonmetal**	**Chemical name**
NaCl	sodium	chlorine	sodium chlor*ide*
HF	hydrogen	fluorine	hydrogen fluor*ide*
FeBr$_3$	iron	bromine	iron brom*ide*
H$_2$O	hydrogen	oxygen	hydrogen ox*ide*
Li$_3$N	lithium	nitrogen	lithium nitr*ide*
Al$_2$S$_3$	aluminum	sulfur	aluminum sulf*ide*

ions (ClO_2); nitrate ions (NO_3); nitrite ions (NO_2); carbonate ions (CO_3); sulfate ions (SO_4); sulfite ions (SO_3); and phosphate ions (PO_4).

These many-atom ions contain the name of the element that is attached to the oxygen. They end with either -*ate* or -*ite.* This ending indicates that they contain oxygen. The -*ite* ion contains one less oxygen atom than the -*ate* ion.

When many-atom ions join with a metal, the name of the compound has two parts. The first part is the name of the metal and the second part is the name of the many-atom ion. Figure 7-4 shows some examples.

Figure 7-4

Compound	Chemical name
$MgSO_4$	magnesium sulf*ate*
$MgSO_3$	magnesium sulf*ite*
$Al(ClO_3)_3$	aluminum chlor*ate*
K_2CO_3	potassium carbon*ate*
$NaNO_3$	sodium nitr*ate*
$Ca_3(PO_4)_2$	calcium phosph*ate*

Compounds with the NH_4 Ion

NH_4 is a many-atom ion that behaves like a metal by joining with a nonmetal. It is called the **ammonium ion.** Compounds starting with (NH_4) have ammonium as their first name. Figure 7-5 shows some examples.

Figure 7-5

Compound	Chemical name
NH_4Cl	ammonium chloride
NH_4ClO_3	ammonium chlorate
NH_4ClO_2	ammonium chlorite
$(NH_4)_2S$	ammonium sulfide
$(NH_4)_3PO_4$	ammonium phosphate

■ Section 7-1 Review ■

Write the definitions for the following terms in your own words.

1. **salt**
2. **ammonium ion**

Answer these questions.

3. Name these salts:
   NaBr H_2S
   BeO CCl_4
   Mg_3N_2 NH_4I
4. Name these compounds:
   $LiClO_3$ $Al(ClO_2)_3$
   $CaSO_3$ Na_3PO_4
   $(NH_4)_2SO_4$ NH_4NO_3
5. NH_3 is always written NH_3 and is always called ammonia. The formula should be reversed to H_3N so that the metal part comes first, but it never is reversed. From the formula H_3N, however, what would be the chemical name for ammonia?

6. Answer questions a through d by analyzing the chemical name.
   a. What two elements are in boron phosphide?
   b. What three elements are in zinc bromate?
   c. What three elements are in ammonium nitrate?
   d. What four elements are in ammonium fluorite?

7-2 Direct Union and Decomposition

■ *Objectives*

☐ *Illustrate direct-union and decomposition reactions.*

☐ *Describe the special features of oxidation, burning, and electrolysis reactions.*

☐ *Tell how to prepare oxygen, test for hydrogen, and decompose water.*

☐ *List some health problems related to burning.*

Have you ever seen coal or charcoal slowly burn with a red glow? Can you recall a beautiful display of fireworks? Did you know that both coal and fireworks burn by the same type of chemical reaction? Both require a union of elements with oxygen in the air. The atoms join to build larger molecules.

The reverse of this is occurring right now inside your body. Complicated molecules in the food you ate are breaking apart into simpler molecules, and even into atoms.

Direct Union with Oxygen

In a **direct-union reaction,** two elements combine to form a single compound. This is also called a synthesis reaction, for *synthesis* means "to put together." An example of direct union is an element joining with oxygen to form an oxide, which is called an **oxidation** reaction. Many oxidation reactions are slow. If the oxidation is rapid, and heat and light are released, you call the reaction **burning.**

Coal and charcoal burn. They consist of carbon. Depending on how much oxygen is available for burning, two different oxides of carbon form.

$$C + O_2 \longrightarrow CO_2 \qquad + \text{ heat } + \text{ light}$$

| sufficient oxygen | carbon dioxide | "burning" |

$$2C + O_2 \longrightarrow 2CO \qquad + \text{ heat } + \text{ light}$$

| insufficient oxygen | carbon monoxide |

Caution Carbon monoxide (CO) is a deadly poisonous gas. Carbon dioxide (CO_2) is not. When you burn charcoal, or any fuel, the fire and the room should be ventilated. Make sure that plenty of air is moving around. When a fire burns poorly because little air is getting to it, there is a danger that carbon monoxide will form. With more air, the fire will produce carbon dioxide, which is safer. Even though CO_2 gas is not poisonous, however, do not let it build up in a room. It is not healthy to breathe air containing too much CO_2. Ventilation prevents this problem.

Metals and Oxygen

Magnesium metal is the main ingredient in fireworks. Once you light it, magnesium burns rapidly and produces a bright light. Even iron can burn in pure oxygen. Below are the equations for these oxidation reactions.

$$2 \text{ Mg} + O_2 \rightarrow 2 \text{ MgO}$$
$$4 \text{ Fe} + 3 \text{ } O_2 \rightarrow 2 \text{ Fe}_2O_3$$

Caution When a piece of magnesium metal burns, the light it gives off is very bright and contains harmful ultraviolet rays. You should avoid staring directly at the flame. Instead, glance at it quickly and sideways. The same warning applies to sparklers, which contain magnesium metal.

Not all oxidations are fast. For example, iron slowly rusts in moist air. Rusting is slow oxidation. Aluminum, magnesium, and the active metals of Groups 1 and 2 on the Periodic Table also join with oxygen slowly, over time. Below are some equations of slow oxidation.

$$4 \text{ Fe} + 3 \text{ } O_2 \rightarrow 2 \text{ Fe}_2O_3 \textbf{ (slow rusting)}$$
$$4 \text{ Na} + O_2 \rightarrow 2 \text{ Na}_2O$$
$$2 \text{ Ca} + O_2 \rightarrow 2 \text{ CaO}$$

Hydrogen gas rapidly joins with oxygen. In fact, a hydrogen-oxygen mixture is very dangerous and will explode with a spark or flame. Examine this equation.

$$2 \text{ H}_2 + O_2 \rightarrow 2 \text{ H}_2O$$

This reaction happens in the test for the presence of hydrogen. When collected and ignited in a test tube, hydrogen "pops" as it rapidly combines with oxygen. This reaction allows a worker to use a hydrogen torch to cut and weld metal. Look at Figure 7-6. At the tip of the torch, pure hydrogen and oxygen gas meet and burn with a very hot, clean flame. As you saw in the equation above, water is the only product. It disappears as steam, because the torch is so hot.

Nonmetals and Oxygen

Nonmetal elements also can combine with oxygen but in a different way. Metal atoms *transfer* electrons to oxygen atoms, but nonmetal atoms *share* electrons with oxygen atoms.

Sulfur, a nonmetal, can burn, and can form two different oxides. Look at the following equations:

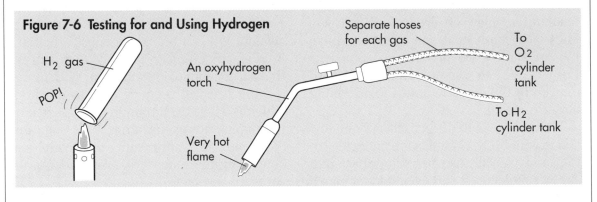

Figure 7-6 Testing for and Using Hydrogen

H₂ gas

POP!

An oxyhydrogen torch

Very hot flame

Separate hoses for each gas

To O₂ cylinder tank

To H₂ cylinder tank

$$S + O_2 \rightarrow SO_2 \text{ (sulfur dioxide)}$$
$$2\ S + 3\ O_2 \rightarrow 2\ SO_3 \text{ (sulfur trioxide)}$$

These direct-union reactions occur when scientists make sulfuric acid, an important industrial chemical.

Nitrogen, another nonmetal, can form a series of oxide compounds, depending on the reaction conditions. Compare these equations.

$$N_2 + O_2 \rightarrow 2\ NO$$
$$N_2 + 2\ O_2 \rightarrow 2\ NO_2$$
$$N_2 + 3\ O_2 \rightarrow 2\ NO_3$$
$$2\ N_2 + 5\ O_2 \rightarrow 2\ N_2O_5$$

These direct-union reactions occur when scientists make nitric acid, another important industrial chemical.

Other Direct-Union Reactions

Direct-union reactions can occur between metals and the halogen elements of Group 17 in the periodic table. For example, iron wool can burn in pure chlorine. In another interesting case, the powder form of the metal antimony (Sb) can be sprinkled into a bottle of chlorine. It sparkles and leaves tiny trails of smoke. Compare the reactions in Figure 7-7.

The equations for these direct-union reactions appear below.

$$2\ Fe + 3\ Cl_2 \rightarrow 2\ FeCl_3 \text{ (iron chloride)}$$
grey solid + green gas $\rightarrow$ red solid
$$2\ Sb + 3\ Cl_2 \rightarrow 2\ SbCl_3 \text{ (antimony chloride)}$$

When a mixture of iron filings and sulfur powder are heated, a new compound forms by direct union. Iron filings are magnetic, which means they attract each other. However, the new material, iron sulfide, is not magnetic and has other properties that differ from those of its "parents," iron and sulfur. Here is the equation.

$$2\ Fe + 3\ S \rightarrow Fe_2S_3 \text{ (iron sulfide)}$$

Decomposition Reactions

Certain reactions are the opposite of direct-union reactions. Recall that in a direct-union

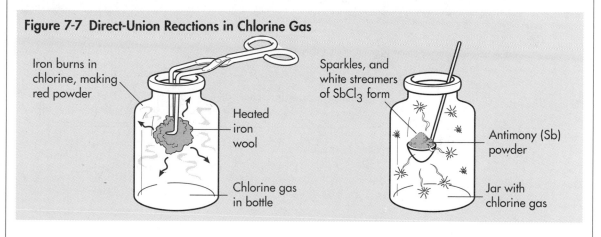

Figure 7-7 Direct-Union Reactions in Chlorine Gas

Iron burns in chlorine, making red powder

Heated iron wool

Chlorine gas in bottle

Sparkles, and white streamers of $SbCl_3$ form

Antimony (Sb) powder

Jar with chlorine gas

reaction, elements combine to create a new material. In a **decomposition reaction,** a compound breaks apart into its elements, into simpler compounds, or into both.

The English chemist Joseph Priestley used a famous decomposition reaction in 1774 to prepare pure oxygen gas. He first used a lens, or a curved piece of glass, to focus sunlight on mercury oxide, HgO. Figure 7-8 shows how the heat and light broke the compound into its elements, mercury and oxygen. Here is the equation.

$$2\ HgO \rightarrow\ 2\ Hg\ +\ O_2$$
red solid → silver liquid + colorless gas

Caution Mercury metal and its compounds are poisonous. Today, researchers use safer methods of preparing oxygen. These methods involve the decomposition of hydrogen peroxide, H_2O_2, or potassium chlorate, $KClO_3$. Preparing oxygen also involves using a catalyst, which is a substance that speeds up a reaction. A catalyst affects a chemical reaction, but is not changed by it. A catalyst of manganese dioxide, MnO_2, is now used to speed up the decomposition reactions that produce oxygen. Look at these equations.

$$2H_2O_2 \xrightarrow{\text{catalyst}} 2H_2O\ +\ O_2$$

$$2KClO_3 \xrightarrow{\text{catalyst}} 2KCl\ +\ 3O_2$$

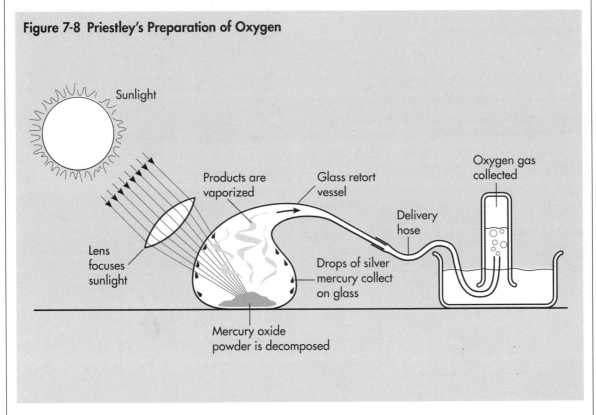

Figure 7-8 Priestley's Preparation of Oxygen

Sunlight

Lens focuses sunlight

Products are vaporized

Glass retort vessel

Delivery hose

Oxygen gas collected

Drops of silver mercury collect on glass

Mercury oxide powder is decomposed

In these decomposition reactions, one compound is breaking apart into a simpler compound and an element. Notice that the word *catalyst* is written above the arrow. When you write a chemical equation, special conditions are shown on the arrow. This includes catalyst, electricity, and pressure, if needed.

Decomposition over Time, and by Light

Many compounds, especially in foods, decompose naturally over time. The warmth of the air is enough to decompose these compounds. This is why foods "go bad." The products of their decomposition usually have foul odors.

Some compounds decompose more quickly in light. To prevent this, they are put in dark bottles or in sealed boxes. Hydrogen peroxide is a hair bleach and an antiseptic for skin cuts. Mercurochrome is an even stronger antiseptic medicine. Both materials are kept in dark bottles so that they will not lose their effectiveness, for they easily decompose in light.

Electrolysis of Water

You can decompose water into its elements by passing electricity through the water. A decomposition reaction that uses electricity is called **electrolysis**. Look at this equation.

$$2H_2O \xrightarrow{\text{electricity}} 2H_2 + O_2$$

The added energy, written above the arrow, is electricity. Electrolysis of water is usually performed in a piece of laboratory equipment called the Hoffman apparatus, which is illustrated in Figure 7-9.

Figure 7-9 Electrolysis of Water in a Hoffman Apparatus

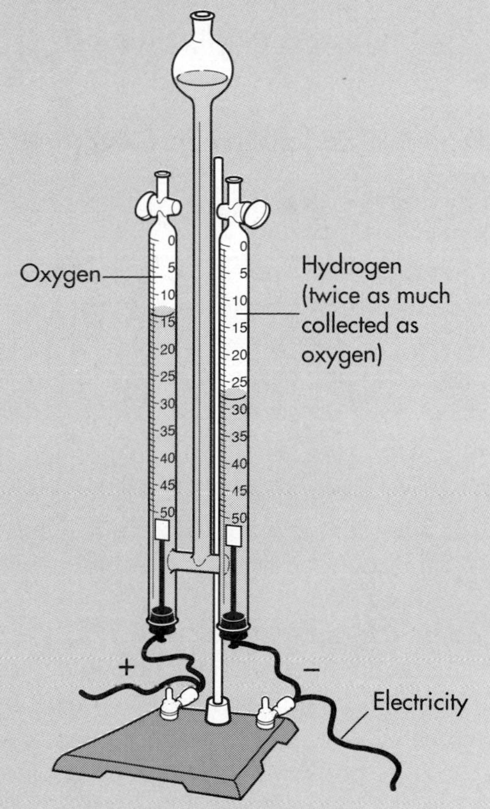

Oxygen

Hydrogen (twice as much collected as oxygen)

+

−

Electricity

Energy in Decomposition Reactions

Energy usually is released when elements form by direct-union reactions. This happens in oxidation, for example. Decomposition is the opposite of direct union. As you might guess, decomposition reactions usually require energy in the form of light, heat, or electricity. Compounds usually do not break down by themselves.

The energy released during the formation of a compound must be added back to make the compound decompose. This is

What Happens in Direct-Union Reactions?

Process Skills observing; interpreting data

Materials fine iron filings or iron powder; sulfur powder; test tubes; test tube holder; Bunsen burner; small glass plate or watchglass; ceramic plate; magnets; copper sulfate crystals

Procedures

1. Mix the iron and sulfur in a ratio of 60 percent iron to 40 percent sulfur. Fill about one third of a test tube with the mixture.

2. Using a test tube holder, carefully heat the mixture. Gently move the test tube in the flame to heat the mixture evenly. Once the mixture begins to glow from within, remove it from the flame. The mixture will continue to react and glow on its own. Do not heat the mixture so strongly that the sulfur boils up and burns at the mouth of the test tube. Should this occur, simply put the flame out with a flat glass plate or watchglass.

   Caution This experiment should be done in a well-ventilated room because some SO_2 gas is always formed as sulfur burns in the tube.

3. Let the test tube cool and break the test tube to empty the contents onto a glass plate. Inspect the solid material that has formed by comparing it to the original elements for appearance, color, and magnetic property. Write down your observations.

Conclusions

4. What type of reaction is illustrated in this exercise? What happened that made you think a chemical reaction occurred? What safety precautions were used in performing these activities? Why was each needed?

---

how chemical reactions balance each other over time. Another way to say this is that direct-union reactions are usually exothermic, meaning that they release energy. Decomposition reactions are usually endothermic; they absorb energy. This is summarized in the following general equations.

For direct union reactions:
element + element → compound + energy

For decomposition reactions:
compound + energy → element + element
compound + energy → element + a simpler compound
compound + energy → simpler compounds

Section 7-2 Review

Write the definitions for the following terms in your own words.

1. **direct-union reaction**
2. **oxidation**
3. **burning**
4. **decomposition reaction**
5. **electrolysis**

Answer these questions.

6. The tip of a match contains phosphorus. When you strike the match, heat makes the following reaction occur.

$$4\ P + 5\ O_2 \rightarrow 2\ P_2O_5$$

Why is this reaction an example of
a. direct union? b. oxidation?
c. burning?

7. How would you classify each of these reactions?

a. $3H_2 + N_2 \rightarrow 2NH_3$

b. $2NaCl \xrightarrow{\text{electricity}} 2Na + Cl_2$

c. $Mg(OH) \rightarrow MgO + H_2O$

d. $4K + O_2 \rightarrow 2K_2O$

e. $4HNO_3 \rightarrow 4NO_2 + 2H_2O + O_2$

8. How does the decomposition of HgO illustrate the property changes that happen in chemical changes?
9. What safety precautions should you take when burning coal or any other fuel?
10. a. Why are some medicines kept in dark bottles?
    b. Why are some medicines best stored in a refrigerator?

7-3 Single and Double Replacements

■ *Objectives*

☐ *Identify single- and double-replacement reactions.*

☐ *Use a chemical activity series to predict whether one element can replace another in a compound.*

☐ *List the products that make a double-replacement/ion-exchange reaction happen.*

Have you ever played or watched tennis? Imagine that, in a friendly game of tennis doubles, the four players agree to switch partners, just for a change. Maybe you have experienced such a switch in tennis, or when friends "cut in," or change partners, at a dance. Similar exchanges can occur among atoms and molecules in certain chemical reactions. In this section, you will learn how and why these exchanges take place.

Single-Replacement Reactions

In a **single-replacement reaction,** one element replaces another element in a compound. There are two types of single-replacement reactions. In Type 1, a metal replaces another metal in a compound. For example, copper may replace silver. Look at this equation.

$$Cu + 2\ AgNO_3 \rightarrow 2\ Ag + Cu(NO_3)_2$$

copper + silver nitrate → silver + copper nitrate

Now look at Figure 7-10 which shows copper replacing silver in a compound.

Figure 7-10 Copper Replacing Silver in a Compound

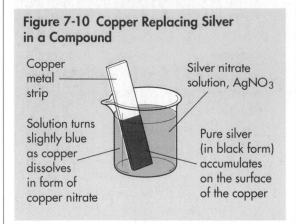

Copper metal strip

Silver nitrate solution, AgNO₃

Solution turns slightly blue as copper dissolves in form of copper nitrate

Pure silver (in black form) accumulates on the surface of the copper

The pattern in Type-1 reactions is that metal B replaces metal A, or "pushes" metal A out of its compound.

$$B + AX \longrightarrow A + BX$$

In Type-2 single-replacement reactions, a nonmetal replaces another nonmetal in a compound. For example, chlorine may replace bromine. Look at this equation, and at Figure 7-11.

**Cl_2 + $MgBr_2$ → Br_2 + $MgCl_2$
chlorine + magnesium bromide →
bromine + magnesium chloride**

The pattern in Type-2 single-replacement reactions is that nonmetal Y replaces nonmetal X, or "pushes" nonmetal X out of its compound.

$$Y + AX \longrightarrow X + AY$$

Figure 7-11 Chlorine Replacing Bromine in a Compound

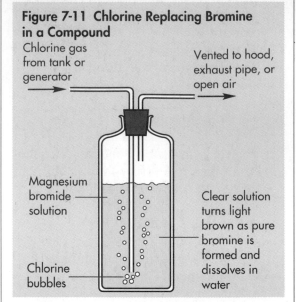

Chlorine gas from tank or generator

Vented to hood, exhaust pipe, or open air

Magnesium bromide solution

Clear solution turns light brown as pure bromine is formed and dissolves in water

Chlorine bubbles

Examples of Single Replacement

If you place a strip of zinc metal (Zn) into a solution of copper sulfate ($CuSO_4$), a single-replacement reaction will occur. Look at Figure 7-12. Zinc metal in the strip will take the place of copper in the copper sulfate solution. Copper metal will "come out" and stick to the zinc strip, making it look orange. Here is the equation for this reaction.

$$Zn + CuSO_4 \rightarrow Cu + ZnSO_4$$

$$zinc + \frac{copper}{sulfate} \rightarrow copper + \frac{zinc}{sulfate}$$

Acids contain hydrogen. In another example of a single-replacement reaction, some metals can replace the hydrogen in an acid. Figure 7-13 shows the standard laboratory procedure for making hydrogen gas. In this case, zinc metal is mixed with hydrochloric acid, which is hydrogen chloride, or HCl, dissolved in water. Here is the equation.

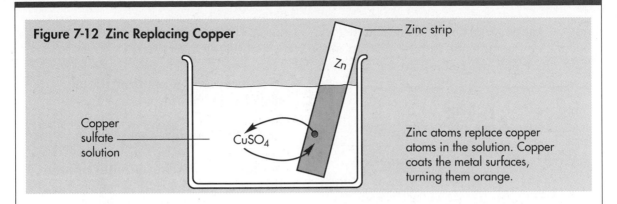

Figure 7-12 Zinc Replacing Copper

Zinc strip

Copper sulfate solution

$CuSO_4$

Zinc atoms replace copper atoms in the solution. Copper coats the metal surfaces, turning them orange.

$$Zn + 2HCl \rightarrow H_2 + ZnCl_2$$

$$zinc + \begin{matrix} \text{hydrogen} \\ \text{chloride} \end{matrix} \rightarrow hydrogen + \begin{matrix} \text{zinc} \\ \text{chloride} \end{matrix}$$

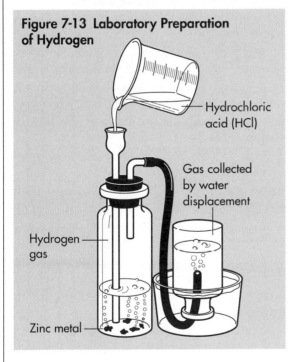

Figure 7-13 Laboratory Preparation of Hydrogen

Hydrochloric acid (HCl)

Gas collected by water displacement

Hydrogen gas

Zinc metal

Chemical Activity

Each element has its own degree of **chemical activity**, which is the ability to replace another element in a compound. An element can only replace elements with a lower chemical activity than its own. A table can show the chemical activity of elements, from most active to the least active. Such a table is called an "activity series."

An element that has a higher place in the series is more chemically active. The most active metal is potassium, and the least active metal is gold. The most active nonmetal is fluorine, and the least active nonmetal is iodine. An element in a compound can only be replaced by an element of higher activity. For example, zinc in a compound can only be replaced by the elements listed above it in the activity series. Study this example.

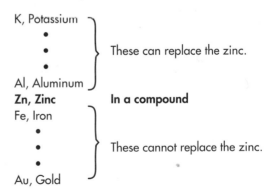

K, Potassium
•
•
•
Al, Aluminum

These can replace the zinc.

Zn, Zinc — **In a compound**

Fe, Iron
•
•
•
Au, Gold

These cannot replace the zinc.

Not all reactions can occur. Using the activity series, you can predict whether or not a single replacement reaction can occur. See the table in Figure 7-14 to see how elements are related to each other.

Figure 7-14 Activity Series (decreasing activity going down)

Metals	Nonmetals
Potassium (K)	Fluorine (F)
Calcium (Ca)	Chlorine (Cl)
Sodium (Na)	Bromine (Br)
Magnesium (Mg)	Iodine (I)
Aluminum (Al)	
Zinc (Zn)	
Iron (Fe)	
Nickel (Ni)	
Tin (Sn)	
Lead (Pb)	
Copper (Cu)	
Hydrogen (H)	
Mercury (Hg)	
Silver (Ag)	
Gold (Au)	

Compare these two equations with the position of the elements in the activity series.

Metal replacing another metal
nickel + tin chloride → tin + nickel chloride
lead + tin chloride → no reaction

Nonmetal replacing another nonmetal
bromine + tin chloride → no reaction
fluorine + tin chloride → chlorine + tin fluoride

Most metals are above hydrogen in the activity series. This means that most metals can replace the hydrogen in acids. Look again at Figure 7-13. Active metals like magnesium, aluminum, and zinc quickly liberate, or replace, the hydrogen in acids. Less-active metals, such as iron and copper, replace hydrogen at a slower rate. Mercury, silver, and gold cannot replace the hydrogen in acids.

Ionic Solutions

Recall that ions are atoms with an electric charge. Ions form when a metal atom gives electrons to a nonmetal atom. The opposite charges on the ions then keep the molecule together in an ionic bond. This type of compound is therefore called an **ionic compound.**

When ionic compounds dissolve in water, the molecules do not stay together. Instead, they break into their ion parts. Water has the ability to separate the ions. The solution that results contains ions, not molecules, so it is called an **ionic solution**. Examine Figure 7-15.

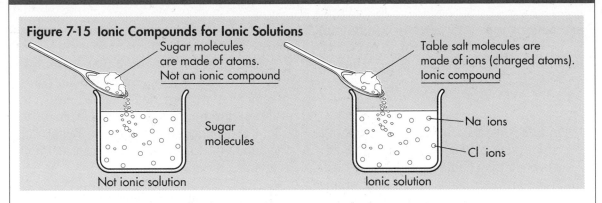

Figure 7-15 Ionic Compounds for Ionic Solutions

Sugar molecules are made of atoms.
Not an ionic compound

Sugar molecules

Not ionic solution

Table salt molecules are made of ions (charged atoms).
Ionic compound

Na ions

Cl ions

Ionic solution

As you can see, sugar is not an ionic compound. This means that its molecules are not made of ions. Therefore, when you dissolve sugar in water, the individual sugar molecules remain.

On the other hand, table salt is an ionic compound. Its molecules consist of ions. Therefore, when you dissolve table salt in water, the individual molecules break apart into their ions. The solution contains Na^+ and Cl^- ions, not NaCl molecules. Salty water is an ionic solution.

Double-Replacement/Ionic-Exchange Reactions

Two compounds may react to form one or more other compounds. For example, think about this equation.

$$NH_3 + HCl \rightarrow NH_4Cl$$

ammonia + $\dfrac{\text{hydrogen}}{\text{chloride}} \rightarrow \dfrac{\text{ammonium}}{\text{chloride}}$

In most cases, however, two compounds exchange their metal parts, or their metal ions, and the reaction produces two new compounds. This is called a **double-replacement reaction**. Look at the following equation, and at Figure 7-16.

$$NaCl + AgNO_3 \rightarrow AgCl + NaNO_3$$

$\dfrac{\text{sodium}}{\text{chloride}} + \dfrac{\text{silver}}{\text{nitrate}} \rightarrow \dfrac{\text{silver}}{\text{chloride}} + \dfrac{\text{sodium}}{\text{nitrate}}$

Notice that in this reaction sodium (Na) and silver (Ag) exchange places. Because a double-replacement reaction involves

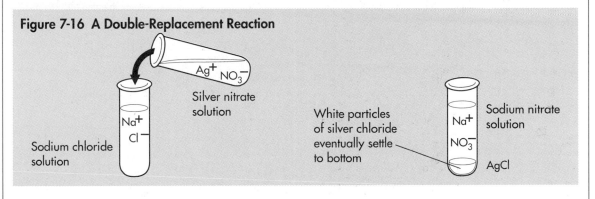

Figure 7-16 A Double-Replacement Reaction

Ag^+ NO_3^-

Silver nitrate solution

Na^+
Cl^-

Sodium chloride solution

White particles of silver chloride eventually settle to bottom

Sodium nitrate solution

Na^+
NO_3^-

AgCl

exchanging ions between two compounds, it is also called an **ionic exchange reaction.** Look at this simple equation.

$$AX + BY \longrightarrow BX + AY$$

Double-replacement or ionic-exchange reactions work best in a solution with water. This is because a compound usually breaks into ions in water. However, if you mix *two* compounds in water, there is no guarantee that a reaction will occur. You may just have a mixture of many ions. Look at Figure 7-17.

In order for an ionic-exchange reaction to occur, a product must be formed that "removes" some of the ions from the solution. There are three such products. Some are precipitates, which are solid particles that settle to the bottom in a solution. You can remember what precipitate means by thinking of precipitation, such as falling rain that lands on the ground. In an equation, precipitates are shown by a downward-pointing arrow ↓.

Other products that cause ionic-exchange reactions are gases. In equations,

these are shown by an upward-pointing arrow ↑. The third special product is water.

You should keep in mind that a **precipitate** is a material of low solubility. This means that it does not dissolve easily in water or another liquid. You will learn more about precipitates in Chapter 8. Below are the equations for these double-replacement reactions.

$$Na_2CO_3 + MgSO_4 \rightarrow MgCO_3\downarrow + Na_2SO_4$$
insoluble
precipitate formed

$$CaO + H_2SO \rightarrow H_2O + CaSO_4$$
water formed

$$CaS + H_2SO \rightarrow H_2S\uparrow + CaSO_4$$
gas formed

The reaction between an acid and a base produces a salt and water. These are ionic-exchange or double-replacement reactions. Study this equation.

$$HCl + NaOH \rightarrow NaCl + H_2O$$
(acid) (base) (a salt) (water)

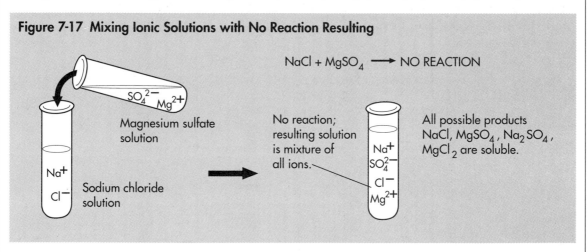

Figure 7-17 Mixing Ionic Solutions with No Reaction Resulting

$NaCl + MgSO_4 \longrightarrow$ NO REACTION

Magnesium sulfate solution

SO_4^{2-} Mg^{2+}

Sodium chloride solution

Na^+
Cl^-

No reaction; resulting solution is mixture of all ions.

Na^+
SO_4^{2-}
Cl^-
Mg^{2+}

All possible products NaCl, $MgSO_4$, Na_2SO_4, $MgCl_2$ are soluble.

You will study acids and bases in detail in Chapter 8. For now, notice that this reaction is forced to happen because water is a product.

How Can You Observe Single and Double Replacements?

Process Skills observing; recording information; interpreting data

Materials saturated copper sulfate solution ($CuSO_4$); diluted zinc nitrate solution; lead, zinc, iron, and magnesium, sanded clean; salt solutions such as copper sulfate, barium chloride, potassium iodide, sodium bromide, aluminum sulfate, potassium dichromate, ferrous (iron II) sulfate, ferric (iron III) sulfate, silver nitrate; diluted sodium hydroxide should be provided to each group.

Part A
1. Place lead, zinc, iron, and magnesium into separate test tubes containing the $CuSO_4$ solution.
2. Note the changes that take place on the surface of each metal at intervals of several minutes. Compare the time it takes for the reaction to occur with each metal. Write down your observations.

3. Repeat steps 1 and 2, using the diluted zinc nitrate solution.
Caution Avoid getting solutions on your hands and body. Wash your hands after each experiment. These solutions are toxic.

Part B
4. Combine various pairs of the salt solutions together in a test tube. Use about 5 mL of each solution and combine just two solutions for each test.
Be sure to include combinations involving the diluted sodium hydroxide (NaOH) solution.
5. Make a table that lists all the materials you have tested along the top and again along the side. Record the results of each combination in the boxes where the materials intersect.

Conclusions
6. What are the implications of the results regarding the activity level of each metal?
7. In what combinations of salt solutions did a reaction happen? How do you know?

Interpreting Data

Look at Figure 7-18. This table tells how well some compounds dissolve in water. You can use the table to figure out many possible reactions. For example, you could find out if a material will be a precipitate in a reaction. In the table, a precipitate is any material marked "i," and is nearly insoluble.

1. Find calcium chloride on the table. It is labeled *s* for soluble. Now find calcium phosphate. It is labeled *i* for insoluble. Calcium phosphate would be a precipitate if it was formed by a reaction.

2. Find calcium oxide on the table. It is labeled *ss* for slightly soluble. Now find calcium sulfide. It is labeled *d,* which means it decomposes when you dissolve it in water. It reacts with the water.

3. What is the solubility of the following materials in water?
   a. aluminum acetate (acetate is a many-atom ion)
   b. copper bromide
   c. zinc carbonate
   d. sodium oxide

4. Determine the solubility of the following compounds.

a. $MgCO_3$
b. KI
c. Ag_3PO_4
d. NH_4Cl
e. $(NH_4)_2SO_4$

5. List the compounds that decompose when placed into water.

6. a. Which metals have compounds that are almost always soluble, with only one or two exceptions?
   b. Which metals have compounds that are almost always insoluble or slightly soluble?

7. a. Which nonmetals or many-atom ions have compounds that are almost always soluble, with only one or two exceptions?
   b. Which nonmetals or many-atom ions have compounds that are almost always insoluble or slightly soluble?

8. Use this table to predict which of these possible ionic exchange reactions will actually occur.
   a. sodium + silver →
      bromide   nitrate
   b. barium + potassium →
      nitrate   phosphate
   c. magnesium + aluminum →
      chloride   nitrate
   d. ammonium + zinc →
      acetate   sulfate

Figure 7-18 Table of Solubilities in Water

	Acetate	Bromide	Carbonate	Chloride	Hydroxide	Iodide	Nitrate	Oxide	Phosphate	Sulfate	Sulfide
Aluminum	s	s	n	s	i	s	s	i	i	s	d
Ammonium	s	s	s	s	s	s	s	n	s	s	s
Barium	s	s	i	s	s	s	s	s	i	i	d
Calcium	s	s	i	s	ss	s	s	ss	i	ss	d
Copper II (cuprous)	s	s	i	s	i	d	s	i	i	s	i
Magnesium	s	s	i	s	i	s	s	i	i	s	d
Potassium	s	s	s	s	s	s	s	s	s	s	s
Silver	ss	i	i	i	n	i	s	i	i	ss	i
Sodium	s	s	s	s	s	s	s	d	s	s	s
Zinc	s	s	i	s	i	s	s	i	i	s	i

i = Nearly insoluble

ss = Slightly soluble

s = Soluble

d = Decomposes

n = Does not exist

Section 7-3 Review

Write the definitions for the following terms in your own words.

1. **single-replacement reaction**
2. **double-replacement or ionic exchange reaction**
3. **ionic compound**
4. **ionic solution**

Answer these questions.

5. Complete these single-replacement word equations.
   a. aluminum + iron nitrate →
   b. magnesium + gold sulfate →
   c. calcium + hydrogen chloride (an acid) →
   d. A (a metal) + BY →
   e. X (a nonmetal) + BY →
6. Complete these double-replacement word equations.
   a. lead nitrate + sodium bromide →
   b. hydrogen iodide + calcium oxide →
   c. AX + BY →
7. What is necessary for an ionic-exchange reaction to occur?

SCIENCE, TECHNOLOGY, & SOCIETY

Air Pollution and Oxides

Most air pollution is the result of oxides produced during burning. For example, cars, trucks, and planes burn fuel, and their exhaust gases contain oxides. In 1983, these forms of transportation produced 66 million tons of pollution.

Carbon monoxide (CO) creates many health problems. This poisonous gas makes up 72 percent of the waste from cars, trucks, and planes. Cars also make nitrogen oxide (NO_2), which causes acid rain.

Figure 7-19

Acid rain damages trees, crops, lakes, and buildings and makes drinking water less pure.

Sulfur dioxide (SO_2) comes from burning sulfur, from volcanos, and from burning coal that contains sulfur. In the air, SO_2 can change to SO_3, or sulfur trioxide, which also causes acid rain.

New technology may improve the situation. Scientists are looking for ways to reduce sulfur from coal. New laws have forced car makers to design engines that give off less pollution. Cars now must have special parts called catalytic converters, which change CO into carbon dioxide (CO_2). Industry must now "scrub" NO_2 and SO_2 from furnace gases.

But cleaning up the oxides caused by vehicles and by burning coal has a high price. Adding catalytic converters to car engines increases the price of cars and decreases the gas mileage, so more gasoline must be used. The Environmental Protection Agency estimates that reducing the sulfur dioxide in air by 40 percent will cost consumers about $3 billion per year. In addition, the enforcement of air-pollution laws could cause some industries to close down, which would mean the loss of thousands of jobs.

People and government must decide whether they are willing to pay the price for cleaner air. That price could be fewer jobs and higher costs for products such as cars.

Follow-up Activities
1. Look in newspapers or magazines for articles related to air pollution. Collect them, summarize what they say, and present a report to your class.
2. Try to find out how your community or city is threatened by air pollution. What groups are involved in trying to improve the situation?

██████████ **KEEPING TRACK** ██████████

Section 7-1 Naming Compounds

1. A salt consists of a metal and a nonmetal. The metal part is named before the nonmetal part.
2. The suffix *-ide* means only two elements are in a compound; *-ate* and *-ite* mean there are three or more elements, one of which is oxygen.

Section 7-2 Direct Union and Decomposition

1. In direct-union reactions, two elements join and form a compound. Oxidation is a direct-union reaction involving oxygen. Burning is rapid oxidation, and it produces light and heat.
2. Many metals and some nonmetals can combine with oxygen.
3. In decomposition reactions, a compound breaks into simpler compounds or elements. Elcctrolysis is a decomposition reaction in which electricity provides energy.
4. Direct-union reactions usually release energy, and decomposition reactions usually rcquire energy.
5. Some compounds decompose naturally over time. Some decompose more readily in light.

Section 7-3 Single and Double Replacements

1. An element can replace an element less active than itself in a compound.
2. Some metals can replace the hydrogen in acids, releasing hydrogen gas.
3. Solutions of ionic compounds consist of free ions, not molecules.

4. Double-replacement, or ionic-exchange, reactions will occur between compounds in solutions if a precipitate, a gas, or water is formed.

██████████ **BUILDING VOCABULARY** ██████████

Write the term from the list that best matches each statement.

burning, chemical activity, electrolysis, ionic solution, oxidation, oxygen, precipitate, salt, direct union

1. compound with a metal and a nonmetal part
2. in all compounds ending with *-ate*
3. combining of a metal element and oxygen
4. term for release of visible light and heat
5. reaction involving electrical energy
6. ability to replace another element in a compound
7. reaction in which two elements form a compound
8. water containing dissolvcd ions, not molecules
9. an insoluble material

Explain the difference between the terms in each pair.

10. *-ide* compounds, *-ate* compounds
11. *-ite* compounds, *-ate* compounds
12. direct union, decomposition
13. oxidation, burning
14. more-active element, less-active element

SUMMARIZING

Write the missing term for each sentence.

1. NaF is called sodium ___.
2. Na_2SO_4 is called sodium ___.
3. $NaClO_2$ is called sodium ___.
4. $(NH_4)_2CO_3$ is called ___ carbonate.
5. An element in a compound can only be replaced by an element that is ___.

INTERPRETING INFORMATION

Figure 7-20 shows some diagrams of chemical reactions. Each shape represents a different kind of atom. Study the figure and answer questions 1-4.

1. What type of reaction is illustrated in each example, direct union, decomposition, single replacement, or double replacement?
2. Match one of the illustrations to each of the following reactions.

$$Cu + S \rightarrow CuS$$
$$CaCO3 \underset{heat}{\rightarrow} CaO + CO2$$

3. Which reaction probably must take place in a water solution?
4. In part a of the figure, why must one of the products shown be a gas or a precipitate?

THINK AND DISCUSS

Use the section number in parentheses to help you find each answer. Write your answers in complete sentences.

1. Name the following compounds.
   a. Li_2O b. $MgCO_3$
   c. HBr   d. $Be(ClO_3)_2$ (7-1)
2. What elements are in each of these many-atom ions?
   a. sulfate   b. phosphate
   c. sulfite   d. ammonium (7-1)

Figure 7-20

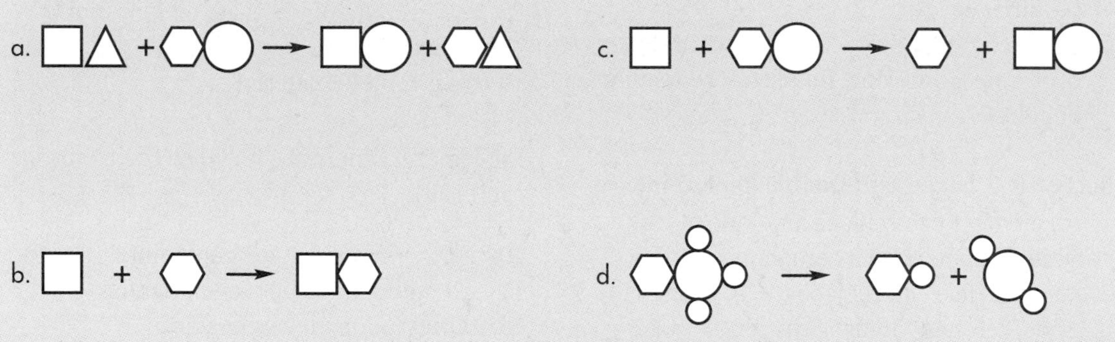

3. How many elements are in
   a. boron nitride?
   b. boron nitrate?
   c. copper chloride?
   d. ammonium phosphate? (7-1)
4. a. Name four elements that may form two or more oxides and give examples of each.
   b. What is the main factor that controls which oxide will form? (7-2)
5. What is the difference between the way a metal combines with oxygen and the way a nonmetal combines with oxygen? (7-2)
6. Describe two ways to prepare oxygen in the laboratory. Make a sketch of each procedure. (7-2)
7. Describe two ways of preparing hydrogen in the laboratory. Make a sketch of each procedure. (7-2, 7-3)
8. Usually, what reactions require energy and what reactions release energy? In what forms might you observe released energy? (7-2)

GOING FURTHER

Many industrial processes make chemicals by using the types of reactions discussed in this chapter. Below is a list of some of these important processes. Your teacher may add others.

Working with a group, investigate some of the processes on this list and prepare a report. Use a chemical textbook or a science encyclopedia. In your report, describe the raw materials with which scientists begin, how the process works, and what types of reactions take place during the process.

- Using a Downs cell to produce sodium or potassium
- Using the Solvay process to produce sodium carbonate
- Getting magnesium from sea water
- Softening "hard" water
- Making plaster of paris and cement
- Producing iron in a blast furnace
- Refining copper
- Producing aluminum from bauxite ore
- Welding thermite
- Using the Haber process to get ammonia

COMPETENCY REVIEW

1. $A + X \rightarrow AX$ is the general equation for
   a. direct-union reactions
   b. decomposition reactions
   c. single-replacement reactions
   d. double-replacement reactions
2. $A + BX \rightarrow B + AX$ is the general equation for
   a. direct-union reactions
   b. decomposition reactions
   c. single-replacement reactions
   d. double-replacement reactions
3. How many elements are in a compound ending with -ide?
   a. one   b. three, with oxygen
   c. two   d. three, without oxygen
4. How many elements are in a compound ending with -ate?
   a. one   b. three, with oxygen
   c. two   d. three, without oxygen
5. The name of CaS is
   a. sulfur calcide   b. calcium sulfate
   c. calcium sulfide   d. calcium sulfite

CHAPTER 8

COMMON CHEMICALS

Do you eat citrus fruits such as oranges and grapefruits? Do you use vinegar on salads? If so, you have been eating common chemicals. In fact, your digestion of food depends on other chemicals that are made in your stomach. Many common chemicals are important in nutrition. Some common chemicals, however, can be dangerous. For example, car batteries contain strong chemicals that are very corrosive. Even using too much salt on your food may sometimes affect your health.

As you learned in Chapter 2, all things are made of chemicals. In your life, you use certain chemicals almost daily. Your body creates chemicals regularly. This chapter will introduce you to the chemistry of some "old friends," the chemicals you use regularly.

Some of the chemicals you are about to study may already be familiar to you. Less familiar, though, may be the reactions that produce them, the reactions they can enter, and the products they can form. You are also about to learn how you can predict the products of some reactions. As you make new discoveries in this chapter, you will deepen your understanding of an idea you have already begun to explore. There is nothing under the sun that does not involve chemicals and chemical reactions.

8-1 Acids and Bases

■ *Objectives*

☐ *List the characteristics of acids and bases.*

☐ *Explain why some acids and bases are strong and others are weak.*

☐ *Outline the pH scale.*

☐ *Define the neutralization reaction.*

There are many types of compounds, such as the salts and oxides. Perhaps the two most important and common compounds are acids and bases. They are usually mentioned and studied together because they

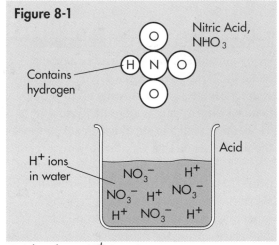

Figure 8-1

Nitric Acid, NHO_3

Contains hydrogen

H^+ ions in water

Acid

Acids release H^+ ions in water.

are similar in some ways and opposite in other ways. In fact, acids and bases can cancel each other out.

Characteristics of Acids

All **acids** contain hydrogen. More exactly, an acid is any compound that contains hydrogen and releases hydrogen ions when the acid is dissolved in water. For example, HNO_3 is an acid. Study Figure 8-1.

A compound that contains hydrogen is not necessarily an acid, however. Water, H_2O, and alcohol (C_2H_5OH) contain hydrogen, but neither is an acid. The hydrogen in an acid must be attached to the rest of the molecule by a weak chemical bond. In water, this bond breaks, and the hydrogen atom comes off as a positively charged hydrogen ion, H^+. Compare the atom to the ion in Figure 8-2.

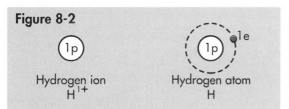

Figure 8-2

Hydrogen ion
H^{1+}

Hydrogen atom
H

Compare a hydrogen atom and a hydrogen ion.

Acids are corrosive. This means they can react with and dissolve away most materials. Diluted or weak acids can slowly dissolve cement, bones, teeth, rocks, and metal surfaces. Concentrated acids are very dangerous to surfaces, clothing, eyes, and skin. **Caution** Avoid tasting strong acids. Do not use a strong acid without your teacher's supervision. Should a strong acid spill on you or on your clothing, rinse with plenty of water. You can also weaken an acid's strength by rinsing with ammonium hydroxide, ammonia water, or sprinkling on sodium bicarbonate (baking soda).

You can find many common acids at school and at home. For example, sulfuric acid (H_2SO_4) is often used in labs. You drink carbonic acid (H_2CO_3) when you have soda water or seltzer. You can find citric acid ($H_3C_6H_5O_7$) in citrus fruits and in vitamin C. The acid in vinegar is acetic acid ($HC_2H_3O_2$).

Acid Strength

The strength of an acid depends on how well it ionizes in water that is, what percentage of the acid's ions are released into the water. Strong acids ionize 100 percent, so all of their hydrogen atoms come off in the water as hydrogen ions. Strong acids react readily with metals and other materials. Weak acids do not ionize readily in water so only a small fraction of the hydrogen atoms come off in water as hydrogen ions. Weak acids react slowly with other materials. Compare the acids in Figure 8-3.

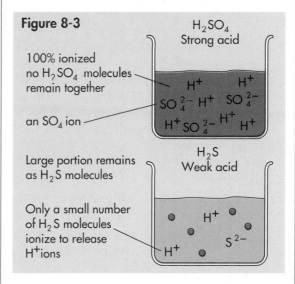

Figure 8-3

H_2SO_4
Strong acid

100% ionized
no H_2SO_4 molecules remain together

an SO_4 ion

H^+ H^+
SO_4^{2-} H^+ SO_4^{2-}
H^+ SO_4^{2-} H^+ H^+

H_2S
Weak acid

Large portion remains as H_2S molecules

Only a small number of H_2S molecules ionize to release H^+ ions

H^+
H^+
S^{2-}
H^+

Characteristics of Bases

A **base** is any compound that releases hydroxide ions in water. All bases contain the hydroxide ion, OH^-. Hydroxide is a many-atom ion with a single negative electric charge, or -1. For example, $NaOH$ and $Mg(OH)_2$ are bases. A compound with OH^- is not necessarily a base, however. Alcohol (C_2H_5OH) has an OH group, but it is not a base. To be a base, the hydroxide ion must be attached to the rest of the molecule by a weak chemical bond. In water, this bond breaks, and the OH^- ion comes off. Compare this explanation of a base to the description of an acid on page 155. Then, look at Figure 8-4, and compare it to Figure 8-1.

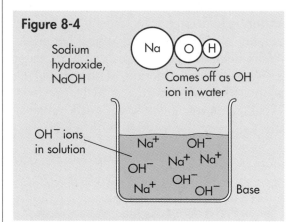

Figure 8-4

Sodium hydroxide, NaOH

Comes off as OH ion in water

OH^- ions in solution

Base

Bases release OH^- ions in water.

Bases are corrosive and can harm your clothing and skin.

Caution Avoid tasting strong bases. Do not use a strong base without your teacher's supervision. Should a strong base spill on you or on your clothes, wash with plenty of water. You can also weaken the base's strength by rinsing the area with vinegar.

You can find many common bases at school and at home. You can use sodium hydroxide (NaOH) to clean clogged pipes or make soap. Sodium hydroxide is more commonly known as lye. Ammonium hydroxide (NH_4OH), which is also known as ammonia water, is used for cleaning. Milk of magnesia, which you might use when you have a stomach ache, is a base called magnesium hydroxide. The formula for magnesium hydroxide is $Mg(OH)_2$.

Base Strength

The strength of a base depends on how well it ionizes in water. A strong base releases large amounts of OH^- ions. A weak base releases a small amount of OH^- ions. Compare the bases in Figure 8-5.

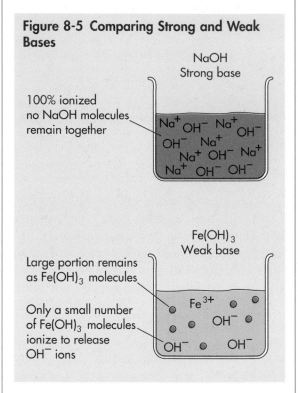

Figure 8-5 Comparing Strong and Weak Bases

NaOH
Strong base

100% ionized no NaOH molecules remain together

$Fe(OH)_3$
Weak base

Large portion remains as $Fe(OH)_3$ molecules

Only a small number of $Fe(OH)_3$ molecules ionize to release OH^- ions

Scientists often call strong bases **alkalis**. Sodium hydroxide, potassium hydroxide, and lithium hydroxide are examples of alkalis. They are very strong and release many hydroxide ions in water. Ammonium hydroxide is of medium strength. Calcium hydroxide and magnesium hydroxide are weak bases. They do not release many hydroxide ions in water.

Indicators

In the lab or at home, you might want to find out whether a substance is an acid or a base. There are some easy ways to do this. Certain chemicals turn different colors in acid or base solutions. They are called **indicators,** and researchers use them to learn whether a solution is an acid or a base. For example, the indicator phenolphthalein is colorless in an acid, but reddish-pink in a base. Litmus is a red indicator in acids but is blue in bases. A litmus indicator is usually put into paper strips called litmus paper. You can use litmus paper to quickly test whether a solution is an acid or a base.

Another indicator, methyl orange, is red in acids and yellow in bases. Finally, with a special type of indicator called a universal indicator, various colors show whether a solution is an acid or a base. The colors also show if the substance is weak or strong.

The pH Scale

The strength of acids and bases is measured on the **pH scale**. The letters *pH* stand for "potential for hydrogen." The scale starts at zero and goes to 14, as you can see in Figure 8-6. Pure water is in the middle of this scale, at pH 7, which is neutral. This means that pure water is neither an acid nor a base.

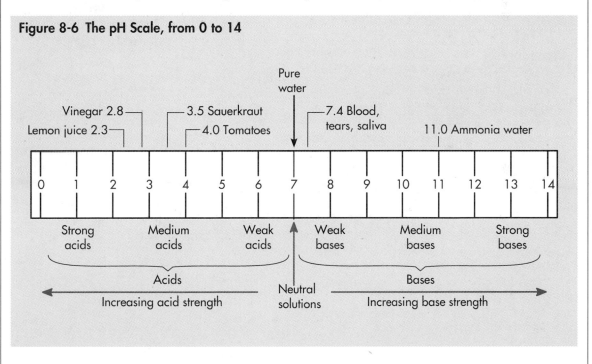

Figure 8-6 The pH Scale, from 0 to 14

Acids have pH values of less than 7, and bases have pH values of more than 7. You may want to refer back to this scale as you continue to read Chapter 8.

Neutralization

Scientists call the reaction of an acid with a base **neutralization**. Neutralization is an ionic exchange reaction that occurs because water is formed. The H^+ ion of acids readily joins with the OH^- ion of bases. This reaction forms water. In this reaction, the electrical charges cancel each other out, so the product is electrically neutral.

A C T I V I T Y

What Are Neutralizations?

Process Skills recording information; interpreting data

Materials 1 M solutions of $NaOH$, KOH, NH_4OH, HCl, HNO_3, and H_2SO_4; (1 M describes the concentrations of these solutions); phenolphthalein; litmus solution; red and blue litmus paper; and other indicators; droppers; stirring rods; and test tubes.

Procedures

1. Use a glass stirring rod to transfer one drop of an acid and one drop of a base to a piece of litmus paper. Observe whether the drops make the paper go from red to blue or blue to red. Write down your observations.
2. Set up test tubes with each available indicator solution. Add three drops of one acid and note the color. Add 10 to 15 drops of one base to each test tube. This completely neutralizes the acid and makes the solution a base. Note the color changes and write them down.

3. Place about ten mL of a base into a small beaker. Add three drops of phenolphthalein. While stirring, add the acid one drop at a time until the indicator turns colorless and remains colorless. When just enough acid is added to neutralize the base, you will reach the end point.
4. Carefully boil away the water from the solution you made in Step 3. Note the crystals of salt that remain. This is the salt produced by the neutralization reaction.
5. Show that you can get the same results in Steps 3 and 4 if you start with ten mL of an acid; place drops of base into it until the solution turns pink and stays pink.

Conclusions

6. Write an equation for each of the reactions that took place in your procedures. State what is occurring to the pH of the solution during the neutralization.

Look at this example:

$$HCl + NaOH \rightarrow NaCl + H_2O$$

hydrochloric + sodium → sodium + water
acid hydroxide chloride

You can see that the acid, hydrochloric acid, releases hydrogen ions that end up as water. The base, sodium hydroxide, releases hydroxide ions that also end up as water. The remaining sodium and chlorine atoms combine and make sodium chloride, which is table salt. The general equation for all neutralizations is

acid + base → a salt + water

$$\text{H}X + B\,\text{OH} \rightarrow BX + HOH$$
$$(\text{H}_2\text{O})$$

Look again at Figure 8-6. During neutralization, the pH value of each product is closer to 7 because water is formed. The strengths of the acid and of the base act against each other and produce a neutral solution. The products of neutralization, a salt and water, are not as dangerous and corrosive as the acid and base reactants.

Neutralization is an important reaction in laboratories. It also occurs in everyday life. For example, if you have an "acid stomach," either your body has made extra stomach acid, or the foods you ate contained a lot of acids. Too much acid causes a burning pain in your stomach, and if you do not do something about it, the condition could lead to painful sores called ulcers. Remedies include milk of magnesia and antacid, or "anti-acid," tablets. These remedies contain bases that neutralize acids.

■■■ **Section 8-1 Review** ■■■

Write the definitions for the following terms in your own words.

1. **acid**   2. **base**
3. **alkali**   4. **pH scale**
5. **neutralization**

Answer these questions.

6. a. Why are some acids and bases strong?
   b. Why are others weak?
7. List three acids and three bases that you could find at home.
8. Complete these neutralizations.
   a. acid + base →
   b. hydrochloric + potassium →
      acid   hydroxide
   c. $HNO_3 + NaOH \rightarrow$
   d. $H_3PO_4 + Al(OH)_3 \rightarrow$

8-2 Chemicals Around Us

■ *Objectives*
☐ *Name some important salts and oxides and explain their uses.*
☐ *Describe hard water and water softeners.*
☐ *Give examples of pure elements and alloys in common use.*

Common Salts

Recall that a salt is a compound with a metal part and a nonmetal part. Salts result when an acid and a base neutralize each other. A salt contains the negative ion from an acid and the positive ion from a base. Look at the examples of common salts in Figure 8-7. Study the following equations, which show how double-replacement reactions produce salts.

$$HX + BOH \rightarrow BX + H_2O$$

acid + base → a salt + water

$$HF + NaOH \rightarrow NaF + H_2O$$

hydrofluoric acid + sodium hydroxide → sodium fluoride + water

In this example, an acid and a base release hydrogen and hydroxide ions. This reaction makes the salt sodium fluoride, as well as water.

Scientists can find most salt compounds in rocks and in sea water. The sea is a great storehouse of chemicals, especially salts. Sea water contains about 2.5 percent dissolved solids by mass. The dissolved solids give sea water a density of 1.025 g/mL. Pure water has a density of 1.000 g/mL.

The sea is not the only place to find salt. In many places around the world, there are deposits with high amounts of one kind of salt. The salt is mined and shipped to other locations by truck, train, and boat. Some of the salts that people mine or produce in laboratories may be familiar to you. For example: table salt (NaCl); sodium fluoride (NaF); and sodium hypochlorite (NaClO), which is the active ingredient in bleach, are three examples.

People use salts for many purposes, such as preserving food, making fertilizers, and making drugs. Salts are also important in explosives and photography. Some salts can be used to change the type of water you use. Calcium sulfate ($CaSO_4$) makes water hard. **Hard water** does not form suds with soap. **Water softeners** make the water soft by removing the calcium ions as a precipitate.

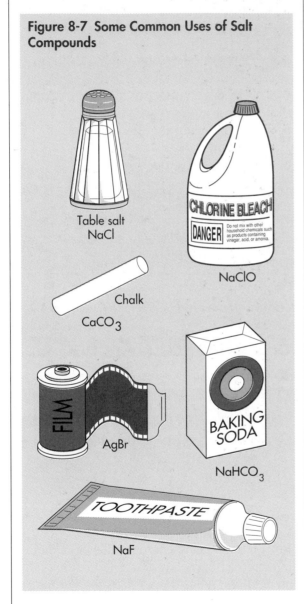

Figure 8-7 Some Common Uses of Salt Compounds

Table salt
NaCl

CHLORINE BLEACH
DANGER Do not mix with other household chemicals such as products containing vinegar, acid, or amonia.
NaClO

Chalk
CaCO₃

FILM
AgBr

BAKING SODA
NaHCO₃

TOOTHPASTE
NaF

Sodium carbonate (Na_2CO_3) also makes water hard. Heating decomposes this compound and softens the water. Look at Figure 8-8 for a comparison of hard water and soft water.

Common Oxides and Sulfides

An **ore** is a rock from which people can extract, or take out, metals. Sand (SiO_2) is the ore from which people extract silicon, which is used in computer chips. Bauxite (Al_2O_3) is the ore from which people produce aluminum.

People mine ores from the earth and then have them refined and chemically treated. Most ores contain oxide and sulfide compounds of metals.

Common Elements

Some elements are common, but many occur only as diatomic molecules or in compounds. You will recognize many things that are made of pure or diatomic elements. You will also recognize many compounds of elements.

For example, oxygen and nitrogen are the elements that make up most of the air you breathe. About 20 percent of air is oxygen. About 78 percent of air is nitrogen. Another element basic to life is carbon. You can find carbon in coal and charcoal. The "lead" in your pencil is carbon, too.

Have you seen neon lights? The element that makes them glow is neon. Aluminum is used in cans, aluminum foil, and house siding. Two of the most familiar elements are nickel and copper, which are used in coins.

Common Alloys

An **alloy** is a mixture of two or more metals. An alloy has special properties—properties that are not present in the individual metal

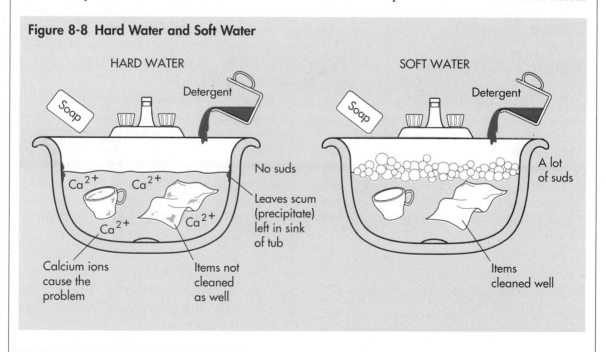

Figure 8-8 Hard Water and Soft Water

HARD WATER

Soap

Detergent

No suds

Leaves scum (precipitate) left in sink of tub

Ca^{2+} Ca^{2+}

Ca^{2+} Ca^{2+}

Calcium ions cause the problem

Items not cleaned as well

SOFT WATER

Soap

Detergent

A lot of suds

Items cleaned well

elements. Steel is an alloy of iron and carbon. People use steel in home building and in many common objects. Cars are made mostly of steel.

Other familiar alloys include bronze, which is made of copper and tin, and brass, which is made of copper and zinc. Two examples of items made of bronze are statues and bells. Brass is commonly used in hardware and electrical fixtures.

You might be familiar with the alloy your dentist uses to fill cavities. This alloy is known as dentist's amalgam, and it is made from silver and mercury.

■■■ Section 8-2 Review ■■■

Write the definitions for the following terms in your own words.

1. **hard water**
2. **water softener**
3. **ore**
4. **alloy**

Answer these questions.

5. List the name, formula, and uses of three common salts and two common oxides.
6. Calcium compounds dissolved in water make the water hard. What does this mean? What can be done to make the water soft?
7. Name three objects found at home that contain a pure element.
8. Name four common alloys and the metals found in each alloy.

8-3 Solubility

■ *Objectives*
☐ *Define solubility.*
☐ *Interpret a solubility table and curve.*
☐ *Distinguish between saturated, unsaturated, and supersaturated solutions.*
☐ *List the factors that affect the solubility of a material and the rate at which it dissolves.*

Experience tells you that when you mix sugar in coffee or tea, the sugar dissolves easily. Experience also tells you that flour dissolves very poorly. As a scientist, you need to explore these reactions more carefully. You will want to find out certain facts about how well a material can dissolve.

The Solubility Table

In the Skill Builder of Chapter 7, you used a table of solubilities. See page 149. Figure 8-9 shows part of that table. The table tells you whether a material is soluble, slightly soluble, or insoluble.

Recall that dissolving is a physical change. After a material dissolves it is still present in the water, either as molecules or ions. You can recover the material from the solution. Some materials, however, will react with the water, and this reaction will result in a chemical change. The table of solubility indicates these materials by d, which stands for decomposes. Sometimes, compounds do not form, as indicated by n. This letter means that the compound does not exist. For example, the table shows that ammonium oxide does not exist.

Figure 8-9 Table of Solubilities in Water

i - Nearly insoluble ss - Slightly soluble s - Soluble d - Decomposes n - Does not exist	Acetate	Bromide	Carbonate	Chloride	Hydroxide	Iodide	Nitrate	Oxide	Phosphate	Sulfate	Sulfide
Ammonium	s	s	s	s	s	s	s	n	s	s	s
Barium	s	s	i	s	s	s	s	s	i	i	d
Calcium	s	s	i	s	ss	s	s	ss	i	ss	d

Measuring Solubility

Solutions are mixtures. Solutions may contain different amounts of solute and solvent. The solute is the substance that is dissolved, and the solvent is the liquid. Water is the most common solvent. Water is also called the **universal solvent**. This means water can dissolve more materials than any other liquid.

Solutions are clear and look the same throughout. You can see through them. The solute is dissolved evenly throughout the solution. Look at Figure 8-10. A given amount of water will continue to dissolve a

Figure 8-10 Characteristics of a Solution

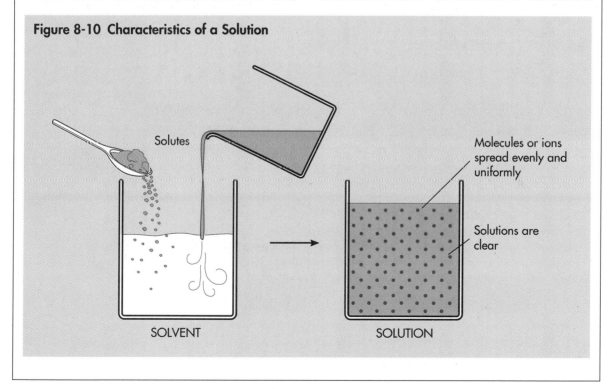

Solutes

Molecules or ions spread evenly and uniformly

Solutions are clear

SOLVENT

SOLUTION

solute up to a certain amount. The maximum, or largest, amount of material that a given quantity of water can dissolve is known as the material's **solubility**. Usually, solubility is defined as the maximum number of grams of solute that can dissolve in 100 grams of water.

For example, suppose seven grams of a substance is the maximum amount of that substance that will dissolve in 20 grams of water. What is the solubility of this substance? To answer this question, divide 100 by the number of grams of water.

100 / 20 = 5

This means that five times as much of the substance will dissolve in 100 grams of water

5 × 7g = 35g

The solubility of the material is 35 grams per 100 grams of water.

Before the maximum amount is dissolved, the solution is **unsaturated.** When the maximum amount is dissolved, the solution is **saturated**. Study Figure 8-11. If you add extra solute to a saturated solution, it will not dissolve but will fall to the bottom. The extra material is under the water, but it is not dissolved into the water. Dissolved materials are part of the liquid phase, but undissolved materials are a separate, solid phase. Any extra solute added to a saturated solution cannot dissolve into the liquid phase. Instead, it stays in a separate, undissolved solid phase.

Certain materials can be "tricked" into dissolving to a greater amount than is normal. The solution is **supersaturated**. It contains more than the maximum amount of

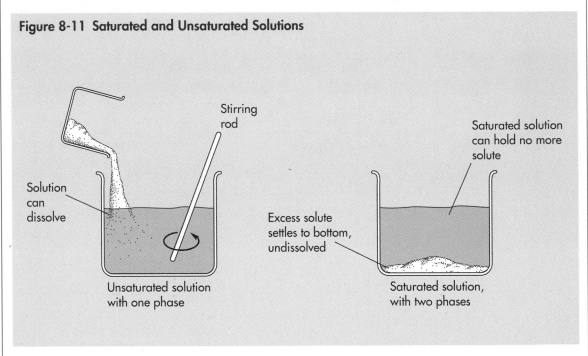

Figure 8-11 Saturated and Unsaturated Solutions

Stirring rod

Solution can dissolve

Unsaturated solution with one phase

Saturated solution can hold no more solute

Excess solute settles to bottom, undissolved

Saturated solution, with two phases

solute usually observed. Look at Figure 8-12. You can prepare supersaturated solutions by dissolving a solute at a high temperature and carefully cooling the solution to a lower temperature. Supersaturated solutions are unstable. A sudden noise, a small shake, or a bit of dirt will cause the extra material to precipitate out as a solid. The remaining solution is saturated.

Factors that Affect Solubility

The maximum amount of a solute that will dissolve in 100 grams of water depends on several factors. These factors include the chemical composition of the solute and the temperature of the water.

The time it takes for a solute to dissolve in 100 grams of water is called the rate of dissolving. The rate of dissolving depends on the chemical makeup of the solute, the temperature of the water, the amount of stirring, and the size of the particles. Small particles dissolve faster than larger ones.

Most solids dissolve faster, and in a greater amount, in hot water than in cold water. The opposite is true for gases. Most gases dissolve slower, and in a lesser amount, in hot water than in cold water.

For example, if you heat soda or seltzer, bubbles of CO_2 quickly appear. The CO_2 gas cannot remain dissolved in the warm soda water.

Solubility Curves

A **solubility curve** is a mathematical picture that shows you how much of a material will dissolve in 100 grams of water at various temperatures. Look at Figure 8-13.

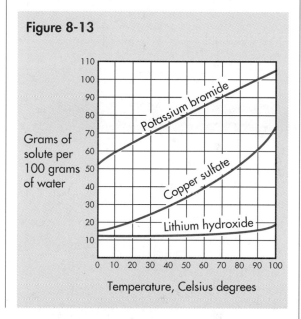

Figure 8-13

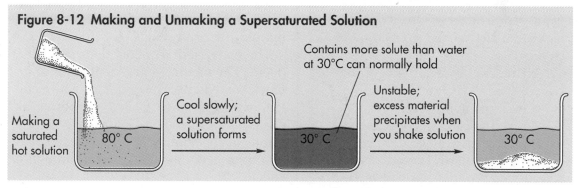

Figure 8-12 Making and Unmaking a Supersaturated Solution

Making a saturated hot solution 80° C

Cool slowly; a supersaturated solution forms

Contains more solute than water at 30°C can normally hold

30° C

Unstable; excess material precipitates when you shake solution

30° C

There are several ways to use solubility curves. For instance, you can use one to find a material's solubility at a certain temperature. Look at Figure 8-14.

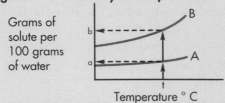
SKILL BUILDER

Interpreting Data

Use the solubility curve in Figure 8-13 to answer these questions.
1. What is the solubility of
   a. potassium bromide at 60° C?
   b. of copper sulfate at 65° C?
2. Between 0° C and 100° C, only one of these compounds will make a saturated solution with 30 g in 100 g of water. What material is this? What temperature is needed to achieve the saturated solution?
3. Which material changes least in solubility between 0° C and 100° C? How much did it increase?
4. A saturated solution of copper sulfate in 100 g of water at 80° C is cooled to 70° C. How much remains dissolved, and how much is no longer dissolved?

Remember that solubility is the maximum amount of a substance that will dissolve in 100 grams of water. When this amount is dissolved, you have a saturated solution. The curves in Figure 8-13 tell you what amounts are soluble for temperatures between 0° C and 100° C.

Suspensions and Colloids

A suspension, unlike a solution, contains undissolved particles that swirl around in the water and eventually settle to the bottom. Pineapple juice is a suspension. A typical can of pineapple juice instructs you to shake the can. If you do not shake the can, the pineapple particles will settle at the bottom. Suspensions can be separated into their parts by filtering because the particles in suspensions are large. The particles that make up solutions are tiny molecules or ions and thus, solutions cannot be separated by filtering. Like all mixtures, suspensions can be made in varying proportions.

A **colloid** is a mixture of a material with water. In some ways the properties of a colloid are similar to those of a solution and in some ways similar to those of a suspension. The particles in a colloid are larger than particles in a solution but smaller than particles in a suspension. Gelatin, milk, and mayonnaise are colloids.

The material in a colloid does not settle down. The particles are large enough to block some light. That explains why a colloid is neither as clear as a solution or as cloudy as a suspension. Colloids are slightly cloudy in appearance. You can see the path of a light beam as light passes through a colloid.

What Factors Affect Rate of Dissolving?

Process Skills observing; recording information; interpreting data

Materials sugar, $KMnO_4$, and $CuSO_4$ in a variety of forms, such as solid crystal, broken crystal, granular, and powder; balance; beakers; cold and hot water

Procedures

1. Prepare identical masses of two forms of one material. Place each into a separate beaker with the same amount of water. Make sure the beakers are at the same temperature. Stir both equally. Which form of the material was the first to dissolve completely? Write down your observations.

2. Prepare two identical masses of the same form of a single material, preferably granular or powder. Then, at the same moment, place one sample into a beaker of cold water and the other sample into a beaker of hot water. Stir the material in both beakers equally. In which beaker of water does the material dissolve first? Write down your answers.

Conclusions

3. In these experiments, why is it essential to keep certain things identical, such as the amount of water or the time of stirring? In any experiments, how many things should be changed at one time?

▬ Section 8-3 Review ▬

Write the definitions for the following terms in your own words.

1. **solubility**   2. **unsaturated**
3. **supersaturated** 4. **solubility curve**
5. **colloid**

Answer these questions.

6. Study the solubility tables given in this chapter and in the Skill Builder for Chapter 7. Name a compound that
   a. is soluble.   b. is slightly soluble.
   c. is insoluble.   d. decomposes

7. Six grams of a material are dissolved in 25 grams of water to make a saturated solution. What is the solubility of the material per 100 grams of water?

8. a. As you begin to heat water, air bubbles appear long before the water is hot enough to boil. If you let a glass of cold water stand at room temperature for an hour, you will notice air bubbles formed on the glass. Why do these events happen?
   b. What can be done to make a cube of sugar dissolve faster in water?

9. In what ways is a colloid like and unlike a solution and a suspension?

SCIENCE, TECHNOLOGY, & SOCIETY

Garbage, a Big, Big Problem

Figure 8-15

Do you know where your garbage goes? Most people hardly think about garbage once it is out of their sight.

Technological advances make modern society possible. However, new technologies often result in more garbage. For example, consider plastic bags. Until recently, people carried their groceries home in brown paper bags. Because of technological advances in the plastics industry, groceries now offer plastic bags, which some people find more convenient. Unlike paper bags, plastic bags cannot be recycled. Therefore, plastic bags result in more garbage.

Garbage is a serious problem in the United States. There is too much garbage and no place to put it. Many garbage dumps are filled up and have been closed.

Some communities and states have passed tough laws controlling the disposal of garbage. These laws involve expensive fees and fines. Many states, like New York, have a plan for the management of solid wastes. Most plans involve four goals.

The top goal is simply to make less garbage. One way is to reduce garbage is to eliminate or cut back on packaging. Think of all the packaging that comes with some foods and home appliances. Small items are packed in large cartons. Styrofoam stuffing and plastic wrapping are used. These packaging items add to our garbage pile. Are they really needed?

The next goal concerns reusing materials. Soda bottles, for example, can be reused. Other materials, such as newspapers, can be used like a raw material for maufacturing.

Another goal is to turn waste into energy. Engineers are looking into ways to produce electricity from garbage, just as coal and oil are used.

The last goal involves landfills, which are dumps. Should more landfills be created? Planners want to avoid opening new landfills. They feel that more landfills would only make people ignore efforts to reduce garbage and reuse waste materials. Community planners also argue that landfills are *not* permanent solutions to the garbage problem.

Follow-up Activities
1. Interview a community leader, a government official, a sanitation engineer, or a city planner about the issue of garbage disposal and reuse.
2. Do a library investigation on the extent of the garbage problem. How much garbage is there? What can be done? Prepare a brief report on your research.

KEEPING TRACK

Section 8-1 Acid and Bases

1. Acids release H^+ ions, and bases release OH^- ions in water. You can find acids and bases in many common materials.

2. The strength of an acid or a base depends on how well it ionizes.

3. The pH scale measures the strength of acids and bases on a scale of zero to 14. Acids have a pH of less than 7, and bases have a pH of more than 7. Neutral solutions and pure water have a pH of 7.

4. You can use indicators to learn whether a solution is acidic or basic.

5. In a neutralization reaction, an acid and a base react and form a salt and water.

Section 8-2 Chemicals Around Us

1. People commonly use many salts, oxides, sulfides, and pure elements for a variety of purposes.

2. Hard water does not form suds. Water softeners remove calcium ions from hard water and make the water soft.

3. Alloys are mixtures of metals. Alloys have unique properties and uses.

Section 8-3 Solubility

1. A solubility table summarizes the solubilities of materials. Solubility curves show the maximum amount of a material that can dissolve at various temperatures.

2. Saturated solutions contain the maximum amount of solute that dissolves at a given temperature. Unsaturated solutions contain less than this amount, and supersaturated solutions contain more than this amount.

3. Solubility is often measured as the maximum number of grams of a material that can dissolve in 100 grams of water.

4. The solubility of a material depends on its chemical properties, and the temperature. The time required to make something dissolve depends on these factors and also on particle size and stirring.

5. Usually solids dissolve more, and gases dissolve less, as the temperature of the water increases.

6. Colloids show some properties of solutions and suspensions.

BUILDING VOCABULARY

Write the term from the list that best completes each sentence.

hard water, base, saturated, alloy, alkalis, water softeners

The presence of hydroxide ions means a solution is a ___1___. Strong bases are known as ___2___. ___3___ remove the calcium ions from ___4___, so that the washing action of soap is improved. Brass is an ___5___ made up of copper and zinc. A solution that contains the most amount of solute it can hold is ___6___.

SUMMARIZING

Write *true* if the statement is true. If the statement is false, change the *italicized* term to make the statement true.

1. Acids release hydrogen *atoms* in water.
2. Litmus paper turns from *blue to red* in acids.
3. Steel is an alloy of *zinc and aluminum*.
4. In air, the element nitrogen *is* more common than the element oxygen.
5. *Stirring* affects how much of a material will dissolve in a given amount of water.
6. *Supersaturated* solutions can have more solute added.

INTERPRETING INFORMATION

Figure 8-16 is a short pH table for some common materials. Use the table to answer the questions that come after it.

1. Which is more acidic, milk or soda? In other words, which has a lower pH number?

Figure 8-16

Acidic*	pH	Basic**
Stomach acid	2.0	
Soda, seltzer	3.0	
Apples	3.1	
Milk	6.5	
	7.4	Blood, tears, saliva
	7.8	Eggs
	8.4	Baking Soda
	11.0	Ammonia water

*Acidic means the material contains acids.
**Basic means the material contains bases.

2. Which is more basic, ammonia water or egg? In other words, which has a higher pH number?
3. An acid stomach can be relieved by drinking a spoonful of baking soda in a glass of water. According to the table, why does this help?

THINK AND DISCUSS

Use the section number in parentheses to help you find each answer. Write your answers in complete sentences.

1. Glycerin, a candy sweetener, has the formula $C_3H_5(OH)_3$. Glycerin contains H atoms and OH groups yet is neither an acid nor a base. Why? (8-1)
2. Why is it wise to rinse metal pots, utensils, and iron knives soon after they are used with foods like lemons, tomatoes, and sauerkraut? (8-1)
3. a. What home material can neutralize a spill of
      (1) strong acid like battery acid?
      (2) strong base like lye?
   b. What other steps should be taken in case of a spill? (8-1)
4. Why are salt deposits, ores, and sea water important in industry and in producing chemicals? (8-2)
5. a. Name four different salts or products containing a salt used in the home. (8-2)
   b. Name four common uses of pure elements. (8-2)
6. At 50° C, the maximum amount of potassium bromide (KBr) that will dissolve in 100 g of water is 80 g. Use this fact to help

you figure out which of the following solutions is saturated, unsaturated, or supersaturated. Explain your answers. (8-3)

Solution A 50 g KBr in 50 g H_2O at 50° C

Solution B 100 g KBr in 200 g H_2O at 50° C

Solution C 8 g KBr in 10 g H_2O at 50° C

GOING FURTHER

1. Expand on the examples of household and industrial chemicals presented in this chapter. Then choose one type of chemical for more study in the library. Learn where the material is found, how it is produced, the tons made each year, and the chemical's importance to industry and the economy. You may decide to investigate acids, bases, salts, salt deposits, ores, refining a metal, making steel, types of steel, and alloys. Other topics, like colloids, may be suggested by the chapter or your teacher. You may decide to focus on only one important compound or element for your report.

2. Most common chemicals and elements have a long and interesting history of their use. Plan a report on this theme. Here are some examples.

   ■ Iodized table salt contains some NaI, or sodium iodide, mixed with the NaCl. Someone discovered that NaI lets the salt flow freely without clumping up in moist weather.

   ■ Copper metal was thought to cure arthritis and aching bones when worn as jewelry.

   ■ Lead is a toxic metal, yet the wealthy people of ancient Rome had their food cooked in lead pots.

3. Find out more about pH from advanced chemistry textbooks. How is the number determined? How much stronger is an acid with pH 1 compared to acids with a pH 4 or 6? What is a pH meter and how is it used? You may want to give some historical, mathematical, or practical information about pH in your report.

COMPETENCY REVIEW

1. Which one of the following could be a base?
   a. HF
   b. H_2O
   c. KOH
   d. KF

2. Acids react with bases to produce
   a. hydrogen gas.
   b. oxygen gas.
   c. oxide compounds.
   d. a salt and water.

3. Ammonia water is an example of a common
   a. acid.
   b. base.
   c. salt.
   d. indicator.

4. Which pH value indicates an acid?
   a. pH of 1
   b. pH of 7
   c. pH of 10
   d. pH of 14

5. Iron and copper come from
   a. salt deposits.
   b. ores.
   c. sand.
   d. sea water.

6. The air is mostly
   a. oxygen.
   b. nitrogen.
   c. hydrogen.
   d. carbon dioxide.

7. The universal solvent is
   a. alcohol.
   b. water.
   c. strong bases.
   d. petroleum.

CHAPTER 9

INTRODUCTION TO ORGANIC CHEMISTRY

You are mostly composed of carbon compounds. So are other living things, including plants. Carbon is a part of the food we eat, many of the medicines we take, the clothes we wear, and the houses we live in. Carbon and carbon compounds make up the coal, natural gas, oil, and gasoline that people use for heating and for running cars, buses, and planes. Almost anything we can think of contains carbon. Many compounds that scientists create contain carbon. Scientists even use an isotope of carbon to tell how old some objects are.

The word *organic* was connected, at first, with living things. The study of compounds of carbon is called **organic chemistry**. Inorganic chemistry, on the other hand, is about the compounds of all the other elements. Why does carbon rate its own special part of chemistry? Because organic compounds, the carbon compounds, far outnumber inorganic compounds. In fact, scientists have found almost 20 times more organic compounds than inorganic compounds.

9-1 Organic Compounds

■ *Objectives*
☐ *Distinguish between organic compounds and inorganic compounds.*
☐ *Compare the properties of organic and inorganic compounds.*
☐ *Interpret a structural formula.*
☐ *Define an isomer.*

Can you imagine having arguments, listening to sermons, or even getting into fights about chemistry? It is odd to even ask such

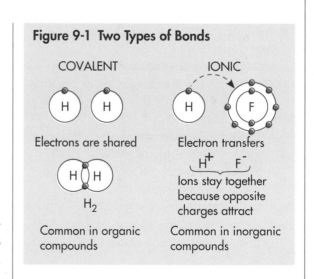

Figure 9-1 Two Types of Bonds

COVALENT

H H

Electrons are shared

H$_2$

Common in organic compounds

IONIC

H F

Electron transfers

H$^+$ F$^-$

Ions stay together because opposite charges attract

Common in inorganic compounds

a question. Yet, 150 years ago, people preached and fought about organic chemistry. Many people felt it was wrong to talk about and study organic materials in the same way as inorganic materials. They argued that organic materials could only come from living things and must, therefore, contain a mysterious "life force."

History of Organic Chemistry

All living things are made of organic compounds, and 150 years ago, all organic materials came only from living things. Milk came from cows, rubber and syrup came from trees, and medicines came from plants. For these reasons, some people thought organic materials were more important than inorganic materials.

The wall between organic and inorganic chemistry was broken forever in 1828. It was then that German chemist Friedrich Wöhler discovered that by mixing two inorganic salts, he could produce urea. Urea is an organic material made by human beings. Today, most organic materials do not come from living things; they are made in laboratories and chemical companies. However, people still study organic compounds and inorganic compounds separately because of the large number of organic compounds, their special properties, and their special reactions.

Organic and Inorganic Compounds

Organic compounds are more easily decomposed by heat than inorganic compounds. They usually **char** when heated, meaning that they get a black coating of carbon. For example, sugar, which is an organic material, melts and burns easily. A sugary marshmallow browns, chars, and catches fire when toasted. Salt is inorganic, and therefore does not burn easily.

Most organic compounds are only slightly soluble in water, but many inorganic compounds are soluble. Organic compounds often react slowly, but inorganic compounds react quickly. An exception is the speed of some organic reactions in living things.

Organic molecules and some inorganic molecules are held together by covalent bonds, in which the atoms join by sharing electrons. Most inorganic compounds are held together by ionic bonds, in which electrons are transferred from one atom to another. Compare the bonds in Figure 9-1.

Bonds within Organic Compounds

Look back at Figure 9-1. You can see that two hydrogen atoms join by a covalent bond. A pair of electrons in a covalent bond may be represented by a dash ($-$) or two dots (:) in a **structural formula**. A structural formula shows the atoms in a molecule, the bonds joining them, and their arrangement. Look at the structural formula for hydrogen, H_2, in Figure 9-2.

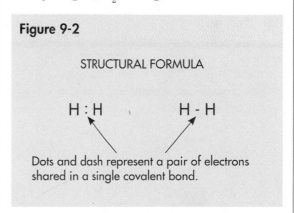

Figure 9-2

STRUCTURAL FORMULA

H : H   H - H

Dots and dash represent a pair of electrons shared in a single covalent bond.

What Decomposes an Organic Compound?

Process Skills *observing; recording information; making predictions*

Materials pond water; beakers; tap water; teaspoon; sugar; test tubes;

Procedures

1. Place about 50 mL of pond water into a beaker. Slowly boil the water away. Continue to heat what is left for two minutes. Note its appearance and write down your observations.
2. Repeat Procedure 1 with water from your tap and compare your results.
3. Place ¼ teaspoon of sugar into a test tube. Heat the sugar gently until it melts. Continue heating it strongly for two minutes. Note the changes. After the leftovers cool, scrape them from the test tube and examine. Repeat this procedure using table salt. Write down your observation.

Conclusions

5. What effect would you say heat generally has on organic compounds?
6. Predict the effect of heating bread, vegetables, and meat for at least 5 minutes in a test tube.

Recall that all organic compounds contain carbon. Carbon tends to form compounds with other atoms by four covalent bonds. For example, look at Figure 9-3 to see how one carbon joins with four hydrogens to make methane, CH_4.

Carbon has an unusual ability. It can form covalent bonds with other carbon atoms. This permits carbon atoms to form long chains and rings. This ability explains why there are so many different organic compounds.

Figure 9-3

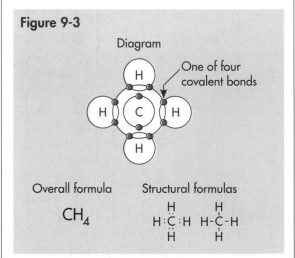

Methane is the simplest organic compound.

Saturated and Unsaturated Compounds

An organic compound is **saturated** (SACH-uh-rayt-ud) if all of its carbon atoms are joined to one another by sharing just *one* pair of electrons. This sharing of one pair of electrons is called a single covalent bond. Carbon atoms may also join together by sharing either two or three pairs of electrons. This sharing results in double and triple covalent bonds between the carbons.

In a double covalent bond between two carbon atoms, the two atoms share two pairs of electrons. This is shown in a structural formula by four dots (::) or two lines (=). In a triple covalent bond, two atoms share three pairs of electrons. This is shown by six dots (:::) or three lines (≡). An organic compound is **unsaturated** if it has one or more double or triple bonds. Look at the bonds in Figure 9-4.

Isomers

Another reason there are so many organic compounds is that the same atoms can be arranged in different patterns. Look at the examples in Figure 9-5. Each molecule has the same chemical formula, C_5H_{12}. You can see, however, that the atoms are arranged differently in each molecule. The molecules have different structural formulas. They are called **isomers** (EYE-su-murz) of one another. The examples in Figure 9-5 are all isomers of C_5H_{12}. Each is a different compound with different properties.

Figure 9-5   Isomers of C_5H_{12}

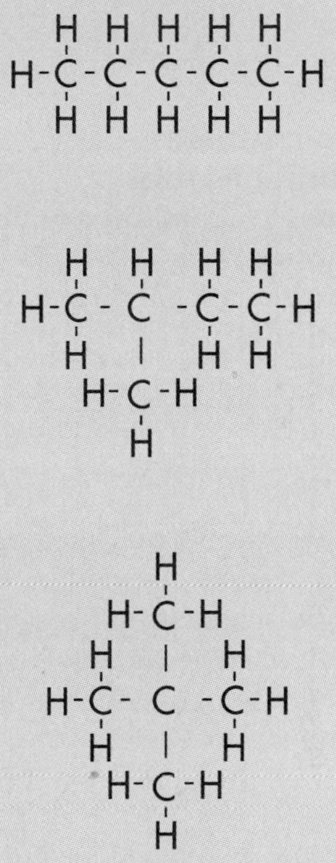

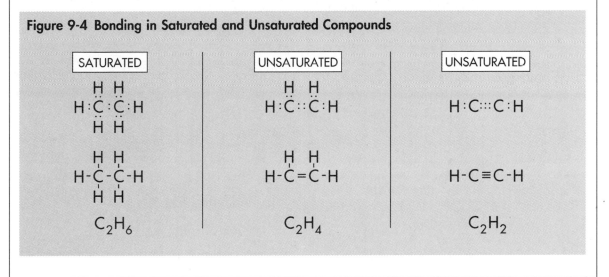

Figure 9-4 Bonding in Saturated and Unsaturated Compounds

Section 9-1 Review

Write the definitions for the following terms in your own words.

1. **organic chemistry**
2. **structural formula**
3. **saturated (regarding organic compounds)**
4. **unsaturated (regarding organic compounds)**
5. **isomers**

Answer these questions.

6. Figure 9-6 shows one molecule of propane, an organic compound.
   a. Why do scientists say propane is held together by covalent bonds? How many covalent bonds are shown in this diagram?
   b. What is the chemical formula for propane?
   c. Draw the structural formulas for propane using dashes and dots.
   d. Explain why this a saturated compound.
7. Name three facts about carbon that explain why there are so many organic compounds.
8. Which compounds in Figure 9-7 are isomers of each other?

Figure 9-6

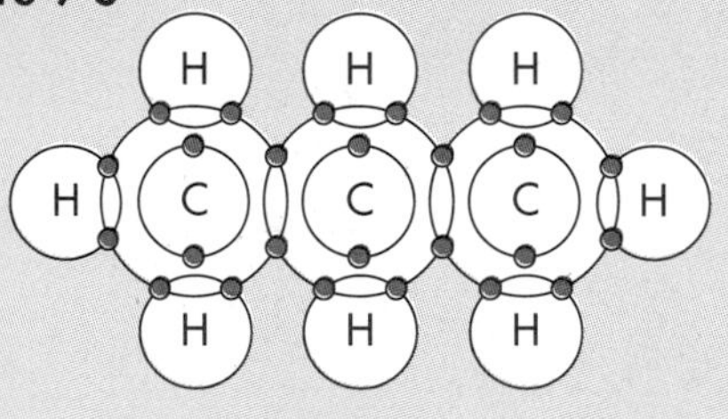

Figure 9-7

a.

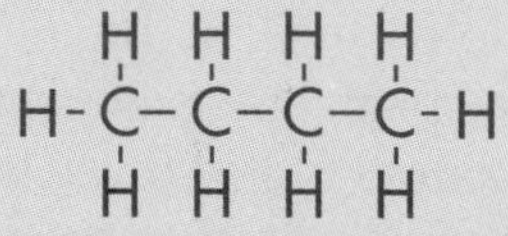

b.

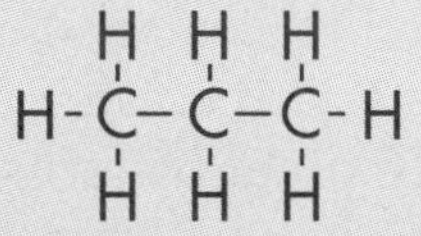

c.

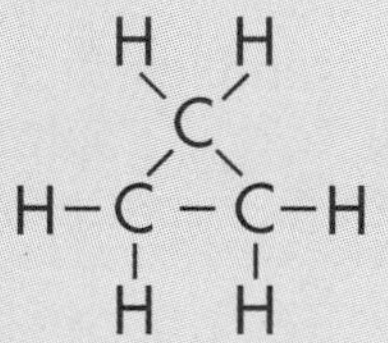

d.

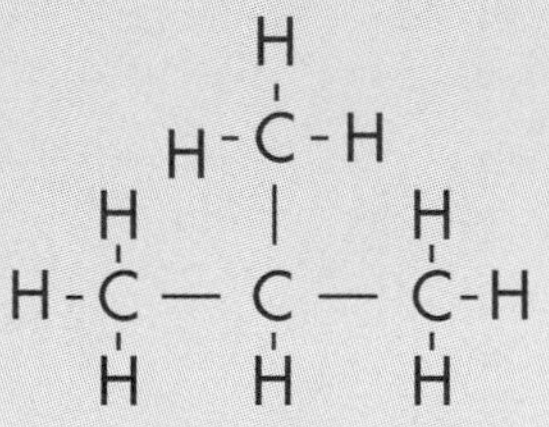

9-2 Types of Organic Compounds

■ *Objectives*

☐ *Recognize hydrocarbons, alcohols, organic acids, esters, and aromatics.*

☐ *List key hydrocarbons obtained from petroleum.*

☐ *Identify some common alcohols and organic acids.*

☐ *Explain why amino acids are vital to life.*

You know that beer, wine, and other alcoholic beverages contain alcohol. There are, however, many different kinds of alcohol. People use wood alcohol to clean up spilled paint. They use rubbing alcohol to work cramps out of muscles. Alcohols are one type of organic compound. You will learn more details about alcohols later in this section.

You can recognize types or classes of organic compounds by the arrangement of atoms in their structural formulas. Each type has its own pattern and **functional group**. A functional group is a group of atoms that is attached to an organic molecule. For example, all alcohols contain one or more groups of -OH atoms. The properties of each type of organic compound depend in part on the functional group.

Hydrocarbons

A **hydrocarbon** contains only carbon and hydrogen atoms. Hydrocarbons may be straight chains, chains with branches, and rings. Some hydrocarbon molecules contain over 50 carbon atoms. Hydrocarbons may be saturated. Saturated hydrocarbons contain only single covalent bonds. Other hydrocarbons are unsaturated, with one or more double or triple covalent bonds. Compare the hydrocarbons in Figure 9-8.

Figure 9-8 A Variety of Hydrocarbons

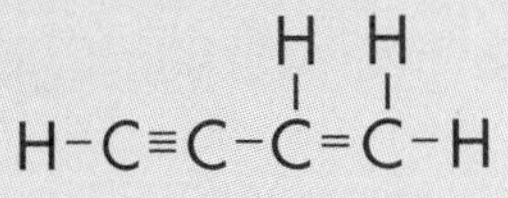

$$H-C\equiv C-C=C-H$$
$$(C_4H_4)$$

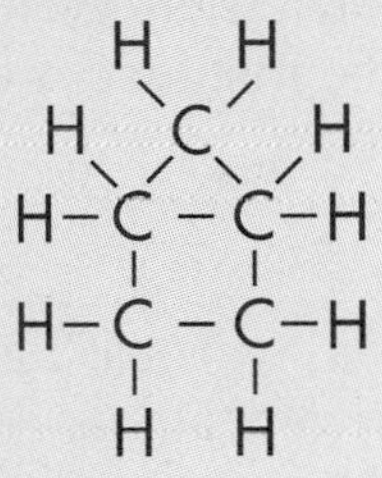

$$(C_5H_{10})$$

Crude oil, which is called **petroleum,** is a mixture of many hydrocarbons. These hydrocarbons have short and long chains of carbon atoms. Oil companies pump petroleum from under the ground, where it was formed by the decay of ancient plants and animals. Workers in oil refineries boil the petroleum and separate each hydrocarbon according to its own boiling/condensation temperature. This method of separation is known as fractional distillation.

People use many hydrocarbons every day. For example, cooking and heating gas contains the hydrocarbon methane, CH_4. The gasoline for cars contains octane, C_8H_{18}. A fuel to light lamps is kerosene, $C_{12}H_{26}$. The grease used to lubricate machines is the hydrocarbon $C_{17}H_{36}$. Candle wax is the hydrocarbon $C_{20}H_{42}$.

Some hydrocarbons are important industrial chemicals. Companies use them to produce many other organic materials, such as plastics and medicines. Other carbon compounds could come from the same hydrocarbons. These compounds are called derivatives, even though they may not now be manufactured in that way. Study Figure 9-9.

Alcohols

An **alcohol** is a hydrocarbon that has one or more hydrogen atoms replaced by an -OH group, the hydroxyl group. The -OH group is the functional group of alcohols. Examine the alcohols in Figure 9-10. Alcohols may be made by oxidizing, or adding oxygen to, hydrocarbons. The general formula for alcohol is R-OH. The R stands for any hydrocarbon minus one H atom.

Figure 9-10 Some Alcohols

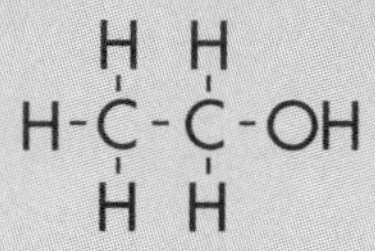

a. C_2H_5OH, ethyl alcohol

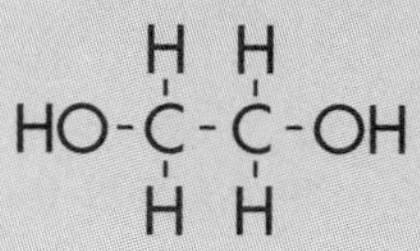

b. $C_2H_4(OH)_2$, glycol

Figure 9-9   Hydrocarbon Derivatives

Hydrocarbon: C_2H_6 Ethane

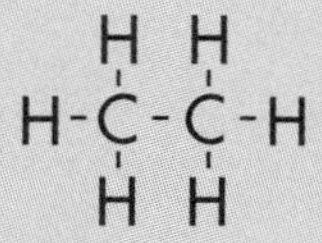

Various derivatives: C_2H_5Br Ethyl bromide

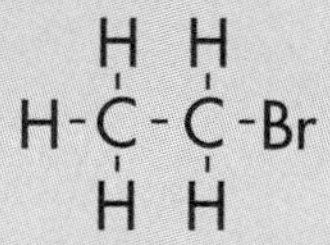

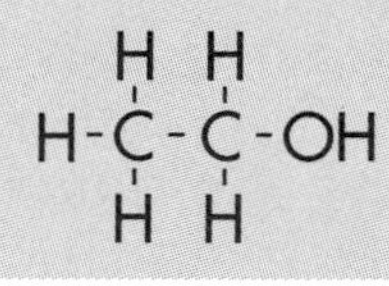

        C_2H_5OH Ethyl alcohol

Interpreting Data

Figure 9-11 is a graph that shows how the boiling points of hydrocarbons change with their formula mass. Use the graph to work through the following example. Then use it to answer the questions.

Figure 9-11

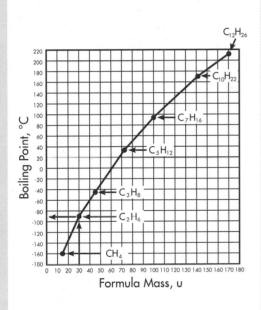

Begin by finding the formula mass and boiling point of C_2H_6. The formula mass of two carbon atoms is 2×12, or 24. The formula mass of six hydrogen atoms is 6×1, or 6. Add these formula masses, and you find that C_2H_6 has a formula mass of 30. Use the graph to find the boiling point. It is about $-90°$ C. You should read the boiling point to the nearest multiple of five.

1. a. Calculate the formula mass of $C_{10}H_{22}$.
   b. Does this agree with the position of the dot on the graph?
   c. What is its boiling point in °C?
2. Determine the boiling points for the hydrocarbons missing from the graph. (Hint: first calculate their formula masses.)
   C_4H_{10} C_6H_{14} C_8H_{18} C_9H_{20}
3. Which hydrocarbon boils
   a. at 215° C?
   b. at $-45°$ C?
   c. close to water's boiling point?
   d. close to water's freezing point?
4. What connection is there between boiling points of hydrocarbons and their formula masses?
5. Which hydrocarbons will be gases at room temperature, about 20° C?

Alcohols have various uses. For example, methyl alcohol, CH_3OH, also called wood alcohol, is a paint solvent. It is also poisonous. Ethyl alcohol, C_2H_5OH, also called grain alcohol, is the alcohol in liquor and gasohol fuel. The antifreeze for car radiators and windshields has the formula $C_2H_4(OH)_2$. It contains two OH groups, and it is called a glycol. Glycerin, $C_3H_5(OH)_3$, is used to make candy, ice cream, and soap. Glycerin is also a part of human body fat.

Organic Acids

The functional group for **organic acids** is -COOH, the carboxyl group. This group attaches itself to the end of a hydrocarbon. The general formula for organic acids is R-COOH. Look at the organic acid and its general formula in Figure 9-12. The reason these compounds are acids is that the hydrogen atom in the -COOH group comes off in water as an H^+ ion.

Figure 9-12 Some Organic Acids

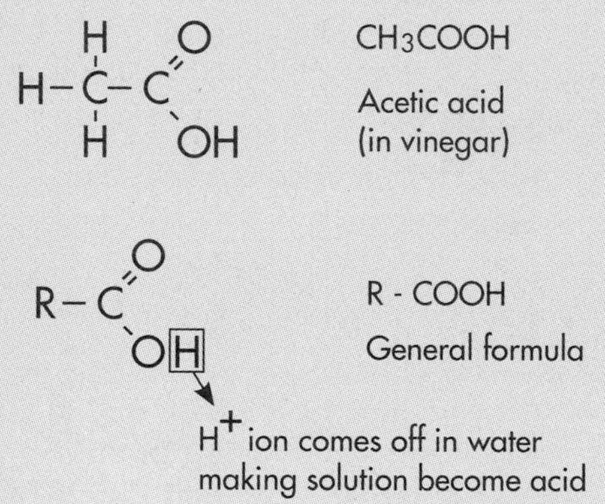

CH3COOH

Acetic acid
(in vinegar)

R - COOH

General formula

H^+ ion comes off in water making solution become acid

Scientists prepare organic acids by oxidizing alcohols. Organic acids made from long-chain hydrocarbons, such as $C_{17}H_{35}COOH$, are called fatty acids. This is because they are the building blocks of fat.

Amino Acids

A very special group of organic acids is made up of **amino acids.** These acids contain the amino group, $-NH_2$, as well as the carboxyl group, -COOH. Examine Figure 9-13.

Amino acids are very important to life. They are the units that make up all proteins. Living things use the amino acids in the food they eat. Amino acids build and repair tissues and body parts.

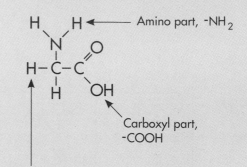

Figure 9-13 The Simplest Amino Acid

Amino part, $-NH_2$

Carboxyl part, -COOH

This H is replaced by R, or longer hydrocarbon parts in more complex amino acids.

Esters

The general formula of an ester is R-COO-R. Animal and vegetable fats are esters. Many esters have strong odors that give flavoring to foods such as bananas, pineapples, and mint. Look at the ester in Figure 9-14.

Esters are formed by the reaction of an organic acid and an alcohol. In some ways, this reaction is similar to the neutralization of an inorganic acid and a base. You studied neutralization in Chapter 8. Compare these two reactions.

inorganic acid + base → a salt + water
organic acid + alcohol → an ester + water

Figure 9-14 An Ester and How to Make One

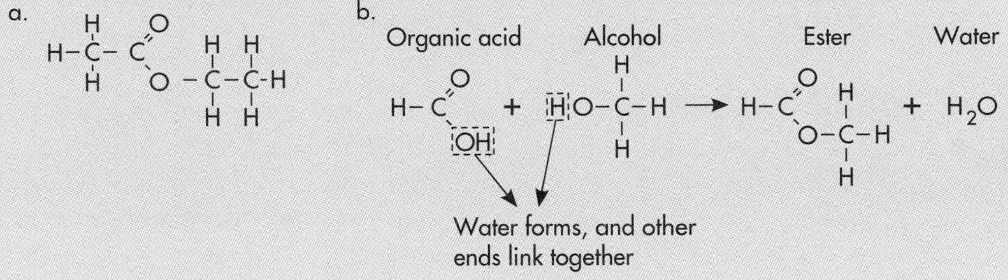

a.

H—C—C with structure showing ester linkage

b.

| Organic acid | Alcohol | → | Ester | Water |

Water forms, and other ends link together

Other Types of Organic Compounds

Other classes of organic compounds are shown in Figure 9-15. An extremely important class are the aromatics. These compounds contain six carbon atoms arranged in a hexagon, or a six-sided ring. In aromatics, the ring is called the **benzene ring**. Most medicines, hormones, and enzymes necessary for life are aromatics. They may contain several benzene rings.

Figure 9-15 Other Types of Organic Compounds

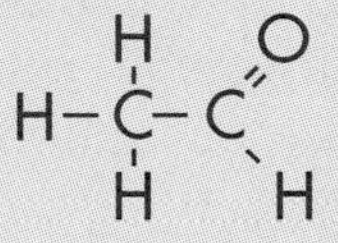

Aldehydes, with
R - CHO structure

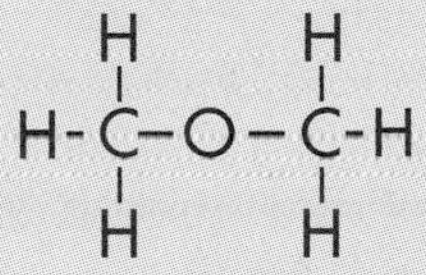

Ethers, with
R - O - R structure

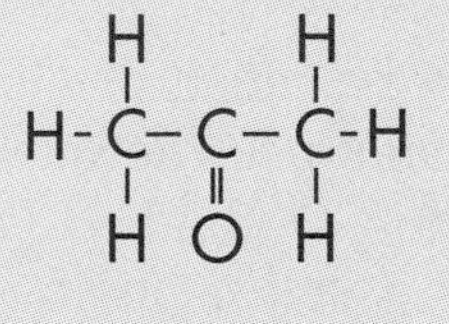

Ketones, with
R - CO - R structure

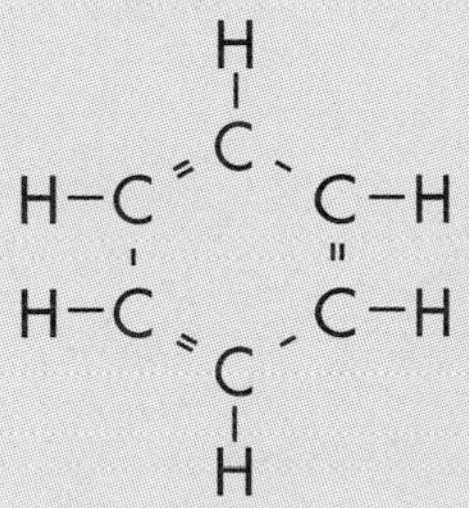

Aromatic, with
benzene ring

Aromatic, a complex
molecule with two
benzene rings

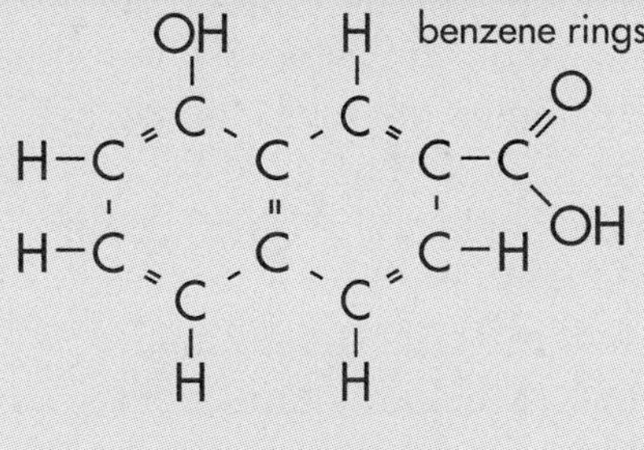

Section 9-2 Review

Write the definitions for the following terms in your own words.

1. **functional group**
2. **hydrocarbon**
3. **alcohols**
4. **amino acids**
5. **benzene ring**

Answer these questions.

6. What type of organic compound is each molecule in Figure 9-16?

Figure 9-16

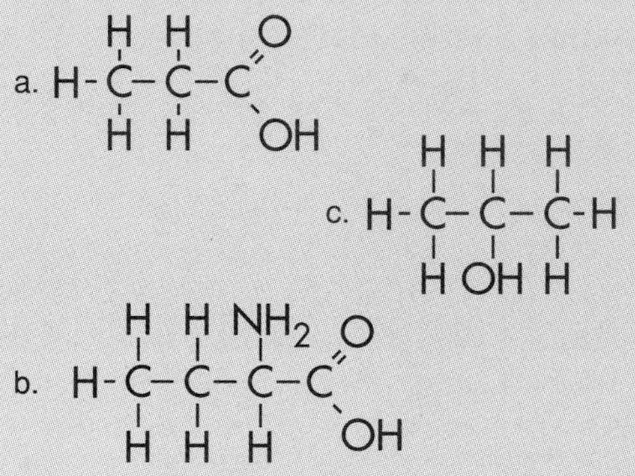

7. What common materials contain alcohols, organic acids, or esters? Give at least two examples of each of these types of organic compounds.
8. Why are amino acids and aromatic compounds essential to life?
9. State how each material can be made:
   a. an alcohol   b. an organic acid
   c. an esther

9-3 Organic Chemicals and Life

■ *Objectives*

☐ *Name some key organic compounds of life.*

☐ *Explain why a balanced diet is important.*

☐ *Contrast proper drug use and drug abuse.*

☐ *Describe the action and effects of drugs.*

Life Processes

Nearly all chemicals involved with life are organic. They contain complex carbon molecules. For example, the organic compound **hemoglobin** is in your blood. It makes your blood red. It carries oxygen from your lungs to all your cells. It also carries carbon dioxide back to your lungs, where you breathe it out. There is an iron (Fe) atom in the center of the hemoglobin molecule. To have healthy blood, you need a good source of iron compounds in the foods you eat.

Many of the organic molecules needed for life come from the foods you eat. A balanced diet will supply you with these molecules. A balanced diet contains meats or protein-rich beans and nuts. It also includes vegetables and fruits, energy-rich foods like bread, and dairy products such as milk and cheese.

All the foods in a balanced diet can be traced to plants. Plants use the energy of sunlight to change CO_2 and H_2O into foods. This process is called photosynthesis, and the organic molecule that makes photosynthesis possible is **chlorophyll**

(KLAWR-uh-fil). Chlorophyll may be the most important organic molecule of all life.

Drug Use and Abuse

Drugs are organic compounds that interfere with the normal chemical reactions that take place in your body. When used properly, drugs can reduce pain and promote healing. Drugs, however, can be abused. Drug abuse or overuse is very dangerous. Many drugs are **addictive**. This means your body and mind can become dependent on the drugs. The best advice regarding medical drugs is to follow your doctor's orders and follow the directions printed on the label of all medicines.

Many drugs come from plants. Some drugs are produced by bacteria and other microorganisms grown in warm, clean tanks. Many drugs are made by scientists, using chemical reactions.

Some drugs work by slowing down or stopping chemical reactions in your body. These drugs are called **depressants**. Some drugs work by speeding up and promoting chemical reactions in your body. They are called **stimulants**. Drugs that dull your senses are called **narcotics**. Drugs that attack infections or prevent illness from germs are called **antibiotics**.

Ethyl alcohol, C_2H_5OH, is the main ingredient in alcoholic drinks. It is a depressant, so it reduces a person's heart rate and blood pressure. It causes a temporary feeling of happiness, followed by confusion and depression. Alcoholic drinks are addictive. They damage body organs, especially the liver. Accidents caused by drunk driving are the biggest killer of teenagers in the United States today. Alcohol abuse is a serious problem for many individuals in society today.

Cigarette smoke contains nicotine, and coffee and tea contain caffeine. Both nicotine and caffeine are stimulants, and both can be addictive. Smoking has so many damaging effects on the body that you should avoid it altogether.

Drugs like cocaine, crack, and heroin cause a "rush," or a brief feeling of happiness. This is followed by a "crash," a feeling of confusion and sadness. These drugs are addictive. Marijuana and LSD alter thinking and feeling. Their effects may be frightening. All of these drugs are illegal. They harm and even kill individuals and create serious problems for families, law enforcement, and society. The President of the United States has declared war against these dangerous drugs.

■■■ Section 9-3 Review ■■■

Write the definitions for the following terms in your own words.

1. **hemoglobin**   2. **chlorophyll**
3. **depressant**   4. **stimulant**

Answer these questions.

5. Name at least six organic materials found in foods or your body.
6. What foods make up a balanced diet?
7. What harmful effects can you expect from the following?
   a. ethyl alcohol
   b. cigarettes, coffee, and tea
   c. illegal drugs

SCIENCE, TECHNOLOGY, & SOCIETY

Some "Most Wanted" Chemicals

Many chemicals help society. Some chemicals, however, are dangerous. Government, industry, and individuals are working to stop the use of dangerous chemicals. Like the FBI's "most wanted" criminals, these chemicals must be arrested and prevented from doing further damage.

For example, a Freon gas, CCl_2F_2, is "wanted" for damaging the ozone (O_3) layer of the atmosphere. The ozone layer protects people from the harmful ultraviolet light from the sun. When the ozone is damaged, more ultraviolet rays reach the ground. More cases of skin cancer are expected to result from this. People, however, use Freon in refrigerators, air conditioners, and industrial processes. Scientists must find a substitute for Freon—one that does not destroy the ozone layer.

"Wanted" for causing lung and stomach cancer is asbestos. Asbestos crumbles into a fine dust that living things can breathe or eat.

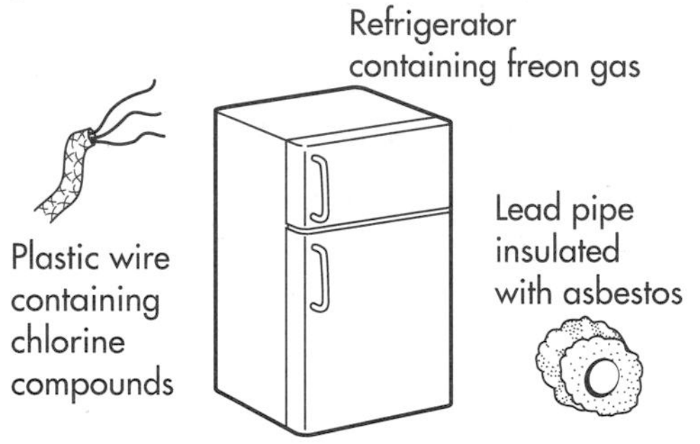

Asbestos used to be woven with other fibers in fire blankets and fire-resistant clothing. Insulation material put around pipes and furnaces, in ovens, and in ceiling tiles used to contain asbestos mixed with plastic and clay. Such insulation is widely present in old homes, schools, and factories.

Asbestos products are no longer made. Old asbestos that is crumbling must be removed by experts at great expense. Old asbestos that is not crumbling is best left in place and coated with a layer of special paint or plastic.

Plastics that contain chlorine are "wanted" for producing toxic fumes as they burn. These plastics are used as building materials and as covering for electric wires.

Follow-up Activities

1. Investigate these and other polluting chemicals in greater detail, using the library or interviews.
2. Methane gas from farm waste and garbage dumps around the world has recently been accused of making the greenhouse effect worse. Investigate what this effect is and why it is a cause for concern.
3. How does this reading illustrate that technology can have unexpected consequences?

Figure 9-17

Refrigerator containing freon gas

Plastic wire containing chlorine compounds

Lead pipe insulated with asbestos

KEEPING TRACK

Section 9-1 Organic Compounds
1. Organic compounds contain carbon.
2. Organic compounds are more numerous than inorganic compounds and have different properties.
3. A structural formula shows how the atoms and the bonds in a molecule are arranged.
4. Carbon atoms link together by saturated or unsaturated covalent bonds. Different arrangements of the same number of atoms result in isomers.

Section 9-2 Types of Organic Compounds
1. Scientists classify organic compounds into types by studying each molecule's structure and functional groups.
2. Hydrocarbons are obtained from petroleum.
3. Many organic chemicals are important in industry, our daily lives and our bodies. These chemicals include alcohols, organic acids, esters, and amino acids.

Section 9-3 Organic Chemicals and Life
1. Life depends upon organic compounds. A balanced diet supplies a healthful amount of each key compound.
2. Drugs can help or hurt people. Drugs include depressants, stimulants, and antibiotics. Some drugs are also addictive, narcotic, and illegal.

BUILDING VOCABULARY
Write the word from the list that best matches each statement.

alcohols, antibiotics, benzene rings, char, functional groups, hemoglobin, organic acids

1. This is the black coating when organic compounds are heated.
2. -OH and -COOH are examples.
3. R-OH is the general formula.
4. R-COOH is the general formula.
5. Aromatic compounds contain these.
6. This is the cause of the red color of blood.
7. These kill bacteria and germs.

Explain the difference between the terms in each pair.

8. organic, inorganic
9. chemical formula, structural formula
10. hydrocarbon, petroleum
11. organic acid, amino acid
12. addictive, narcotic
13. depressant, stimulant

SUMMARIZING
Write the missing word for each sentence.

1. Organic compounds must contain ___.
2. ___ easily decomposes organic compounds.
3. Organic compounds react more ___ than inorganic compounds.
4. Carbon atoms link by ___ electrons.
5. Atoms in hydrocarbons are arranged in ___ and rings.
6. Combining an organic acid and an alcohol produces an ___.
7. A benzene ring has the shape of a ___.
8. ___ are used to form body proteins.
9. The ___ in coffee can be addictive.

CHAPTER REVIEW

INTERPRETING INFORMATION

All hydrocarbons that are saturated and form a chain are known as alkanes. Alkane hydrocarbons have the general formula C_nH_{2n+2}. For example, C_3H_8 follows this formula.

For carbon, n = 3
For hydrogen, 2n + 2 = 2(3) + 2 = 8

1. Supply the missing subscript for H in the following alkane hydrocarbons. Write each formula with the correct subscripts.

    $C_2H_?$ $C_9H_?$ $C_{15}H_?$ $C_{40}H_?$

2. All four formulas below are hydrocarbons, but only one is an alkane hydrocarbon. Which one is it?

    C_2H_2 C_6H_{10} C_7H_{14} C_8H_{18}

THINK AND DISCUSS

Use the section number in parentheses to help you find each answer. Write your answers in complete sentences.

1. Tell what you look for to identify the following compounds.
    a. hydrocarbons   b. organic acids
    c. fatty acids   d. amino acids
    e. alcohols   f. aromatics (9-2)
2. a. What did people 150 years ago believe was true about organic chemicals? What discovery helped to change that view? (9-1)
    b. Why is the study of carbon compounds still a special branch of chemistry today? (9-1)
3. Figure 9-18 shows the structural formulas of four materials.

Figure 9-18

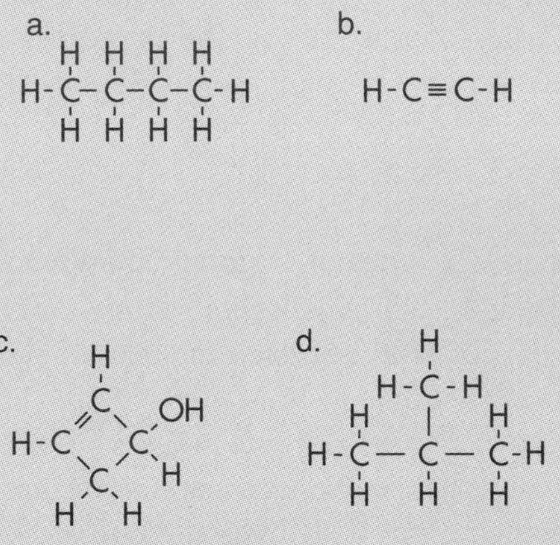

a. Which two are unsaturated?
b. Which two are isomers?
c. Which is an alcohol? (9-1, 9-2)
4. Determine the formula mass of the four molecules in Figure 9-18. (9-1, 9-2)
5. a. What type of organic compounds does your body require for your good nutrition? (9-3)
    b. What type of compounds does your body make from the food you eat? (9-3)
6. What are the dangers in each of the following?
    a. alcoholic drinks
    b. cigarettes

c. excessive coffee or tea

d. illegal drugs (9-3)

7. How are medical drugs made? (9-3)

8. How does fractional distillation work? (9-2)

GOING FURTHER

1. Practice drawing structural formulas for each type of organic compound discussed in this chapter. Make a large chart to show each type in class. Use wooden molecular models to create examples of each type of compound.

2. Using a chemistry textbook or library references, find out more facts about the preparation and uses of hydrocarbons, alcohols, organic acids, or esters.

3. Copy the labels from various medicines. Look for each medicine's ingredients, directions for proper use, and warnings. Report on your findings.

4. Obtain more information about substance abuse, which means misuse of drugs, tobacco, or alcohol. What is the extent of the problem? What biological and chemical reactions occur? How is the public being educated about the dangers? Look for related newspaper and magazine articles. Present your research as a report.

COMPETENCY REVIEW

1. All organic materials contain
   a. oxygen.   b. iron.
   c. carbon.   d. chlorine.

2. Organic compounds usually
   a. are very soluble in water.
   b. come from rocks.

c. break apart when heated.

d. have fast chemical reactions.

3. A special ability of carbon atoms is that they
   a. join to make long chains.
   b. form covalent bonds with hydrogen.
   c. form ionic bonds with hydrogen.
   d. have one electron in the outer shell.

4. All the choices below refer to the simplest of all hydrocarbons. Which choice gives its structural formula?
   a. methane
   b. home cooking and heating gas
   c. CH_4
   d.

Figure 9-19

5. Isomers have the same
   a. properties.   b. structure.
   c. chemical formula.   d. bonds.

6. Which material is not organic?
   a. wood   b. water
   c. food   d. body tissue

7. An important warning about alcohol is that it is
   a. illegal.   b. habit forming.
   c. a stimulant.   d. an antibiotic.

8. Which organic material is highly addictive, narcotic, and illegal?
   a. nicotine   b. cocaine
   c. caffeine   d. alcohol

CHAPTER 10

FORCE, MOTION, AND WORK

You see and feel forces everywhere. When you pull open a door, push a shopping cart, or lift a box, forces are at work. Anything that is moving needed a force to get it started. Forces have several effects on objects. You will learn about these effects in this chapter.

You also experience motion everywhere. You move, cars move, and the wind moves. Sometimes the motion is steady. Many times the motion is changing in speed or direction. In this chapter, you will learn how to detect and measure motion.

Have you used the words *mass* and *weight* and wondered how they differ? You will learn why they are different and why your weight can change, even when your mass does not change.

Finally, science has uncovered three laws of motion. These laws tell about the relationship between forces and motion. The three laws of motion are discussed in Section 3 of this chapter. You will find some of this material very obvious, some not so obvious, and some even surprising!

10-1 Characteristics of Forces

■ *Objectives*
☐ *Describe the effects of force.*
☐ *Distinguish between mass and weight.*
☐ *State the law of universal gravity.*

A **force** is a push or a pull. You cannot see a force, but you can see its effects on objects. When you ride a bicycle, for example, your legs produce a force that makes the pedals start to move. The force is invisible,

Figure 10-1

Invisible forces produce visible results.

but you can see its effect. This same invisible force gets transferred to the bike chain, then to the wheel gears, to the rear tire, and finally to the whole bike. Each starts to move as a result of the force. After you get moving, if you stop pedalling and just coast, air meets you with an invisible force that makes you slow down and stop. Look at Figure 10-1.

The Effects of Forces

Whenever an object changes its motion, a force must be causing the change. Whenever scientists observe a change in the motion of anything, they immediately look for the force that caused that change. For example, a small hammer in a doorbell moves forward and hits a bell by the force of a magnet. It then moves back by the force of a spring. This happens many times each second.

Forces can also change the shape of objects upon which they act. Look at Figure 10-2. A balloon gets squeezed, a car tire flattens out, and a rubber band gets stretched. All of these things happen because forces act on the materials

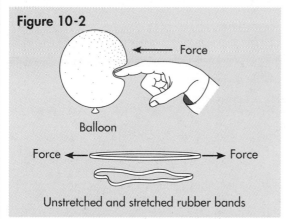

Figure 10-2

Balloon

Unstretched and stretched rubber bands

Forces can change the shape of some objects.

in the objects. Force can change both an object's motion and its shape.

Mass and the Force of Gravity

You may know that there is a force pulling things toward the earth. It is called the force of **gravity**. Did you know that you also have your own gravity force? The force of gravity is an attraction that every object has to all other objects. Gravity force is usually just called gravity. Gravity is a property of matter. The gravity of your body pulls on everything around you. It pulls on the furniture, the walls, nearby trees, and everything else. The gravity force of your body, however, is so tiny that only special equipment could detect it.

The amount of an object's force of gravity depends on its mass. Recall that mass is a measure of the amount of matter in an object. Your mass is small, so your force of gravity is weak. The mass of the earth is very large, so its force of gravity is much stronger. Remember that mass is usually measured in grams (g) or in kilograms (kg). One thousand g = 1 kg. You may wish to look again at Chapter 3, to see how these measurements are related.

The Law of Universal Gravity

All objects have a mass that produces gravity. Therefore, all objects attract other objects. The strength of this attraction force between two objects will increase if the mass of one or both objects increases, or if the distance between the objects decreases. The reverse is also true. The strength of the force of attraction will decrease if the mass of one or both objects decreases, or if the distance between the objects increases.

This is the law of universal gravity. As an example, suppose you stand half way between an elephant and a friend. The attraction of the elephant is stronger because of its larger mass. As you walk toward your friend, however, your friend's force of attraction increases, and the force of attraction of the elephant decreases, because you are farther away from the elephant. Look at Figure 10-3.

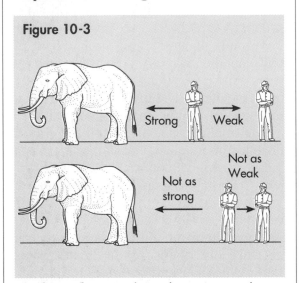

Figure 10-3

The force of gravity depends on mass and on distance.

Force of Gravity and Weight

When you stand on a scale, the weight you read is actually the force of attraction the earth has on you. If you could go to the moon with the scale, you would discover that you weigh less. This is because the moon has less mass than Earth, so it pulls with less gravity force. **Weight** is the force of attraction that a large mass, usually the earth's mass, has on an object of small mass. Your weight would be lower on the moon, but your mass stays the same because your body has the same amount of matter.

Measuring Force and Weight

In the United States, the common unit of force is one **pound (lb)**. Recall that in the metric system, the unit for force is one newton (N). Both of these units are also used to measure the weight of an object because weight is a force.

In the metric system, the weight of an ordinary apple is about one newton. A conversion to remember is that an object with a mass of 454 g has a weight on earth of about one lb, which is about 4.4 N. Look at the table in Figure 10-4.

Figure 10-4

Object	Mass	Approximate weight on Earth in pounds	In newtons
Melon	454 g	1 lb	4.4 N
Cat	2,700 g	6 lb	26.5 N
Person	54,500 g	120 lb	534.0 N

Recording Information; Interpreting Data

Figure 10-5 shows a large rubber band being stretched between two scales. These scales give the force of the pull in newtons. A meter stick is used to measure the unstretched and stretched lengths of the rubber band.

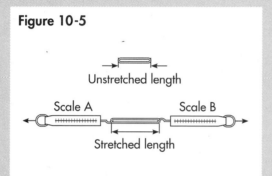

Figure 10-5

Unstretched length

Scale A   Scale B

Stretched length

The length of the unstretched rubber band was 5.0 cm. Examine the results in Figure 10-6. Note that some information is not yet complete.

1. Copy and complete this table. To find the missing changes in length, use this formula

   change in length = stretched length − 5.0 cm

2. Figure 10-7 shows the scales for making a graph from this table. Copy the scales onto your own graph paper.

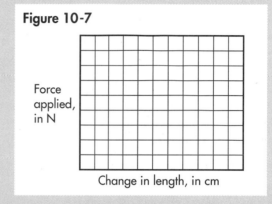

Figure 10-7

Force applied, in N

Change in length, in cm

3. Carefully plot the values on the graph. Connect the points on the graph with a straight line.

Figure 10-6

Force on scale A	Force on scale B	Force applied	Stretched length	Change in length
0 N	0 N	0 N	5.0 cm	0 cm
1 N	1 N	1 N	6.5 cm	1.5 cm
2 N	2 N	2 N	8.0 cm	
3 N	3 N	3 N	9.5 cm	
4 N	4 N	4 N	11.0 cm	
5 N	5 N	5 N	12.5 cm	

■■■ **Section 10-1 Review** ■■■

Write the definitions for the following terms in your own words.

1. **force**   2. **gravity**
3. **weight**   4. **pound**

Answer these questions.

5. Forces cannot be seen, but you can see their effects. Explain this, using two or three examples.
6. Name two ways in which an object's force of gravity can be made stronger.
7. What is the difference between mass and weight?
8. Jupiter is much bigger in size and mass than Earth. What would happen to your mass and to your weight if you could move to Jupiter right now?

10-2 Balanced and Unbalanced Forces

■ *Objectives*

☐ *Explain what happens when balanced and unbalanced forces act on an object.*

☐ *Describe the effect of friction.*

Have you ever played tug-of-war? Sometimes, both teams pull with exactly the same force. Always, one team soon pulls a little harder and wins. These two situations involve forces that are first in balance and then not in balance. The change in balance of forces makes a difference.

Balanced Forces

The force with which one team pulls can be as strong as the force with which the other team pulls. This is called an **equilibrium** (ee-kwuh-LIB-ree-um), a situation in which the forces acting on an object cancel each other out, or balance out. In the case of just two forces, the forces must be equal in strength but must act in opposite directions. During equilibrium, an object does not move or it continues to move at a constant speed and in the same direction. The people holding the rope and the rope itself stay in place. Look at Figure 10-8.

Often, more than two forces may act on an object with just the right combination of strengths and directions that they balance each other. For example, suppose three people are all pulling on a basketball in different directions. If the basketball is not moving, you know that the forces are in balance.

Balanced forces can produce two different results, no motion and unchanging motion. It is no surprise that if balanced forces act on an object that is not moving, the object stays at rest. This is the "no motion" result. However, if balanced forces act on a moving object, that object will continue at a constant speed, in a straight line indefinitely. This is the "unchanging motion" result.

Perhaps you have flown in an airplane. Imagine that as the plane flies west, the air from the west pushes against it with a 10,000-lb force. This force will cause the plane to slow down. However, if the engines of the plane produce a forward force of 10,000 lbs, there will be a balance, or an equilibrium, between both forces. In such

Figure 10-8 Equilibrium of Two Forces

Left force 300 lbs

Right force 300 lbs

a situation, the plane keeps moving in a straight line at a steady speed.

The engines do not produce the motion of the plane once it is moving. Instead, the engines simply produce a forward force to balance the resisting force of the air that would slow down the plane. The plane automatically moves at a steady speed because all forces are balanced. Look at Figure 10-9.

One very clear example of this idea is a rocket ship travelling in deep space. With its engines off, the rocket ship will move at a constant speed in a straight line. It needs no fuel to keep going! The reason is that there are no forces in the vacuum of

space that act against the rocket ship's motion. With the engines off, there is no forward force, either. The total force is zero, which results in equilibrium and unchanging motion.

Unbalanced Forces

Imagine sitting in a car stopped at a red light. When the light turns green, the driver gives more gasoline to the engine to produce a large forward force on the car. Look at Figure 10-10. This large forward force is much greater than the force of air resistance and other forces that act in the opposite direction. There is no equilibrium. These are **unbalanced forces**, meaning that the combined forces do not cancel each other out. As a result, the car picks up speed.

Figure 10-9 Constant Speed Because of Equilibrium

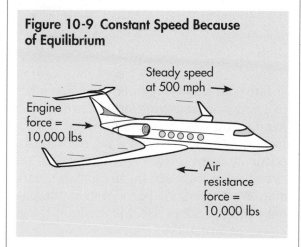

Steady speed at 500 mph →

Engine force = 10,000 lbs →

← Air resistance force = 10,000 lbs

Figure 10-10

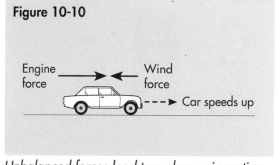

Engine force

Wind force

- - - → Car speeds up

Unbalanced forces lead to a change in motion.

When the forces on an object are not in balance, a change in motion must occur. A change of shape may also occur, as when someone kicks a soccer ball. The change in motion might mean gaining speed, losing speed, or changing the direction of the motion. These changes can result only if forces are unbalanced. For example, a car's brakes create a large, unbalanced force that counteracts the way the car is moving. This unbalanced force makes the car's speed decrease, perhaps even to zero.

Unbalanced forces can also change the direction of motion. Have you ever swung a ball or a yo-yo in a circle around your body? You must keep pulling in on the string. Look at Figure 10-11. This inward, unbalanced force pulls on the ball so it changes direction and goes in a circle rather than in a straight line. Even when you walk or run around a corner, this rule is at work. You make one leg exert more force than the other leg to produce a change in the direction of your motion.

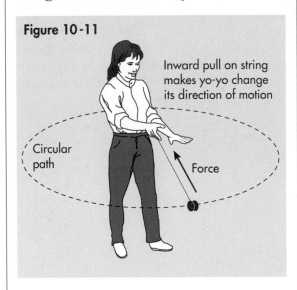

Figure 10-11

Inward pull on string makes yo-yo change its direction of motion

Circular path

Force

In summary, when no force acts on an object, or when the forces present are in balance, the result will be no motion or unchanged motion. When the forces acting on an object are not balanced, the result will be a change in motion. This means a change in speed, in direction, or in both.

Friction

The force created when two surfaces rub against each other is called **friction.** Friction always opposes the motion, which means it acts in the opposite direction of the motion. Look at Figure 10-12. Friction always produces heat.

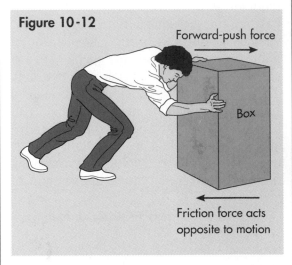

Figure 10-12

Forward-push force

Box

Friction force acts opposite to motion

Rub your hands together. You must push to overcome the friction force between your hands. You can also feel the heat that is made. On a bicycle or in a car, friction can occur in many places. It can happen between gears, between the tire and the road, and between engine parts such as the pistons and cylinders. Heat is made in these places as a result.

The brakes of a bike or car are made to squeeze against the turning wheel to create a large friction force. You use brakes to stop the motion of the wheels with friction. Sometimes brakes and wheels get so hot from friction that they start to smoke or burn.

■■■ Section 10-2 Review ■■■

Write the definitions for the following terms in your own words.

1. **equilibrium**
2. **unbalanced forces**
3. **friction**

Answer these questions.

4. What must be true about two forces that are in equilibrium? How is it possible for more than two forces to be in equilibrium?
5. a. What are two possible effects on motion when no forces act on it or if the forces are balanced?
   b. What are two possible effects on motion when the forces are unbalanced?

10-3 Measuring Motion

■ *Objectives*
☐ *Calculate speed and acceleration*
☐ *State the three laws of motion*

Imagine being in a spaceship traveling at 20,000 miles per hour. Would you be thrilled at moving so fast? Do you think it would be as exciting as amusement park rides, or train rides, or even skating and skiing? Probably not! A spaceship ride will feel pretty quiet, even dull. Why?

Detecting Motion

To detect, experience, and measure motion, you must be able to see changes in distance from a chosen **reference point**, a spot from which distance is measured. In fact, **motion** is defined as the change in the position of an object when it is compared to a reference point. Look at Figure 10-13. When you walk or run, you judge your motion by looking at buildings, trees, or marks on the ground. In an airplane, you may even forget that you are moving, unless you watch the roads and cities slowly go by down below. You might also

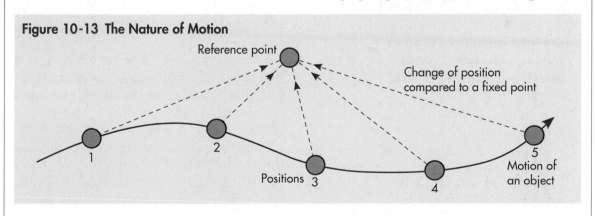

Figure 10-13 The Nature of Motion

Reference point

Change of position compared to a fixed point

Motion of an object

Positions 1 2 3 4 5

feel motion when you see clouds zip past with great speed.

In a spaceship, however, there would be no objects to look at to detect that you really are moving. In deep space, far away from the earth, you could not detect your motion even at enormous speeds. You would feel that you were not moving, because to detect motion you must have a way to measure your change of position from a reference point.

Measuring Speed

Motion can be measured by the speed of an object. **Speed is** the distance traveled by an object in one unit of time. For example, you could find speed by asking how many miles a girl biked in one hour, or how many meters a boy ran in one minute, or how many kilometers light travels in one second. You would then use this formula.

speed = distance moved/time required
or **speed = distance/time**

For example, a helicopter goes 30 km in 15 minutes. You want to find its speed in km/min and in km/hr. The distance moved is 30 km, and the time required is 15 minutes.

speed = distance/time
speed = 30 km/15 min
speed = 2 km per 1 min

Now, change minutes to hours. Since one hr = 60 min, divide 15 by 60.

15/60 = 0.25
speed = 30 km/0.25 hr
speed = 120 km per hr

Study these two extra examples.

1. Sound can travel at about 330 meters per second. About how far will it travel in 8 seconds?

distance = speed × time
distance = 330 m/sec × 8 sec =
2640 m

2. A car is moving at 30 miles per hour. What time is required to go 75 miles at this speed?

time = distance/speed
time = 75 miles/30 miles/hr =
2.5 hr

Measuring Acceleration

Every day, you start to move, stop, change your speed, and change your direction. These are **accelerations** (ik-sel-uh-RAY-shunz). Acceleration is a change in the speed of an object or in the direction of an object's motion. By this definition, an object is accelerating when it starts moving, stops moving, speeds up, slows down, or changes its direction of motion. When an object slows down, we say it has **deceleration** (de-sel-uh-RAY-shun), or negative acceleration.

Suppose you are in a car moving at a steady speed of 55 miles per hour. What will your acceleration be? It will be zero. Acceleration only exists if your speed changes, not if it is steady.

Sometimes car companies will compare their car's "pick-up" power to that of others by stating their car's rate of acceleration. For example, one company claims its car

How Can You Learn the Speed of an Object?

Process Skills performing calculations; recording observations

Materials a small moving toy, wind-up or battery, that moves in a straight line; a meter stick; a watch with a second hand

Procedures

1. Mark a starting line on the table. Place the front end of the toy on the starting line.
2. Allow the toy to move forward. As each second passes, one member of the group should shout "1, 2, 3, . . ." Another member of the group should mark the location of the front of the toy at each second. Be careful not to interrupt the motion of the toy when making this mark.
3. After the last reading, use the meter stick to measure the total distance the toy moved from the starting line to each mark. Record this on a data table like the one in Figure 10-14.
4. Calculate the average speed of the object after each second that passed.

Conclusions

5. Did the speed of the toy remain steady during this experiment, or did it begin to slow down? Why does it slow down, in terms of the concepts of this chapter?

Figure 10-14

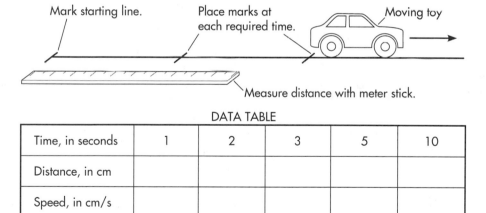

Mark starting line.

Place marks at each required time.

Moving toy

Measure distance with meter stick.

DATA TABLE

Time, in seconds	1	2	3	5	10
Distance, in cm					
Speed, in cm/s					

Interpreting Data

Graphs are an excellent way to present and understand data. Remember that data are the measurements obtained from an experiment. Figure 10-15 shows the speeds of a motorcycle in km/hr during an experiment that lasted 12 minutes. The dots are the data points, or the speed of the bike at each minute. Answer the following questions by studying the graph. On a sheet of paper write the answers or the letter, A, B, C, D, that best answers each question.

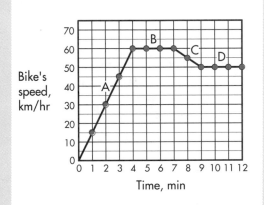

Figure 10-15 Speed of a Motorcycle during an Experiment

1. Which part of the graph shows the motorcycle accelerating?
2. What was the highest speed the motorcycle reached?
3. How many minutes did it take to reach the highest speed?
4. Where was there a deceleration?
5. In what parts of the graph must the forces acting on the bike be
   a. in equilibrium?
   b. unbalanced?
6. Describe what type of motion occurs
   a. from 0 to 4 minutes.
   b. from 4 to 7 minutes.
   c. from 7 to 9 minutes.

can go from 0 to 60 miles per hour in just five seconds. This means that its speed increases by 60/5 or 12 miles per hour during each second. The acceleration rate could be advertised as 12 miles per hour per second.

The Laws of Motion

In Section 10-2 you learned that motion only changes if there is an unbalanced force acting on an object. In other words, unbalanced forces cause acceleration. This idea is also known as the second law of motion. The first law of motion simply says that with no force, or with balanced forces, there will be no acceleration. Figure 10-16 summarizes these laws.

Inertia

In everyday words, inertia means to act or move slowly. In science, **inertia** (in-UR-shuh) means the resistance of an object to changing its motion. The first law of motion states that an object at rest tends to stay at rest, and an object that is

Figure 10-16

	First law of motion	Second law of motion
Forces on object	None or balanced	Unbalanced
Acceleration?	No	Yes
Motion of object	Stays at rest or at constant speed	Changes speed or direction

moving tends to keep moving at the same speed and in the same direction.

The second law of motion says that when a force is applied to an object, it accelerates. In addition, it says that a larger force produces a larger acceleration rate. If the mass of the object becomes larger, however, its rate of acceleration will be smaller. For example, if a car is loaded with people and luggage, it will accelerate more slowly. This example is also an illustration of inertia. A large mass has greater inertia than a small mass, so a large mass resists changing its motion and will not accelerate as quickly as a small object. Compare these two laws by looking at Figure 10-17.

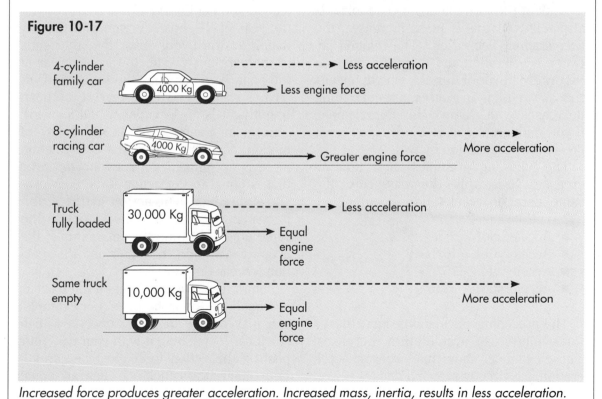

Figure 10-17

Increased force produces greater acceleration. Increased mass, inertia, results in less acceleration.

A more complete statement of the second law of motion is that greater force results in greater acceleration, but greater mass results in less acceleration. The law can also be written as a simple formula.

acceleration = force/mass

According to this formula, acceleration changes with force and mass. If force increases, the acceleration must also increase by the same factor. This relation between acceleration and force is **directly proportional**, which just means that as one thing increases, a second thing also increases by the same factor. For example, as you multiply the force by two, you also multiply the acceleration by two.

Look at the formula again. The formula also tells you that if mass increases, the acceleration decreases. The relation between acceleration and mass, however, is **inversely proportional**, which means that as one thing increases, a second thing decreases by the same factor. For example, if you multiply the *mass* by two, you must *divide* the acceleration by two.

The following examples show how increasing force while the mass stays the same increases acceleration.

- Acceleration = 5/5 = 1
- Acceleration = 10/5 = 2
- Acceleration = 15/5 = 3
- Acceleration = 45/5 = 9

The preceding calculations show that as force increases, acceleration increases. These examples show that force and acceleration are directly proportional.

Notice, what happens, however, if the mass increases while the force stays the same.

- Acceleration = 60/1 = 60
- Acceleration = 60/2 = 30
- Acceleration = 60/3 = 20
- Acceleration = 60/4 = 15

These examples show that as mass increases, the acceleration must decrease. They show that mass and acceleration are inversely proportional. It is harder to make objects with more mass change their motion.

Action and Reaction

Imagine how the oars work when you row a boat. Look at Figure 10-18. You sit facing the rear of the boat. As you pull the oar handles toward your chest, the other ends of the oars push water toward the rear of the boat. A motor does the same thing. It pushes water toward the rear of the boat. In both cases, the boat is pushed forward. This is an example of the third law of motion. This law states that every action force is accompanied by a reaction force that is equal to the action force but opposite in direction. The **action** and **reaction** forces act on different objects, not on the same object. In the rowboat example, the action force is on the water and the reaction force is on the boat.

Action and reaction forces always occur in equal but opposite pairs. For example, you may decide to help someone with a stuck car by pushing it with your car. Your push is the action force. Just be careful! Remember, the stuck car has an equal

Figure 10-18 Action/Reaction Pairs of Forces

Motion of boat

Front of boat

Reaction:
Oar pushes boat
forward at pivot point.

Action:
Oar pushes water to rear.

Rear of boat

reaction force. The stuck car could dent the fender of your car.

Think about another example. When you let go of a balloon, the air inside rushes out with action force in one direction, while the reaction force sends the balloon flying in the opposite direction. In Figure 10-6, the forces on Scale A, Scale B, and the rubber band are all equal. The principle of action and reaction is also at work in jet engines and rockets.

Without the third law, motion would not be possible. For example, as you walk, your shoes exert the action force against the ground. At the same instant, your legs move forward by the reaction force from the ground. The friction between your shoes and the ground causes this pair of forces. Just think how hard walking is when there is too little friction. Sliding on ice or slipping on an oil slick are perfect examples.

Figure 10-19 shows an experiment that demonstrates action and reaction. The string that pulls on the rubber band is burned. When the string breaks and the rubber band snaps forward, the slingshot action throws the steel ball in one direction. Meanwhile, the rubber band and the wooden platform flies in the opposite

Figure 10-19 A Demonstration of Action/Reaction/Recoil

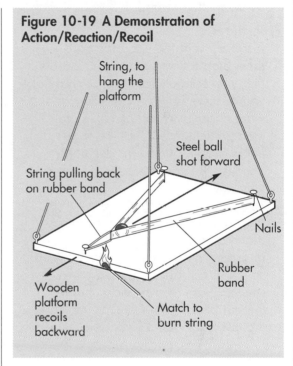

String, to
hang the
platform

Steel ball
shot forward

String pulling back
on rubber band

Nails

Wooden
platform
recoils
backward

Rubber
band

Match to
burn string

direction. Their movements show action and reaction. The movement backward is sometimes called **recoil**. A gun recoils at the instant the bullet flies forward through the gun's barrel. The forward movement of the bullet and backward movement of the gun also show the principle of action and reaction.

How Can Inertia Be Observed?

Process Skills observing; determining relationships from data; interpreting data

Materials one uncooked egg; a heavy object such as a one-kg mass with a hook; some thin thread

Procedures

1. Carefully spin the egg on a hard surface. Stop the egg from spinning but then immediately release it. What do you observe? How might you explain what happens as an example of inertia? Remember that the egg is liquid inside. Do the insides stop because the egg is stopped? Why or why not?
2. Attach the thread to the heavy mass. Very gently begin to drag the mass along a hard surface. It should begin to move very slowly and then pick up speed. If the thread breaks, use a smaller mass or a stronger thread.
3. Repeat Step 2, but this time yank on the thread. What do you see happen?

Conclusions

4. Why does the mass resist moving? Why does slow acceleration go smoothly, but a fast acceleration, the yank, breaks the thread? Where in the thread experiment is an action and reaction force? Name the objects that feel the action and the reaction forces.

Here is a riddle teachers often ask about this topic. A horse pulls a wagon. Since the action and reaction forces are equal and opposite, they will be in equilibrium. Therefore, the wagon cannot move! What is wrong with this argument? Action and reaction forces do not act on the same object, like the wagon in this riddle. They each act on different objects. The action force of the horse acts on the wagon, but the reaction force acts on the horse, not on the wagon.

Applications of the Laws of Motion

You probably have been in a car or aboard a bus that came to a sudden stop. What happened? You were thrown forward in your seat, or you bumped people standing in front of you on the bus. An accident like this in a fast-moving car is especially dangerous. At high speeds, people can be thrown against the windshield, or even through it, with deadly force.

Nothing really throws you forward in such situations. What really happens is that your body obeys the law of inertia. The car or bus stops, but your body tends to keep moving forward at the same speed as before. Look at Figure 10-20.

The opposite occurs when a car or bus suddenly jerks forward when starting. At that moment, you feel as if a force is pushing you backward, but there is really no force. Your body was at rest and is ready to stay that way. The bus starts moving forward, but your body does not.

To protect car passengers from these effects of the laws of motion, people use seat belts and child-restraint seats. Many states have laws requiring all passengers

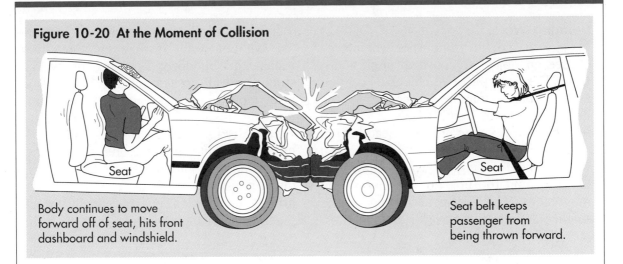

Figure 10-20 At the Moment of Collision

Body continues to move forward off of seat, hits front dashboard and windshield.

Seat belt keeps passenger from being thrown forward.

to use these. You have probably read a sign that says, "Seat Belts Save Lives." Read the Science, Technology, and Society feature at the end of this chapter to learn more about seat belts.

■ Section 10-3 Review ■

Write the definitions for the following terms in your own words.

1. **reference point**
2. **deceleration**
3. **inertia**
4. **action and reaction**
5. **recoil**

Answer these questions.

6. What is required in order to detect and measure motion?
7. a. What is the speed of an ant that runs 72 centimeters in eight seconds?
   b. What is the speed of a car that travelled 90 miles in one hour and 30 minutes?

8. What is the cause of acceleration? What types of changes in motion are accelerations?
9. State the three laws of motion in your own words.
10. Describe two situations that illustrate action and reaction.

10-4 Understanding Work

■ *Objective*
☐ *Define work.*

Who do you think does more work, a furniture mover working for two hours or a bus driver working for eight hours? You can only answer this question by using science. In science, the word work has a very special meaning. A scientist would probably say the furniture mover does more work. You will learn why in this section.

Defining Work

In science, **work** is done whenever a force causes an object to move through a distance. For example, if you lift up a box, you are working. You apply a force to raise the box to a certain height. However, if you now hold the box above your head for ten minutes, your muscles may ache but you will not be working. You are applying a force to hold the box up, but there is no movement. Therefore, according to the science definition, no work is being done.

Imagine a bulldozer pushing against an old building wall. If the wall does not break, the bulldozer is doing no work. Once the wall breaks and the bulldozer's force moves the bricks and dirt through a distance, physical work is being done.

Look at Figure 10-21. Because the furniture mover lifts, pushes, and pulls many heavy objects onto trucks and up stairs, much physical work is done. Because bus drivers usually do not personally exert much force or move objects, they perform

Figure 10-21

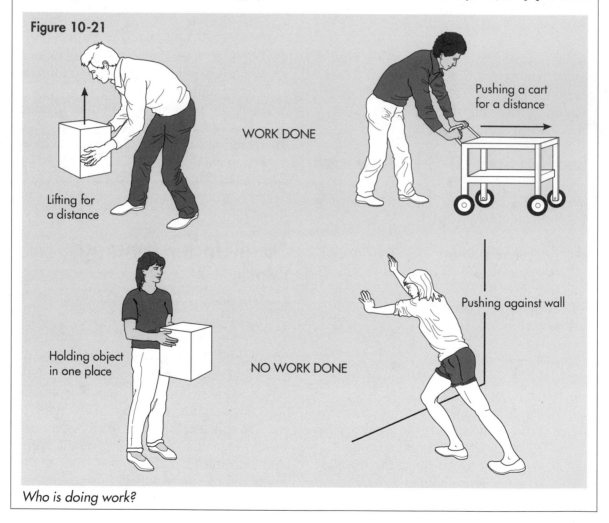

Lifting for a distance

Pushing a cart for a distance

WORK DONE

Holding object in one place

NO WORK DONE

Pushing against wall

Who is doing work?

less physical work. Of course, they work very hard in terms of hours and tasks.

Measuring Work

You can find the amount of work done in moving an object by using this equation.

work = force applied × distance moved

or

work = force × distance

Study this example. A girl pushes a car, using a force of 120 pounds, for a distance of 15 feet. What amount of work is being done?

work = force × distance
work = 120 lbs × 15 ft = 1800 ft-lbs

This unit, **foot-pounds (ft-lbs)**, is the unit of work used in industry and engineering in the United States.

For another example, imagine a boy lifts an eight-newton box from the floor to a shelf two meters up. What amount of work is required? The weight of the box is eight newtons, or 8 N. Therefore, the boy must use at least 8 N of force to do the lifting.

work = force × distance
work = 8 N × 2 m = 16 N-m

This unit, newton-meter (N-m), is the unit of work in the metric system. In fact, it is used so often that scientists have renamed it as the **joule (J)** (JOOL). One joule of work is done when a force of one newton acts through a distance of one meter. For the example of the boy, work = 16 N-m = 16 joules or 16 J.

The basic formula tells you two important facts about work. If no force is applied, then work is not done, even if an object is moving. For example, a spaceship may be coasting quickly in space with no engines on. Since the force is zero, the work is zero. Also, if a force is applied to an object but it does not move, then no work is done, because the distance moved is zero.

■■■ Section 10-4 Review ■■■

Write the definitions for the following terms in your own words.

1. **work**
2. **foot pounds**
3. **joule**

Answer these questions.

4. Why does it take more work to climb a hill than to walk on level ground?
5. How much work is done when a 150-lb person climbs 20 feet up a ladder?
6. Which requires more work, pushing with a force of six newtons over six meters, or pulling with a force of seven newtons over five meters? Calculate your answers in joules.
7. Under what conditions is no work done on an object?
8. Where does a typist do more work, all day at the office, or after work at the gym? Explain your answer.

SCIENCE, TECHNOLOGY, & SOCIETY

Seat Belts and the Law

Many states have passed laws requiring the driver and passengers in the front seat of a car to wear seat belts. Imagine that such a law is being debated in your state legislature.

The governor is in favor of such a law, claiming that it could save more than 1,000 lives each year.

Insurance agents talk about the rising costs of treating injuries from head-on collisions. The costs often go beyond a family's insurance coverage. The state, directly or indirectly, pays for the balance of these costs.

A group of scientists point out that many people think they can react fast enough to stop an accident. However, they say, human reaction time is about 0.75 second, and during that time a car moving at 60 mph travels about 30 feet. The force of impact is more than people imagine, too. At 30 mph, it takes 1,000 lbs of force to brace yourself against flying forward, far more than the body can stand.

Some citizens say that seat belts are a bother and may even be dangerous if you have to exit a car quickly. Others want to pass a law requiring air bags that inflate and cushion people in an accident.

Senator Zirofsky says, "Seat belts are the best trade-off between costs, other technologies, and saving lives."

Car manufacturers report they are trying one new idea involving loose belts that lock only during rapid braking. Some cars have automatic belts that lock by themselves. These expensive changes may make seat belt technology more acceptable to the public.

A university professor says she thinks the long-term benefits in lives and medical costs are worth it. The governor pointed out that the new law would be a compromise between doing nothing and trying expensive safety technologies.

Follow-up Activity

By interviewing relatives and friends and using library resources, draw up a list of pros and cons regarding seat belts. Then prepare a senator's position on the seat-belt law. Be ready to advise the senator on how to respond to pressures that will come from voters in the district.

Figure 10-22

CHAPTER REVIEW

KEEPING TRACK

Section 10-1 Characteristics of Force

1. Force causes objects to change their motion and/or shape.
2. Mass is the amount of matter in an object.
3. Every object has a force called gravity that increases with greater mass and decreases with greater distance.
4. Weight is the earth's force of gravity acting on an object. Weight can change, depending on height above the ground; mass does not change.
5. Force, including weight, is measured in pounds (lbs) or in newtons (N).

Section 10-2 Balanced and Unbalanced Forces

1. Equilibrium exists when all the forces acting on an object balance each other. If only two forces are in equilibrium, they must be equal but opposite.
2. Balanced forces have the same effects as no force; the object affected has either no motion or unchanging motion.
3. Unbalanced forces have the same effect as a single force; the object affected has a change in motion.
4. Friction force is always produced during motion. It acts opposite the direction of motion and produces heat.

Section 10-3 Measuring Motion

1. Motion is measured from fixed reference points.
2. Speed is the distance traveled in a unit of time. Speed = distance / time.

3. Acceleration is any change in speed or in direction of motion.
4. The first law of motion says that balanced forces and zero force cause no acceleration, which could mean no motion or motion at a constant speed.
5. The second law of motion says that unbalanced forces and a single force cause acceleration. The rate of the acceleration produced is directly proportional to the force applied and inversely proportional to the mass of the object.
6. Inertia is the resistance of an object to changing its motion, and it depends on mass.
7. The third law of motion says every action force produces a reaction force, sometimes called recoil.

Section 10-4 Understanding Work

1. Work is a force that moves an object through a distance. Work = force × distance.
2. Work is measured in foot-pounds or newton-meters. The newton-meter is also called a joule (J).

BUILDING VOCABULARY

Write the term from the list that best completes each sentence.

acceleration, equilibrium, gravity, speed, unbalanced force, weight

The cause of changes in an object's motion is always an ___1___. Two forces may be opposite yet equal, thereby producing ___2___. In this situation, an object either does not move or keeps moving at a constant ___3___. ___4___ is an attraction force that every object has because of its matter. Earth's attraction force on an object is called the object's ___5___ . Unbalanced forces produce ___6___.

SUMMARIZING

Write *true* if the statement is true. If the statement is false, change the *italicized* term to make the statement true.

1. A change in motion is *acceleration*.
2. *Constant speed* is created by unbalanced forces.
3. Constant speed is created by *unbalanced* forces.
4. The gravity of an object *increases* with an increase in its mass.
5. On the moon, an object's *mass* will be less.
6. A larger mass has *more* inertia.

INTERPRETING INFORMATION

Figure 10-23 gives data about the speeds of five identical objects taken at intervals of one second.

Which objects, A, B, C, D, or E, best match the following statements? More than one are possible.

1. It is moving at a constant speed.
2. It is accelerating.
3. It is decelerating.
4. It is not accelerating.
5. Unbalanced forces are acting.

Figure 10-23					
Time (seconds)	**0**	**1**	**2**	**3**	**4**
Speeds (meter/sec)					
of A	0	0	0	0	0
of B	0	3	6	9	12
of C	2	4	6	8	10
of D	15	15	15	15	15
of E	7	6	5	4	3

THINK AND DISCUSS

Use the section number in parentheses to help you find each answer. Write your answers in complete sentences.

1. A magician claims she can bend a spoon just by her thoughts. What would a scientist look for if this were done? (10-1, 10-4)
2. A 100-lb person would weigh 16 lbs on the moon and 9,500 lbs on Jupiter. What can you conclude from these facts? (10-1)
3. When a sky diver first jumps from a plane, the diver's falling speed increases rapidly. Once the parachute opens, however, the diver's speed stays constant. Using information gained in this chapter, explain what is happening. (10-2)
4. Which situations result from an unbalanced force acting on an object? (10-2)

a. being at rest
b. starting to move from rest
c. changing speed
d. changing direction of motion

5. If the inertia of moving objects tends to keep them moving forever, then why does everything slow down and eventually stop on the earth? (10-2, 10-3)

6. a. How much work is done as a 100-lb person climbs 25 feet up a ladder?
   b. Which requires more work, pushing with eight newtons for six meters, or pulling with seven newtons for seven meters? Calculate both in joules. (10-4)

GOING FURTHER

1. Earth's gravity pulls on a truck with a force of 80,000 newtons. Calculate
   a. the truck's weight, in N
   b. the truck's weight, in lbs
   c. the truck's mass, in grams
   d. the truck's mass, in kilograms

2. A "hot rod" fan buys a car and decides to increase its rate of acceleration. She cleans and oils the engine and gears. She strips the car of everything it does not need, such as the rear seat. Why will these changes improve acceleration?

COMPETENCY REVIEW

1. Which makes an object change its motion?
   a. zero force
   b. balanced forces
   c. forces in equilibrium
   d. unbalanced forces

2. The mass of an object can be given in

a. grams.   b. pounds.
c. newtons.   d. joules.

3. As an object is sent farther from the earth, its mass
   a. decreases.
   b. increases.
   c. remains the same.
   d. changes shape.

4. Angus walks 20 kilometers in four hours. What is his speed?
   a. 5 km/hr   b. 0.2 km/hr
   c. 24 km/hr   d. 80 km/hr

5. Which combination results in the greatest acceleration?
   a. zero force, small mass
   b. small force, small mass
   c. small force, large mass
   d. large force, small mass

6. Which situation requires the most work?
   a. zero force for a large distance
   b. large force for zero distance
   c. small force for a small distance
   d. large force for a large distance

7. Objects resist changes in motion according to how much mass they have. This is called
   a. action.   b. reaction.
   c. inertia.   d. work.

8. Which is not true about friction?
   a. It is sometimes caused by rubbing between surfaces.
   b. It is always equal to the applied force.
   c. It acts against the direction of motion.
   d. It produces heat.

CHAPTER 11

AN INTRODUCTION TO ENERGY

"Oh, sorry, Mom, but I used up all my energy studying last night. I cannot do any work today. I will just stay in bed until I get some energy again. Okay? I can clean my room tomorrow. Thanks, Mom. Please wake me for lunch."

Does this sound familiar? While this teenager might be making excuses for laziness, there is some scientific truth in what the teenager said. Work is only possible if energy is present. In this chapter, you will learn about energy. What is it? How is it measured? You will also see that there are just two types of energy, but there are many forms of energy.

It is possible to transform one form of energy into another. This occurs continually all around us in nature and in human-made machines. During such transformations of energy, certain laws of science are followed. These laws will be discussed in this chapter.

This chapter's introduction to energy sets the stage for the rest of this book, which is all about the forms of energy, the uses of energy, and the sources of energy in our society.

11-1 The Concept of Energy

■ *Objectives*

☐ *Define energy.*

☐ *Distinguish between energy and work.*

☐ *List standard units for measuring energy.*

In Chapter 10 you learned that work is a force acting over a distance. When you push, pull, or lift something for any

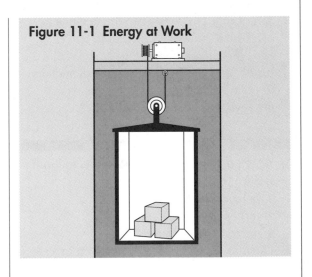

Figure 11-1 Energy at Work

distance, you are doing work. You are able to do this work because you eat food. A machine is able to do work because it is given electricity or fuel. Food and gasoline are examples of materials that contain energy. **Energy** is the ability to do work. This energy might be in a single object, like a machine, or in an organism, like you.

For example, look at Figure 11-1. An elevator motor pulls an elevator up with large force over a large distance. This is work. Energy in the form of electricity had to exist for the motor to do this work.

Windmills were once used to grind grain, like wheat, into flour. They were also used to pump water up from below ground. Windmills are now used to turn electric generators. To do this work, there must by energy somewhere. The energy of the moving wind makes the windmill turn.

Energy and Work

An object has energy if it can do work. Study Figure 11-2. It shows a pile driver banging steel posts into the ground. When the hammer is high up, it has energy because of its height. When it strikes the steel post, the hammer does work by pushing the post into the ground a little more deeply. You can also see that when work is done by the motor to lift the hammer, the hammer gains energy. Perhaps you have used a bow and arrow. As you pull on the string, you are doing work on the bow. That work increases the energy in the bow because the bow is bent. When you release this energy, it returns as work to send the arrow flying.

Work and energy are like two sides of a coin. They must exist together. Work is the result of releasing energy, and energy is

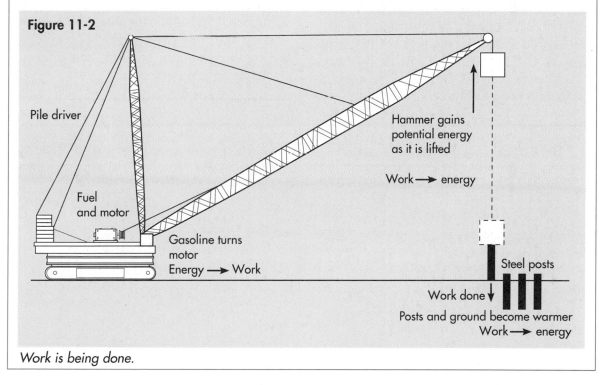

Figure 11-2

Pile driver

Fuel and motor

Gasoline turns motor
Energy ➞ Work

Hammer gains potential energy as it is lifted

Work ➞ energy

Steel posts

Work done

Posts and ground become warmer
Work ➞ energy

Work is being done.

the ability to do work. The pile driver in Figure 11-2 shows several of the back-and-forth changes between energy and work. Find the changes illustrated in the figure.

Measuring Energy

Work can be changed into energy, and energy can be changed into work. For this reason, energy is measured in exactly the same units as work. These units are foot-pounds (ft-lb) in United States engineering and industry, and joules (J) in the metric system. Recall that one joule is one newton-meter.

Heat is a form of energy that increases as the temperature of an object increases. Special units are used to measure heat energy. For instance, **calories** (cal) and kilocalories (kcal) are used in the metric system. The **BTU**, or British Thermal Unit, is used in United States engineering and industry. The energy of electricity is also measured in special units called **kilowatt-hours** (KIL-uh-waht) (kWh). It is always possible to convert these energy units into joules. Look at Figure 11-3.

■■■ Section 11-1 Review ■■■

Write the definitions for the following terms in your own words.

1. **energy**   2. **heat**
3. **BTU**

Answer these questions.
4. Name the unit of energy used
   a. in the metric system.
   b. in United States industry.
   c. for heat energy in the metric system.
   d. for heat energy in United States industry.
   e. for electrical energy.
5. Why are the units for work and energy identical?
6. What is the difference between work and energy?
7. Work is done on a rubber band as you stretch it. What is true about the rubber band after it is stretched?
8. Give an example of work and energy changing into each other.

Figure 11-3

Example	Energy	Energy in joules
50-lb rock, 5 ft off ground	250 ft-lb	340 J
10-N object, 8 m off ground	80 N-m	80 J
Heat from 1 lb of coal	12,000 BTU	13 million J
Heat from bowl of cereal	150 kcal	630,000 J
Electricity for 1 hr of TV	0.1 kWh	360,000 J

11-2 Kinds of Energy

■ *Objectives*
☐ *Describe the two types of energy.*
☐ *Distinguish among the forms of energy.*

Energy can appear in many forms. Food, fuel, batteries, a moving car, the wind, a magnet, and atoms all contain different forms of energy. Although there are many forms of energy, there are just two main types of energy.

Two Types of Energy

Potential energy comes from the position of an object or from its chemical composition and structure. For example, a diver standing at the end of a diving board has potential energy. **Kinetic energy** comes from the motion of an object. When the diver pushes off from the diving board, the motion is kinetic energy.

Objects have potential energy because of their position above the ground. As you climb a ladder or walk up stairs, your potential energy increases. The amount of potential energy in an object depends on its height and its weight.

The higher up an object is, the more potential energy it has. Also, the greater the weight of an object, the more potential energy it has. For example, a heavier person has more potential energy than a lighter person when the two are side by side on a platform above the ground. Study Figure 11-4.

Objects also have potential energy because of their chemical composition and structure. Gasoline is made of molecules that can combine with oxygen to release heat. The potential energy in gasoline comes from its chemical composition. This is true of other fuels, of food, and of the chemicals in a battery. You can think of

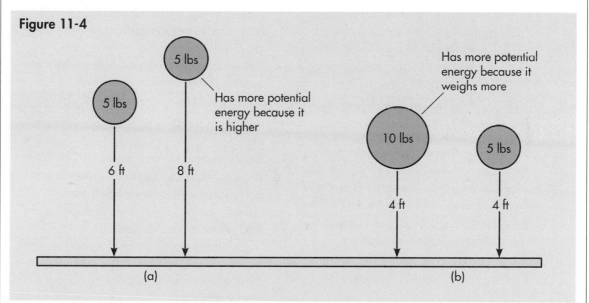

Figure 11-4

5 lbs

5 lbs
Has more potential energy because it is higher

Has more potential energy because it weighs more

10 lbs

5 lbs

6 ft   8 ft

4 ft   4 ft

(a)   (b)

Potential energy depends on height and weight.

How Can You Observe the Effect of Weight and Height on Potential Energy?

Process Skills *gathering and recording data; analyzing; stating conclusions*

Materials three glass tubes of different lengths, through which cylinders can fall; meter stick; scale; soft clay; set of cylinders made of different metals, identical in size but with different weights

Procedures

1. Mark a starting line on each glass tube. Measure the distance, in cm, from the far end of the tube to the starting line. Record the distance for each tube. Then weigh each metal cylinder in newtons. Record the weights. If your scale measures mass in grams, multiply the measurement by 0.01 to find the weight in newtons.
2. Flatten the top of the clay and position the shortest glass tube so that it just touches the surface of the clay. Look at Figure 11-5.
3. Using only the shortest glass tube, drop the cylinders from the starting mark, through the tube, onto the surface of the clay. The hole that the cylinder makes in the clay is called an indentation. Each time, measure the depth of the indentation, in cm. Between each trial, make the clay flat and smooth again. Record your observations.

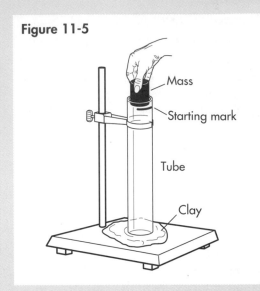

Figure 11-5

Mass

Starting mark

Tube

Clay

4. Repeat this procedure, using only the heaviest cylinder. Drop that cylinder through each of the three tubes. Measure the depth of the indentation after each drop, and then smooth the clay. What does this tell you about how dropping distance influences the indentation made in the clay?

Conclusions

5. In Procedure 3, which variables were kept constant? Which variables were kept constant in Procedure 4?
6. Explain how the depth of the indentation made in the clay shows the amount of potential energy the object had when it was dropped.
7. Study the data of weight, height, and depth of indentation in the clay. What conclusions can you state about potential energy?

Analyzing; Stating Conclusions

Figure 11-6 shows a spool being knocked with a rubber mallet along a string. It is later hit with mallets of different masses and speeds. The distance the spool moves along the string is shown in Figure 11-7.

1. Why do these qualify as well-designed, controlled experiments?
2. How is the kinetic energy of the mallet shown by the distance the spool moves?

3. What conclusions can you state about the effect of speed and mass on the kinetic energy of an object?

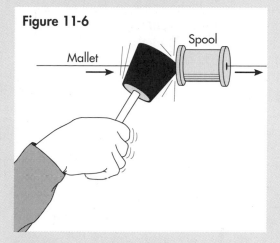

Figure 11-6

Figure 11-7 Experiment 1 The spool is struck with the same mallet at different speeds.

Mallet mass	Hitting speed	Distance spool moves
100 g	Slow	38 cm
100 g	Medium	153 cm
100 g	Fast	604 cm

Experiment 2 The spool is struck with different mallets at the same speed.

Mallet mass	Hitting speed	Distance spool moves
100 g	Medium	158 cm
200 g	Medium	322 cm
300 g	Medium	444 cm

potential energy as energy stored in an object. It is stored by the position of the object or by the chemicals of the object.

All moving objects have kinetic energy. The amount of kinetic energy in an object depends on the speed and mass of the object. As an object's speed increases, so does the kinetic energy. For example, a car moving at 60 miles per hour has much more kinetic energy than when it moved at 30 miles per hour. In fact, although the speed doubles, the kinetic energy is multi-plied by four. Also, a larger mass will have more kinetic energy than a smaller mass if both move at the same speed. Look at Figure 11-8.

Forms of Energy

There are many **forms of energy**, or ways in which energy may exist. Each form of energy may consist of just potential ener-gy, just kinetic energy, or a combination of both. For example, the heat energy re-leased when burning gasoline comes from

Figure 11-8

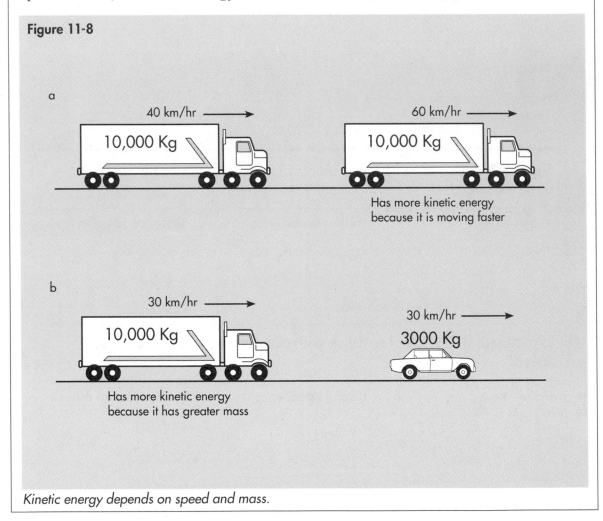

Kinetic energy depends on speed and mass.

the potential and kinetic energy of its molecules. You will recognize many of the important forms of energy.

Mechanical Energy Whenever you talk about how a machine works, you are referring to **mechanical energy.** Mechanical energy is the sum of the potential energy resulting from the height of an object and the kinetic energy of the object. It is the energy of position plus the energy of motion. For example, an airplane's mechanical energy comes from its height above the ground and its speed. Machines such as car engines, pulleys, and pencil sharpeners run on mechanical energy.

Mechanical energy is also increased when a spring is wound up or a rubber band is stretched. For example, the mechanical energy in a wind-up clock comes from the wound up spring.

Chemical Energy The energy present in the chemicals in an object is called **chemical energy**. Chemical energy is stored in the bonds between the atoms and molecules of an object. Food, wood, and gasoline have high amounts of chemical energy. All batteries operate on chemical energy. Explosives such as gunpowder and dynamite contain a large amount of chemical energy that is easily and rapidly released.

Heat Energy The motion of molecules in a material causes **heat energy.** The amount of heat energy in an object is determined by its temperature, mass, and composition. An object with a lot of heat energy is usually hot because its molecules are moving fast. An object with less heat energy is usually cooler because its molecules are moving more slowly. For example, the water in a hot bath contains more heat energy than water in a cup of cold water.

Electromagnetic Energy X-rays and radio waves are examples of **electromagnetic energy** (ih-lek-troh-mag-NET-ik)**.** This energy is contained in electromagnetic waves that exist everywhere. Such energy exists in light waves, microwaves, TV waves, ultraviolet waves, and infrared waves. The electromagnetic energy in microwaves cooks food; in radio waves, electromagnetic energy helps produce sound; and in X-rays, electromagnetic energy goes through your body and forms pictures.

Light Energy The electromagnetic energy in light waves is **light energy.** The light from a firefly, a car's headlight, or lightning contains light energy. The sun is a vast source of light energy and of other kinds of electromagnetic energy. The energy of the sun pours into our planet. Without this energy, life would not be possible.

Sound Energy The squeezing of air molecules when objects vibrate creates **sound energy.** For example, strumming a guitar squeezes the air molecules near the strings as they vibrate. When you speak, air molecules are squeezed in your throat.

Electrical Energy The energy in electricity and in any objects that are electrically charged is **electrical energy.** Electricity is actually a flow of trillions of electrons each second through a wire. This flow creates the energy of electricity. You have seen objects that contain an electrical charge, such as a comb that sparks in your hair, or a balloon that clings by "static" when rubbed. Lightning is just a giant spark formed by the electrical energy in storm clouds.

Magnetic Energy   The energy found in magnets is known as **magnetic energy.** Some magnets are natural and some are made in laboratories. Many magnets are produced by electricity; door buzzers and loud speakers are two examples. The magnetic energy in magnets can pull and move objects made of iron. Recall that this is energy doing work, as introduced in Chapter 10 and Section 11-1.

Nuclear Energy   Locked in the nucleus of every atom is **nuclear energy** (NU-klee-ur). It is tremendously powerful when compared to all other forms of energy, and it is also extremely dangerous. It is the energy of the atom bomb, of nuclear reactors that make electricity, and of the sun itself.

■ Section 11-2 Review ■

Write the definitions for the following terms in your own words.

1. **potential energy**
2. **kinetic energy**
3. **mechanical energy**
4. **light energy**
5. **sound energy**

Answer these questions.

6. Which of the following has the greater amount of potential energy? Explain each answer.
   a. a six-newton object or a four-newton object, if both are three meters from the ground
   b. a 50-lb object when it is 10 feet off the ground or when its 20 feet off the ground
7. Which has the greater amount of kinetic energy? Explain your answer.
   a. a car or a bike moving at the same speed
   b. a baseball thrown slow or fast
8. List eight forms of energy and give one illustration from your own life in which each form is present.

11-3 The Transformations of Energy

■ *Objectives*

☐ *Give examples of energy transformations.*

☐ *Use the law of conservation of energy.*

☐ *Explain two laws related to transforming energy.*

Did you have a good fill of chemical energy this morning? How was your breakfast? The energy in your food can be traced back to light energy from the sun. Sunlight enables plants to grow. You eat food from plants, such as grains, fruits, and vegetables. You might also eat products from animals that eat plants. Animal products include meat, eggs, and milk.

Some chemical energy now in your body will be changed to heat energy to keep your temperature at 98.6° F. Some chemical energy will increase your potential energy and your kinetic energy as you move. Some chemical energy will be converted into work as you push, pull, and lift things. In all three examples, energy is being changed from one form to another.

Transforming Energy

You can always use energy to do work by moving something. This is what defines energy. Energy can also be changed from one form into other forms. This is **energy transformation.** Figure 11-9 shows examples of energy transformations.

Many natural events include energy transformations. The electrical energy in storm clouds is transformed into light energy, or lightning, and sound energy, or thunder. Light energy from the sun is changed into chemical energy in plants and into heat energy that warms the air and the ground. In stars, including our sun, nuclear energy becomes heat energy, which becomes electromagnetic energy that travels through space. Most human inventions also include energy transformations.

Conservation of Energy

Although form of energy can be changed, the total amount of energy cannot be changed. The amount of energy is conserved, which means that it stays the same. This is the law of **conservation of energy**. When energy is transformed, the total amount of energy always stays the same.

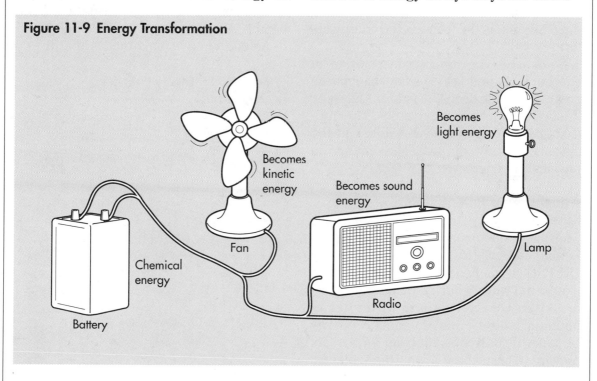

Figure 11-9 Energy Transformation

Figure 11-10	
Before explosion	
Chemical energy in firecracker	8000 joules
After explosion	
Sound energy of the boom	2000 joules
Light energy of the flash	500 joules
Kinetic energy of flying pieces	1500 joules
Work for ripping the cover, pushing air	1000 joules
Heat energy heating the air	3000 joules
Total	8000 joules

Figure 11-10 shows the energy transformations that take place when a firecracker explodes. The total energy after the explosion is the same as the energy before, 8000 J.

Sometimes the law of conservation of energy is stated as "energy cannot be created or destroyed." It is also known as the first law of thermodynamics. **Thermodynamics** (thur-moh-dy-NAM-iks) is the advanced study of heat and energy in physics and engineering.

Transforming Potential and Kinetic Energy

The back-and-forth changes between potential energy (PE) and kinetic energy (KE) occur so often that you can find them almost anywhere. For instance, study Figure 11-11. On the board, the diver has all potential energy and no kinetic energy because the diver is high up and is not moving. During the dive, the potential

energy decreases and the kinetic energy increases. As the diver hits the water, there is no potential energy, because there is no height. All the energy has been changed to kinetic energy. This means the diver's

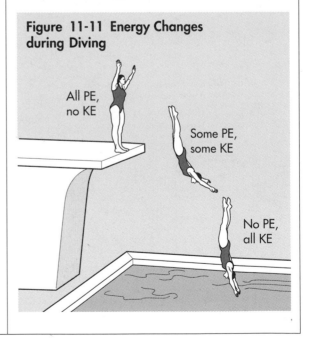

Figure 11-11 Energy Changes during Diving

All PE, no KE

Some PE, some KE

No PE, all KE

speed is greatest just upon hitting the water. For example, suppose the potential energy of the diver standing on the diving board is 3500 joules. When the diver hits the water, the potential energy is zero and the kinetic energy is 3500 joules.

Think about a ride on a roller coaster. A cable pulls the cart up the first big hill. At the top, the cart has a large amount of potential energy but almost no kinetic energy. Now examine Figure 11-12. You know that as the cart rolls downward it picks up speed, so it gains kinetic energy. Where did this kinetic energy come from? It came from the loss of potential energy as the cart's height above the ground decreased. At the bottom, the cart is moving at great speed, with all kinetic energy and no potential energy. As it starts to go up the next hill, it loses some kinetic energy and slows down, but it gains potential energy as it climbs up. The entire ride is a conversion of potential energy into kinetic energy, and kinetic energy into potential energy.

In these examples, the total of the potential energy plus the kinetic energy stays the same. Recall that the mechanical energy of an object is the sum of its potential energy due to height and its kinetic energy.

mechanical energy = potential energy
+ kinetic energy

In general, if there were no friction, the mechanical energy of an object would remain the same.

Figure 11-12

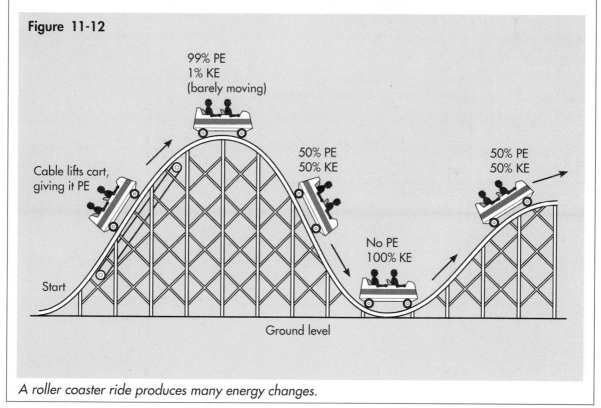

A roller coaster ride produces many energy changes.

How Does Energy Change as a Pendulum Moves?

Process Skills observing; analyzing

Materials a pendulum made from a string and an attached mass; a heavy metal or wooden ball

Procedures

1. Start the pendulum swinging by pulling it up to a certain height above the table. Observe its motion carefully. Find where it moves fastest and note when it stops for a moment to turn around.
2. Allow the pendulum to keep swinging on its own. What begins to happen?
3. Now place a ball on the table so that the pendulum will hit it at the bottom of its swing. Look at Figure 11-13. Record how far the ball rolls along the table. Repeat this for different starting heights of the pendulum above the table.

Conclusions

4. What energy transformations occur? Where does the pendulum have only potential energy, maximum kinetic energy, and a combination of both? How does the kinetic energy change with a higher starting position of the pendulum? Observe this from the distances the ball is sent rolling.

Figure 11-13

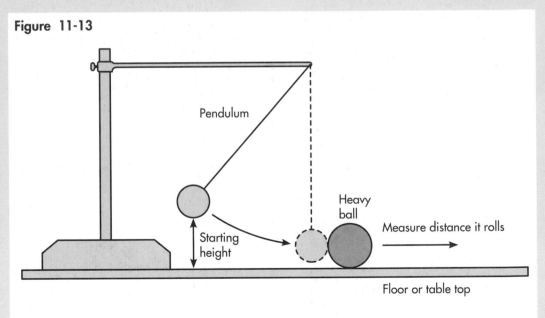

Decrease in Useful Energy

Heat is produced in every transformation of energy. This is the second law of thermodynamics. In the firecracker example in Figure 11-10, 3000 J of the 8000 J was changed into heat energy. When you climb stairs, you are changing the chemical energy of food into potential energy. A tiny amount of the energy turns into heat because of the friction in your joints and chemical reactions in your muscles.

Did you know that 95 percent of the electrical energy used in a light bulb turns to heat, and only five percent becomes light energy? When the light falls on your room, your books, or you, it also changes into heat. Almost all the electrical energy is "lost" as heat energy. The word *lost* emphasizes that heat is not a very useful form of energy. Machines cannot work by heat energy alone. They need fuel. Living things cannot stay alive by heat energy alone. They need food.

Useful energy is easily and almost fully convertible into useful work. Electricity and gasoline contain very useful energy. Heat is a less useful form of energy because it cannot be fully changed into work.

Experts know that a large percentage of a car's gasoline never is used to move the car but instead turns into heat energy in the engine, gears, tires, brakes, and air. In this way, energy is wasted. Reducing the amount of this wasted energy is a major goal of scientists and engineers.

The laws of thermodynamics are the two major laws of energy transformations. The first law of thermodynamics states that the total energy remains constant. The amount of energy never decreases or increases. The Second Law of Thermodynamics states that heat is produced during every energy transformation. Heat energy is not a useful form of energy, because it cannot be fully changed into work. This means that the amount of useful energy decreases.

▰▰ Section 11-3 Review ▰▰

Write the definitions for the following terms in your own words.

1. **energy transformation**
2. **conservation of energy**
3. **thermodynamics**
4. **useful energy**

Answer these questions.

5. When energy is transformed, some is lost because of friction. How can we say, then, that energy is always conserved?
6. The potential energy of an airplane, because of its height, is 50 million ft-lbs. The plane's kinetic energy, because of its speed, is 20 million ft-lbs. What is the mechanical energy of the plane?
7. What form of energy is present in a roller coaster cart at the top of a hill? What is the energy change that takes place as the cart moves down the hill? What energy change takes place when the cart begins to climb the next hill?
8. Why is heat energy sometimes called lost energy? When is energy considered useful?

4. Kinetic energy increases with an object's ___ and mass.
5. Heat can be measured either in ____ or ____.
6. In a toaster, ___ energy is transformed into ___ energy.
7. Only a small percentage of what a light bulb gives off is ____ energy.
8. The fact that the amount of energy available always stays the same is known as the ____ of energy.
9. Energy that can produce work is known as ____.
10. After any energy change, the useful energy is always ___ because some energy is lost as _____.

INTERPRETING INFORMATION

1. Look at Figure 11-15. Which object has the least mechanical energy? Which has the greatest mechanical energy?
2. An experiment was done for one minute on an electric fan. Here are the results:
   Electrical energy used by fan = 12,000 J
   Kinetic energy of air it pushed = 8000 J
   Kinetic energy of spinning fan blades = 400 J
   Energy in noise made by fan = 300 J
   Heat made in motor's wires = 3300 J
   How do these results illustrate the two laws of thermodynamics?

THINK AND DISCUSS

Use the section numbers in parentheses to help you find each answer. Write your answers in complete sentences.

Figure 11-15

Object	Potential energy	Kinetic energy
K	20 J	20 J
L	10 J	25 J
M	5 J	60 J
N	30 J	30 J

1. You wind up a spring inside a toy. (11-1, 11-2)
   a. Why is this called work?
   b. What happens to the work as you wind up the spring?
   c. What happens as you turn the toy on?
2. List at least four forms of energy that may be involved when the television is on. (11-2)
3. A farmer uses an electric pump to lift water from below the ground and spray it onto crops. Electrical energy turns the motor. The work done on the water moves and lifts it above ground and gives it potential energy and kinetic energy. As drops of water fall on the crops, mechanical energy is transformed into heat. What does the law of energy conservation insist must be true in this example? (11-3)

GOING FURTHER

1. The formula for potential energy is
   PE = weight × height above ground

a. Calculate the potential energy of a 150-lb boy who has climbed to the top of a 40-ft tree. Answer in ft-lbs.

b. Calculate the potential energy of a 300-N girl who is painting 6 m up a ladder. Answer in N-m and in J.

2. The formula for kinetic energy is

KE = 1/2 × mass × speed × speed

When the mass is in kilograms (kg) and the speed is in meters per second (m/sec), then the kinetic energy will be in joules (J).

a. Calculate the kinetic energy of a runner with a mass of 60 kg and a speed of 5 m/sec.

b. Calculate the kinetic energy of this runner if his or her speed is doubled.

3. Here are the facts about a roller coaster at one spot along its track. Determine its mechanical energy in Joules at that spot.

mass = 1500 kg
weight = 15,000 N
height above the ground = 30 m
speed = 20 m/sec

Calculate potential energy and the kinetic energy, using the formulas in Exercises 1 and 2.

COMPETENCY REVIEW

1. The ability of something to do work is called
   a. power.   b. force.
   c. energy.   d. speed.

2. What is *not* true as you start to pedal a bicycle?
   a. Work is done on the bike.
   b. Work is done by the bike.
   c. The bike gains energy.
   d. You lose energy.

3. The chief metric unit for electrical energy is
   a. foot-pounds.
   b. BTU.
   c. kilowatt-hour.
   d. joule.

4. The mechanical energy of an object depends on all these things except
   a. speed.
   b. position.
   c. weight.
   d. chemical composition.

5. Gasoline is used as a fuel because it contains
   a. electromagnetic energy.
   b. mechanical energy.
   c. chemical energy.
   d. nuclear energy.

6. Which transforms light energy into chemical energy?
   a. a battery   b. a flashlight
   c. a stove   d. a plant

7. The energy involved in any change always
   a. decreases in amount.
   b. increases in amount.
   c. remains the same.
   d. is used up.

8. During any energy transformation, what is always produced?
   a. heat energy
   b. mechanical energy
   c. work
   d. electric energy

CHAPTER 12

MECHANICAL ENERGY AND MACHINES

Can you imagine life without machines? Early in the history of the human race, there were no machines. There was not even such a thing as a wheel. Yet today machines surround you at home. Cars, washing machines, typewriters, and food mixers are a few. Machines are the muscles of industry. Presses, drills, and robots are just a few of the heavy-duty machines that make the things you use. Farmers use machines such as tractors, combines, and automatic milkers. The printers who made this book used large, complicated printing presses.

Machines are the reason for many jobs too. Millions of people design, build, and sell machines. Others operate machines or fix them. Think of how many times families may have had to repair or replace a broken machine such as a vacuum cleaner, a bicycle, or an electric lawn mower.

In this chapter, you will learn the basics about machines. You will see that machines transfer mechanical energy to do a certain job. You will also learn that some machines do this job better than others.

12-1 The Characteristics of Machines

■ *Objectives*

☐ *State the characteristics of all machines.*

☐ *List the simple machines.*

You may think of machines as complicated things with many parts, like car engines. In science, though, even a nutcracker and a sidewalk ramp are called machines. Whether machines are complicated or simple, all have some features in common.

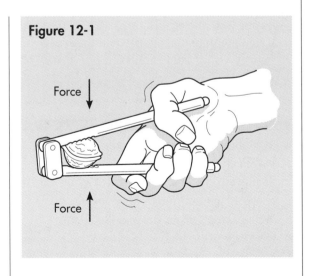

Figure 12-1

Force ↓

Force ↑

Nature of Machines

A **machine** is any device that transfers a force from one place to another. In other words, machines transfer mechanical energy from one place to another. Look at Figure 12-1. A nutcracker is a machine because it transfers the squeezing force of your hand to the nut you want to crack open. A ramp is a machine because it can transfer your push on a cart to a force that lifts the cart higher. A car's engine is a machine because it transfers the force of exploding gasoline to make the tires go around. Even the most complicated machine has the one simple purpose of transferring a force from one place to another.

All complicated machines contain many small, **simple machines**. Simple machines are the basic devices for transferring mechanical forces in all machinery. Figure 12-2 illustrates some simple machines.

Action of Machines

The force you apply to a machine is called the **effort force**. The effort force is the force put into a machine. The distance the effort force acts through is known as the **effort distance**. Have you ever played on a seesaw? Imagine that you are on a seesaw, up in the air. The effort force as you lower yourself to the ground is your own weight. The effort distance is the space between yourself and the ground.

The force the machine applies to some object is called the **resistance force**. This is the force that comes out of a machine. A useful example of this is a seesaw. The resistance force is in the end opposite the effort force. The distance through which the resistance force acts is the **resistance distance**. It is the distance your partner on the seesaw would rise as you move downward.

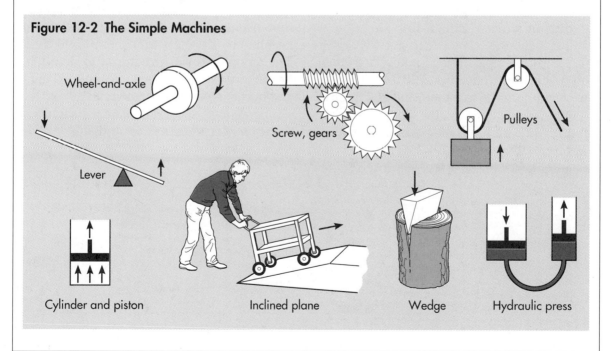

Figure 12-2 The Simple Machines

Wheel-and-axle

Screw, gears

Pulleys

Lever

Cylinder and piston

Inclined plane

Wedge

Hydraulic press

You can calculate the work put into the machine and the work that comes out. The **input work** is the effort force multiplied by the effort distance. See Figure 12-3.

Input work = effort force ×
effort distance

For example,

input work = 50 lbs × 4 ft
input work = 200 ft-lbs

The **output work** is the resistance force multiplied by the resistance distance.

output work = resistance force ×
resistance distance

For example,

output work = 100 lbs × 2 ft
output work = 200 ft-lbs

These examples reveal a key fact about all machines. Notice that 200 ft-lbs of work were put into the machine, and 200 ft-lbs of work were done by the machine. The input work and output work are equal. It is important to remember that machines cannot multiply the amount of work put into them. They can multiply the force or the distance applied to them, but they can never multiply the total work or energy put into them.

This is the **law of machines**. Under ideal conditions, the amount of work put into a machine is equal to the amount of work the machine does. The ideal conditions mean that no heat is made by friction or air resistance when the machine is used.

Real machines have friction, however. A real machine always gives out less energy than it takes in. This is because some mechanical energy is wasted as heat. You can write this as output work < input work, where the symbol < means "is less than."

Purposes of Machines

If machines do not increase the work put into them, what do they do? One answer is that machines make work seem easier and faster. Why do they only seem to do this? The amount of work required to do a job, or the force times distance, cannot change just because you use a machine.

There are several reasons machines make work seem easier and faster. For

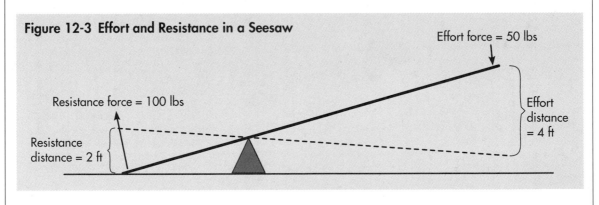

Figure 12-3 Effort and Resistance in a Seesaw

Effort force = 50 lbs

Resistance force = 100 lbs

Resistance distance = 2 ft

Effort distance = 4 ft

example, machines can multiply a force. This is the major purpose of most machines. On the seesaw you studied in Figure 12-3, a 50-lb force was multiplied to become a 100-lb force.

Machines can also change the direction of a force. On the seesaw, the downward effort force became the upward resistance force. Study the force illustrated in Figure 12-4. The pulleys shown in Figure 12-4 enable you to pull downward or sideways in order to pull an object upward. This makes the work seem easier to do.

Machines can also multiply distance. A large effort force only has to be moved a short distance to produce a small resistance force that moves a long distance.

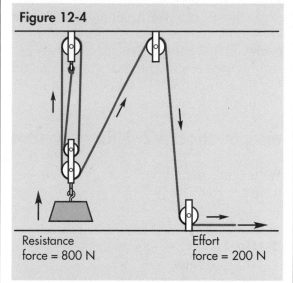

Figure 12-4

Resistance force = 800 N

Effort force = 200 N

Pulleys change the direction and amount of force.

S K I L L   B U I L D E R

Stating a Hypothesis

A research hypothesis is a special way of stating a hypothesis. For example, you may think that a loose chain will make a bicycle slow down. You could create a simple hypothesis, such as, "When a bicycle chain is loose, the bicycle will go slower." In a research hypothesis, however, you would say, "If the chain on a bicycle is loosened by three links, then the bicycle will go three miles per hour slower over a one-mile distance."

A good research hypothesis has four features. It has an "If…, then…" form. It briefly describes an experiment you could use to test the hypothesis, and it tells exactly what to measure or observe. Finally, a good research hypothesis predicts an outcome, which may turn out

to be wrong. The four features of a hypothesis are labeled in Figure 12-5.

1. Write a research hypothesis for the following ideas and label their four features.
   Idea 1: The loudness of a radio affects how far away it can be heard.
   Idea 2: Tire pressure affects a car's gas mileage.

Figure 12-5

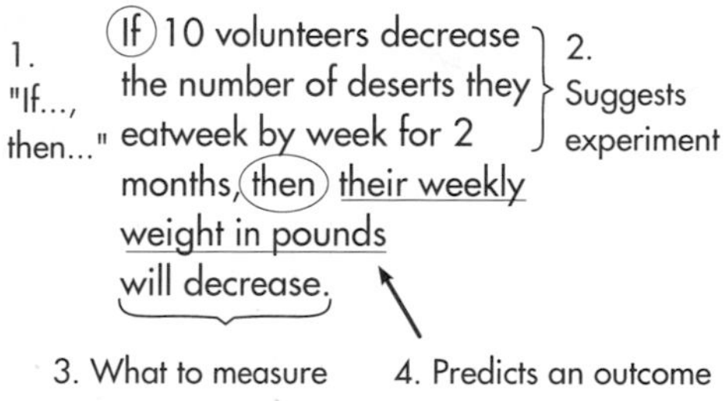

1.
"If…, then…"

(If) 10 volunteers decrease the number of deserts they eatweek by week for 2 months,(then) their weekly weight in pounds will decrease.

2.
Suggests experiment

3. What to measure   4. Predicts an outcome

Machines can also multiply speed. You know that on a ten-speed bicycle in high gear, you move the pedals rather slowly but with a large force. The gears and chain on the bike multiply this speed and the rear tire turns much faster.

■ Section 12-1 Review ■

Write the definitions for the following terms in your own words.

1. **machine**
2. **effort force**
3. **effort distance**
4. **resistance force**
5. **resistance distance**

Answer these questions.

6. Figure 12-6 shows an ideal car jack used to lift a car. The numbers are for just one push down on the handle. You would move the handle up and down many times to raise a car just high enough to change a tire. Recall that when you multiply newtons by meters, you get joules. Determine the following amounts.
   a. Effort force
   b. Effort distance
   c. Resistance force
   d. Resistance distance
   e. Input work
   f. Output work
7. How is the law of machines illustrated in the numbers for the ideal car jack in question 6?

Figure 12-6 Ideal Car Jack

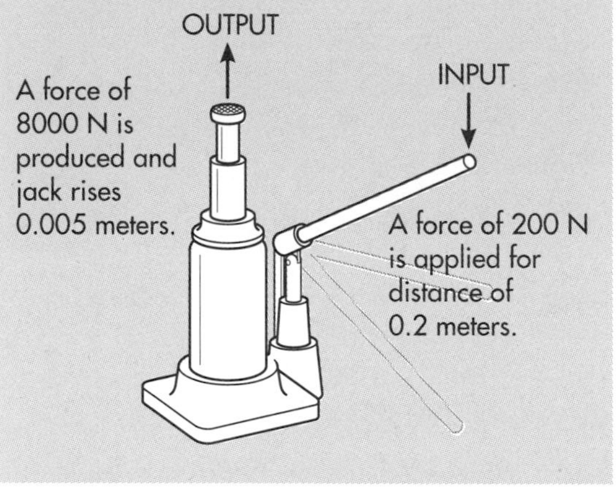

OUTPUT

A force of 8000 N is produced and jack rises 0.005 meters.

INPUT

A force of 200 N is applied for distance of 0.2 meters.

12-2 Efficiency and Mechanical Advantage

■ *Objectives*

☐ *Calculate efficiency and wasted energy.*

☐ *Compare ideal and actual mechanical advantage.*

☐ *Explain how to make machines more efficient.*

☐ *Describe the effects of friction.*

Have you ever tried to ride a bicycle that was in poor condition? The tires were too soft, and they rubbed against the frame. The chain was stiff and dirty. You felt how difficult the bicycle was to ride. This was because a lot of your pedalling force went to overcome friction, not to do useful work. In contrast, a bike in good condition rides like a dream. It takes very little energy for you to get it moving and keep it moving.

Any machine can be in good condition or bad condition. When a machine is in bad condition, it usually takes more energy to get it to work. When a machine is in good condition, it is almost ideal. The output work almost equals the input work.

Determining Efficiency

The **efficiency** of a machine is how well it transfers mechanical energy. Efficiency equals the output work divided by the input work. Usually, the fraction that results is multiplied by 100 so that you can express the efficiency as a percentage. Study the following example.

The input, 8000 joules of work, is supplied to a machine to do a task that required only 6000 joules, the output. What is the machine's efficiency?

> **efficiency = output work / input work**
> **efficiency = 6000 J / 8000 J = 0.75**
> **0.75 × 100 = 75 percent**

The **wasted energy** is the difference between the work put in and the work given out. In other words, the wasted energy is the difference between the work done to a machine and the work done by the machine. The wasted energy becomes heat.

> **wasted energy = input − output**
> **wasted energy = 8000 J − 6000 J**
> **wasted energy = 2000 J**

The percentage of the energy that is wasted is equal to 100 percent minus the machine's efficiency. In this example, the percentage wasted is 25 percent.

Ideal and Real Machines

An **ideal machine** has no wasted energy because it produces no friction. The output work and input work are equal. The efficiency of an ideal machine, therefore, is 100 percent. In **real machines**, however, friction changes some energy to heat, so the output work is always less than the input work. Therefore, the efficiency of a real machine is always less than 100 percent.

Increasing Efficiency

No real machine can ever be 100-percent efficient. It is possible, however, to increase the efficiency of a machine by making it run better and waste less. One way to increase efficiency in a machine is to **lubricate** its moving parts. Lubricating means putting slippery materials, like oil and grease, between any surfaces that rub together. This reduces friction. Most complicated machines need regular oilings. Even locks on doors are usually kept oiled.

Another way to increase efficiency is to make surfaces smooth. Smooth surfaces produce less friction. Metal surfaces are ground smooth and kept polished. Wooden surfaces are sanded smooth with sandpaper, and then they are kept waxed.

Machines that contain **ball bearings**, small steel balls that roll between surfaces like wheels, are often more efficient than machines that do not have ball bearings. Roller skates and bicycle pedals turn easily because they contain ball bearings. Try this experiment. Drag a box along the floor. Then, put it on many marbles and pull it. Which activity is easier? There is less friction when two surfaces can roll over each other instead of sliding over each other.

Machines made with lightweight materials are often more efficient than machines using heavy materials. Planes and some cars are now being made with lightweight alloys instead of steel. These materials are less dense than steel, so the objects and machine parts weigh less. Less weight helps reduce friction because lightweight moving surfaces will not rub together as tightly as surfaces made of heavier materials.

Engineers can improve the efficiency of machines by using **aerodynamic designs** (ar-oh-dy-NAM-ik). Aerodynamic design means shaping an object so that it meets less resistance and less friction as it moves through air or water. Look at Figure 12-7. Engineers have changed the shapes of cars, planes, and boats so that they move through the air and water more easily.

To make a machine more efficient means to increase its efficiency percentage. A machine is called **efficient** if it has a high efficiency percentage, perhaps over 90 percent. A high efficiency percentage results when little energy is wasted because of friction. A machine is called **inefficient** if it has a low efficiency percentage, perhaps under 70 percent. In such cases, a lot of energy is lost because of friction.

Mechanical Advantage

In Section 12-1, you learned that machines can multiply force. The **mechanical advantage (MA)** of a machine is the number of times the machine multiplies the effort force when it produces the resistance force. The formula for finding MA is

$$\text{MA} = \text{resistance force} / \text{effort force}$$

Look back at Figures 12-3, 12-4, and 12-6. You can easily calculate the mechanical advantage of each of those simple machines

MA of the seesaw in Figure 12-3 = 100 lb / 50 lb = 2

MA of the pulley in Figure 12-4 = 800 N / 200 N = 4

MA of the car jack in Figure 12-6 = 8000 N / 200 N = 40

You can find the mechanical advantage for a machine under ideal conditions and under real conditions. The **ideal mechanical advantage (IMA)** is the number of times an ideal machine should multiply the effort force, according to the structure of the machine. The **actual mechanical advantage (AMA)** is how many times a real machine multiplies the effort force. You can only find an ideal mechanical advantage in

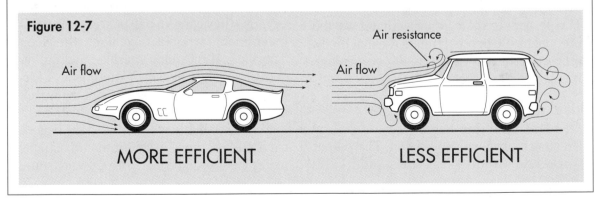

Figure 12-7

Air flow

Air resistance

Air flow

MORE EFFICIENT

LESS EFFICIENT

theory. You can find the actual mechanical advantage by conducting an experiment.

Remember the distinction between an ideal machine and a real machine. It should not surprise you that the ideal mechanical advantage for any machine is always greater than the actual mechanical advantage. Look at the pulleys in Figure 12-8, and try to locate the information used to calculate the ideal and actual mechanical advantages.

Figure 12-8 Mechanical Advantages of a Set of Pulleys

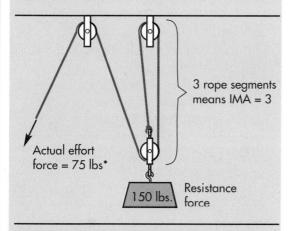

3 rope segments means IMA = 3

Actual effort force = 75 lbs*

150 lbs.

Resistance force

*Ideal effort force should be 50 lbs.

Ideal Mechanical Advantage   Look again at Figure 12-8. The IMA for any set of pulleys is automatically fixed by the number of rope segments that connect the pulleys. This machine has three rope segments connecting the pulleys, so its IMA must be 3. Without friction, these pulleys would triple any effort force you applied.

You could also calculate the IMA from the ideal effort force predicted by theory. This is the effort force if the machine has no friction. Use this formula

IMA = resistance force/ ideal effort force

Using the numbers in Figure 12-8, you can find the IMA for these pulleys.

IMA = 150 lbs/ 50 lbs
IMA = 3

To sum up, you can calculate the IMA by two methods. You can use the structure of the machine or the ideal effort force. Both of these methods are used in this chapter. Most of the time, however, scientists calculate the IMA by analyzing the machine's structure.

Actual Mechanical Advantage   The AMA for the set of pulleys in Figure 12-8, and for all machines, is determined by what actually occurs when the machine is used. The actual effort force was measured at 75 lbs, more than the ideal prediction of 50 lbs. The extra effort force is needed to overcome friction. Use this formula to find the actual mechanical advantage.

AMA = resistance force/ actual effort force

For the example in Figure 12-8

AMA = 150 lbs/ 75 lbs
AMA = 2

This set of pulleys can actually double, not triple, any effort force you apply to the rope. Its IMA is three, and its AMA is two. For all machines, IMA is greater than AMA. The symbol > means "is greater than," and the symbol < means "is less than." You could say that IMA > AMA.

Efficiency versus Mechanical Advantage

The efficiency of a machine is a measurement comparing the work out to the work in. Remember that work includes both the amount of force and the distance through which it acts. The formula for efficiency is output work/input work.

To compare forces, you would measure mechanical advantage. Using mechanical advantage, compare the force going in and out, or the effort force to the resistance force. The formula for mechanical advantage is resistance force/effort force.

Friction reduces efficiency in real machines. This fact makes the following mathematical statements true. These statements are true for all real machines.

- The heat energy produced > 0.
- Work out < work in.
- Efficiency < 100 percent.
- IMA > AMA.

Section 12-2 Review

Write the definitions for the following terms in your own words.

1. **ideal machine**
2. **real machine**
3. **mechanical advantage (MA)**
4. **ideal mechanical advantage (IMA)**
5. **actual mechanical advantage (AMA)**

Answer these questions.

6. Give two examples from your own experience of machines that are not very efficient.
7. Figure 12-9 shows a lever with its forces and distances.
   Determine the following.
   a. the input work, in joules
   b. the output work, in joules
   c. the lever's efficiency
   d. the lever's actual mechanical advantage (AMA)
8. How can machines be made more efficient? Give new examples for each method.

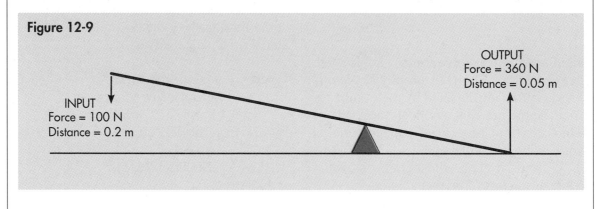

Figure 12-9

OUTPUT
Force = 360 N
Distance = 0.05 m

INPUT
Force = 100 N
Distance = 0.2 m

12-3 How Simple Machines Work

■ *Objectives*

□ *Solve problems concerning levers and inclined planes.*

□ *Describe other simple and complex machines.*

Look back at the simple machines in Figure 12-2 on page 229. You come into contact with a variety of these simple machines each day. For example, a door is a lever. The door knob is a wheel and axle. A sidewalk ramp is an inclined plane. Clocks work with gears, and pianos operate by levers. Clotheslines have pulleys, and so do cranes. See whether you can spot three simple machines near you right now.

Each simple machine can be designed to do a particular job and to have a particular ideal mechanical advantage (IMA). The structure of a simple machine determines its ideal mechanical advantage, which you can figure out with pencil and paper. You cannot calculate the actual mechanical advantage (AMA) until you know the friction forces in the real machine.

Levers

A **lever** is a rigid bar that rotates around a fixed point. Its basic form is illustrated in Figure 12-10. The point around which a lever turns is called the **fulcrum**. The distance between the fulcrum and the point where the effort force is applied is called the **effort arm**. The distance between the fulcrum and the point where the resistance force is applied is called the **resistance arm.**

The Ideal Mechanical Advantage of a Lever

The ideal mechanical advantage of a lever is set by the length of the effort and resistance arms. You can find the IMA by using the following formula.

**IMA = length of effort arm/
length of resistance arm**

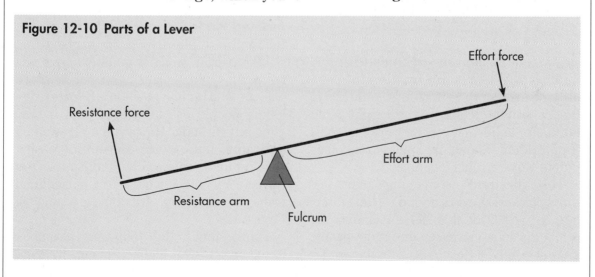

Figure 12-10 Parts of a Lever

Effort force

Resistance force

Effort arm

Resistance arm

Fulcrum

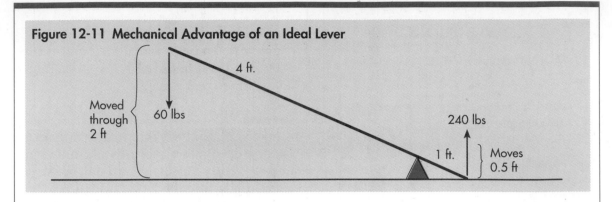

Figure 12-11 Mechanical Advantage of an Ideal Lever

The lever in Figure 12-11 has an effort arm four feet long, an ideal effort force of 60 lbs, and an effort distance of two feet. It has a resistance arm one foot long, a resistance force of 240 lbs, and a resistance distance of half a foot. What is the IMA for this lever?

IMA = 4 ft/ 1 ft
IMA = 4

You can confirm this IMA by using the forces given in Figure 12-11.

IMA = 240 lbs / 60 lbs
IMA = 4

An IMA of four means that this lever will multiply any effort force by 4. An IMA of 4 also means that the effort force only has to be one fourth the size of the weight being lifted. For an ideal lever the work in equals the work out. Show that input work equals output work for this example.

This illustrates some important features of levers. Levers multiply effort force, but the "cost" of this benefit is that the effort distance increases too. Most people cannot apply a force of 240 lbs to lift something, even for 0.5 ft. A lever lets you do the job with only 60 lbs of force. The 60 lbs, however, must be applied for a longer distance, four feet. No work is saved by using the lever, but the effort force is lower.

Friction and Levers Most levers produce very little friction because the only moving part is at the fulcrum. As a result, levers have a high efficiency. This means that, for levers, the input work and output work are nearly equal. The actual mechanical advantage is very close to the ideal mechanical advantage.

Inclined Planes

An **inclined plane** is a flat surface that slants up at an angle; a good example is a ramp. An inclined plane helps people move objects to a higher position. It does so by letting people use less force than if they lifted up directly. Have you ever walked up a ramp? If so, which do you find easier, walking up a ramp or climbing stairs? You can see the main features of an inclined plane in the ideal frictionless ramp in Figure 12-12.

Think about the following example, based on an ideal inclined plane. Imagine

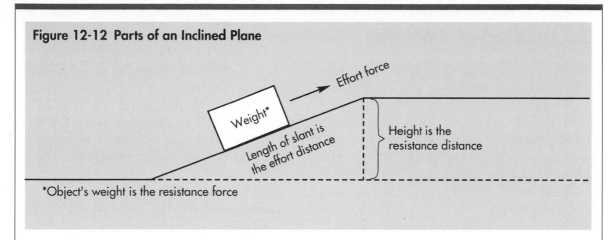

Figure 12-12 Parts of an Inclined Plane

Effort force

Weight*

Length of slant is the effort distance

Height is the resistance distance

*Object's weight is the resistance force

trying to lift a 600 N box from the ground to a platform that is 2 m high. That would require 600 N × 2 m = 1200 N-m, or 1200 J, of work. A 10 m ramp can help you do this work. Remember that work is force times distance. By increasing the distance it takes to do the work, the ramp reduces the force you have to apply.

1. output work = resistance force × height of the platform
   output work = 600 N × 2 m
   output work = 1200 N-m, or 1200 J

Note that the resistance force equals the weight of the object being lifted. The input work must equal the output work, for this is a frictionless ideal inclined plane. The input work is the effort force times the effort distance.

Begin with input work of 1200 J, the equal of the output work. To do this job, you must apply an effort force along the whole length of the ramp.

2. input work = ideal effort force × length of ramp

1200 J = ideal effort force × 10 m
ideal effort force = 1200 N-m / 10 m = 120 N

You can see that, even though the work needed is still 1200 J, the effort force of 120 N is less than the resistance force of 600 N. The ramp lowers the effort force needed to produce the same amount of work.

The Ideal Mechanical Advantage of an Inclined Plane The ideal mechanical advantage of an inclined plane is set by its height and its length. Think about the example above, and consider this formula.

IMA = length / height
IMA = 10 m / 2 m
IMA = 5

You can confirm this by calculating the forces.

IMA = resistance force / ideal effort force
IMA = 600 N / 120 N
IMA = 5

The Actual Mechanical Advantage of an Inclined Plane The actual mechanical advantage of an inclined plane is much less than its ideal mechanical advantage. This is because dragging an object up the length of the plane produces a lot of friction, even if you use a dolly, or platform, with wheels. Figure 12-13 gives the data for an actual inclined plane.

What is the IMA of the inclined plane in Figure 12-13?

IMA = length / height
IMA = 6 m / 1 m
IMA = 6

What is the AMA of this inclined plane?

AMA = resistance force /
**   actual effort force**
AMA = 1200 N / 400 N
AMA = 3

Friction has reduced the mechanical advantage of the inclined plane. An IMA of 6 means that an ideal effort force of 200 N would raise the 1200-N object. Because of friction, however, you will need an effort force of 400 N, resulting in an AMA of 3.

Actual inclined planes are not highly efficient. The work put in is always much more than the work coming out. The AMA is always less than the IMA. Study the following calculation of efficiency for the actual inclined plane in Figure 12-13.

input work = 400 N × 6 m
input work = 2400 N-m or 2400 J

output work = 1200 N × 1 m
output work = 1200 N-m or 1200 J

efficiency = 1200 J / 2400 J
efficiency = 0.5, or 50 percent

You could improve this efficiency by smoothing the surface of the ramp and using ball bearing wheels. Even with these improvements, however, the inclined plane is not very efficient. Why is it used? People use inclined planes because they make it easier to roll heavy weights, such as shopping carts, to a higher position with a reasonable force.

Other Simple Machines

Most simple machines are variations of the lever and the inclined plane. For example, pulleys and gears are really two kinds of levers. A screw is an inclined plane twisted into a spiral shape, and a wedge is two inclined planes placed back to back.

Look back at Figure 12-2 on page 229. Examine the cylinder and piston. In a hydraulic press, fluid moves between two cylinders, each with a piston. The effort force is applied to the smaller piston. This

Figure 12-13

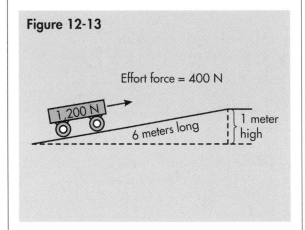

Effort force = 400 N

1,200 N

6 meters long

1 meter high

Why Use an Inclined Plane?

Process Skills *stating a hypothesis; stating conclusions*

Materials inclined plane; lab cart or skateboard; meter stick; spring scale in newtons; string. If only gram scales are available, divide the readings in grams by 100 to get the approximate force in newtons. For examples, 50 g = 50/100 = 0.5 N

1. Form a research hypothesis about the height of an inclined plane and the force needed to pull an object up it.
2. Set up the inclined plane as shown in Figure 12-14.
3. Record the weight of skateboard. This is the resistance force.

4. Record the length and height of the ramp.
5. Gently pull the skateboard up the ramp, keeping the spring scale in line with the ramp. Record the pulling force. This is the effort force.
6. Using your data, calculate the ideal mechanical advantage, the actual mechanical advantage, the output work, the input work, and the efficiency of this inclined plane.
7. Repeat steps 2-6 two more times with steeper inclined planes.

Conclusions

8. Present your data and calculations in a neat, organized laboratory report. How do the data support, or fail to support, your hypothesis about inclined planes?

Figure 12-14

Length of plane

Height of plane

How Do Fixed and Movable Pulleys Compare?

Materials two fixed pulleys and one moveable pulley; an object to lift; spring scale in newtons; two meter sticks; string.

Procedures

1. Attach the object to the movable pulley and record their combined weight.
2. Set up the pulleys as shown in arrangement A. Gently pull up on the spring scale thereby lifting the object and the moveable pulley. Record the force needed to lift them. The moveable pulley is lifted along with the object so that the weight lifted is the same as in arrangement B.
3. Lift the object up a certain length, for example, from 25 to 50 cm. Measure how many meters the spring scale must be pulled up to do this task.
4. Now set up the movable pulley as shown in arrangement B. Record the force needed to lift the moveable pulley and object in this arrangement.
5. Using arrangement B, lift the object the same distance used in Procedure 3. Measure how many meters up the spring scale must be pulled.
6. Present your results in a table.

Conclusions

7. Compare the force needed to lift the object and moveable pulley in each arrangement.

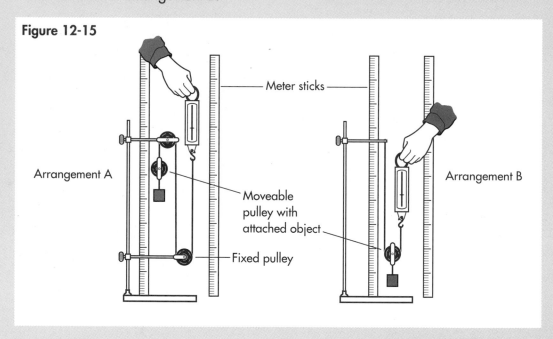

Figure 12-15

Meter sticks

Arrangement A

Moveable pulley with attached object

Fixed pulley

Arrangement B

produces a larger resistance force in the bigger piston. Car brakes work because of such hydraulic presses.

Complex Machines

A **complex machine** consists of many simple machines that interact with one another. The mechanical energy is passed from one part to another. For example, in a wristwatch, gears, levers, and wheel and axles move energy from the spring or battery to the moving hands. In a car, hundreds of simple machines work together. These include cylinders and pistons in the engine and levers in the valves. Levers also make up the hand brake, the turn-indicator stick, the electrical switches, and the door handles. The car's transmission contains gears; steering works because of wheels and axles. There are pulleys in the fan belt and electric generator belt.

Until the 1700s, most machines were not too complex. They ran on the energy from people, animals, wind, and steam. Today, most machines are complex. They are driven by engines that use electricity, oil, gasoline, or coal for energy.

■■ Section 12-3 Review ■■

Write the definitions for the following terms in your own words.

1. **fulcrum**   2. **inclined plane**
3. **effort arm**   4. **complex machine**
5. **resistance arm**

Answer these questions.

6. How can you calculate the ideal and actual mechanical advantages of a lever?
7. How can you calculate the ideal and actual mechanical advantages of an inclined plane?
8. What are the advantages and disadvantages of using a ramp to lift objects?
9. Name the simple machines in Figure 12-16 as you tell how this contraption works. Start on the left.

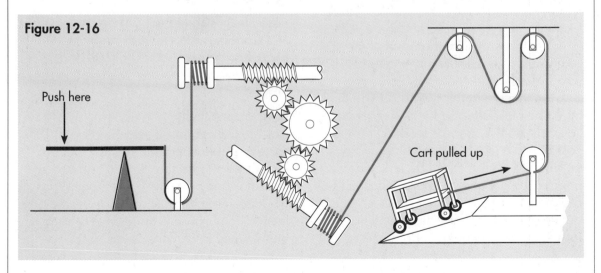

Figure 12-16

Push here

Cart pulled up

SCIENCE, TECHNOLOGY, & SOCIETY

FAX Machines and Technology Assessment

Figure 12-17

Have you seen a FAX machine? FAX machines send printed material and pictures over telephone lines. Within 20 years, 20 percent of all families and 100 percent of all businesses in the United States could have FAX machines. No one can tell today what impact the machines will have on society.

The country's Postal Service might lose money as people mail less and FAX more. Students may do research at home with FAX machines. Just as there is junk mail, businesses may send "junk" material through FAX machines.

From these examples, who do you think might be most affected by FAX machines? You can see why people working for the Post Office and for phone companies will be affected. Banks might be affected, if people begin to pay bills or send checks by FAX. Think of the possibilities.

Such questions form the beginnings of a **technological assessment** of the FAX.

Technological assessment means trying to predict the consequences of some new system, process, or device. Business and government use technological assessment to anticipate problems, make better and earlier decisions, and plan for the changes a new technology may cause.

In conducting a technology assessment, researchers study the new technology in detail and interview experts. They brainstorm with informed citizens to figure out possible consequences. Researchers try to identify the people who may be affected most by the new invention. Technology assessments help society prepare for the future.

Special agencies advise the United States Congress about the possible long-term consequences of new technologies. For example, the Office of Technology Assessment (OTA) might someday suggest that Congress pass laws to regulate the use of FAX machines. Society already has similar laws that regulate the use of telephones.

Follow-up Activity

Hold a brainstorming session in class. List some possible effects of the new idea for selling telephones that let you talk and see the other person at the same time. Write all the ideas on the board. Do not debate them. The purpose is to develop a list of many ideas to be considered.

Section 12-1 The Characteristics of Machines

1. Machines transfer mechanical energy from one object to another.
2. The amount of work done by a machine, or the output, is always less than the amount of work put into the machine, or the input.
3. Machines may make work seem easier.
4. Machines can change the amount, direction, distance, and speed of a force.

Section 12-2 Efficiency and Mechanical Advantage

1. Efficiency = output work/input work
2. A machine can be made more efficient by reducing friction by means of lubrication, smoothing surfaces, and ball bearings.
3. Mechanical advantage (MA) = resistance force/effort force

Section 12-3 How Simple Machines Work

1. Simple machines include the lever and the inclined plane.
2. The ideal mechanical advantage (IMA) for
   a. levers = effort arm/resistance arm
   b. inclined plane = length/height
3. Complex machines consist of two or more simple machines.

Write the term from the list that best completes each sentence.

input, law, simple, effort force, lubricate, resistance distance, output, efficiency

All complicated machines are really made up of ___1___ machines. The ___2___ multiplied by the effort distance gives the ___3___ work done to a machine. The distance an object is pulled or pushed by a machine is the ___4___. This times the resistance force is the ___5___ work done by the machine. In ideal machines, the output work equals the input work. This is known as the ___6___ of machines.

The measurement that tells how the output work really compares to the input work is called the machine's ___7___. To reduce friction, you could ___8___ the parts with oil or grease. You can also use wheels that contain ___9___, or work with an object that has an ___10___ design.

Write *true* if the statement is true. If the statement is false, change the *italicized* term to make the statement true.

1. In *ideal* machines the input work and output work are equal.
2. In real machines the output work is always *equal to* the input work.
3. The weight lifted by a machine is called the *effort* force.
4. The *ideal* mechanical advantage of a machine partly depends on the amount of friction it makes.
5. It *is possible* to have machines with 100% efficiency.
6. Smooth waxed surfaces have *less* friction than rougher surfaces.
7. A truck that is fully loaded has *more* friction than when it is empty.

INTERPRETING INFORMATION

Tomás and María do an experiment using a large pulley system in a car-repair shop. They attach different objects to the pulleys and measure the force it takes to pull them up. Figure 12-18 is a graph of their results. Read the graph to fill in the missing items in Figure 12-19.

Figure 12-18

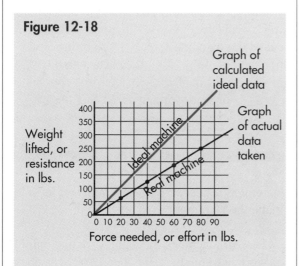

Graph of calculated ideal data

Graph of actual data taken

Weight lifted, or resistance in lbs.

Force needed, or effort in lbs.

THINK AND DISCUSS

Use the section number in parentheses to help you find each answer. Write your answers in complete sentences.

1. What are the four purposes of machines? Give an example that illustrates each. (12-1)
2. Which student is making an impossible claim? Adam says, "My machine produces 100 joules from 25 joules." Eve says, "My machine produces 100 newtons from 25 newtons." (12-1)
3. Why does improved aerodynamic design increase a motorcycle's efficiency? (12-2)
4. A complex machine contains gears that transform 100 N applied over 3 meters into 400 N that moves an object 0.6 meters. Determine its actual mechanical advantage, and its efficiency. First get the work in and work out in J. (12-2)

GOING FURTHER

1. The newest racing bicycles have metal or plastic discs on the inside of each wheel that

Figure 12-19

Weight of object (resistance)	Ideal pulling force (effort)	Actual pulling force (effort)
100 lbs	a	b
150 lbs	c	d
e	f	40 lbs
g	40 lbs	h

completely cover the spokes. Research the reasons for this change in design. What might be the advantage of doing this? What is one disadvantage?

2. Choose a machine that you can examine in detail, perhaps one that you can take apart. You might use some electric tool, some kitchen or garden appliance, or a pencil sharpener. List the parts you can see and name the kind of simple machine in each part. **Caution** Unplug the electric plug. Watch for any sharp edges. Avoid touching moving parts. Ask for adult assistance.

COMPETENCY REVIEW

1. The key energy transformation in machines is
   a. mechanical to mechanical.
   b. mechanical to sound.
   c. chemical to electrical.
   d. heat to chemical.

2. Simple machines transfer
   a. mechanical energy.
   b. nuclear energy.
   c. magnetic energy.
   d. electric energy.

3. When performing a task, machines cannot change
   a. the amount of force needed.
   b. the speed of the force.
   c. the direction of the force.
   d. the amount of work needed.

4. The law of machines says that, under ideal conditions, the work given out by a machine
   a. is less than the work put in.
   b. is equal to the work put in.
   c. is more than the work put in.
   d. lessens over time.

5. The work given out by real machines
   a. is less than the work put in.
   b. is equal to the work put in.
   c. is more than the work put in.
   d. increases over time.

6. Where does the "missing" work go when a machine is used?
   a. to mechanical energy
   b. to kinetic energy
   c. to heat energy
   d. to potential energy

7. What measurement tells how well a machine uses energy to do work?
   a. total energy used per hour
   b. total work done per hour
   c. efficiency
   d. speed of operation

8. Waxing, polishing, and oiling are ways to make a machine
   a. 100 percent efficient.
   b. more efficient.
   c. less efficient.
   d. increase friction forces.

9. Friction in a machine increases if it
   a. uses heavier replacement parts.
   b. is lubricated.
   c. is sanded.
   d. uses ball bearings.

10. Which is *not* a simple machine?
    a. a ramp   b. a wristwatch
    c. a seesaw   d. a steering wheel

CHAPTER 13

SOUND ENERGY

You have heard sounds since you were born. In fact, doctors believe you probably heard sounds before you were born. Sounds are a constant part of your life, always present, like the air. There are sounds of nature, human activity, speech, music, and noise. Some sounds are pleasant; others are very unpleasant. Fortunately, your brain ignores most of the sounds that enter your ears. You are unaware of them.

Sound is a form of energy. If a sound is loud enough, like a blast, it can do work by knocking down a wall or trees. Most sounds have a small amount of energy, just enough to move your ear drums back and forth. That is how your ears detect a sound and send electric signals to your brain. As you have perhaps guessed so far from this introduction, sound involves kinetic energy. There are, however, a number of other interesting things to discover about sound as well.

Sound can work both for us and against us. Some sounds can clean dishes. Other sounds are used by doctors to detect and treat certain medical problems. Sound waves are used to take a picture of an unborn baby inside its mother. On the other hand, some kinds of sound at certain ranges of loudness can be harmful to our physical health, our mental health, or both.

13-1 The Nature of Sound

■ *Objectives*

☐ *Describe sound as vibrations, energy, and waves.*

☐ *Illustrate the four sources of sounds.*

☐ *Explain how sound travels.*

Feel your throat while you speak or sing. Do you feel a slight trembling inside your throat? This is caused by two thin bands in your voice box. They move back and

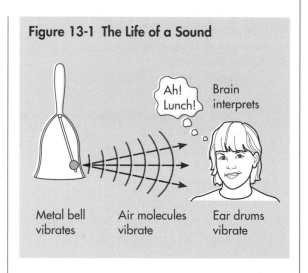

Figure 13-1 The Life of a Sound

Ah! Lunch!

Brain interprets

Metal bell vibrates

Air molecules vibrate

Ear drums vibrate

forth rapidly as air passes through them. This is the origin of your voice.

Sound Is Vibrations

Sound is produced by **vibrating** matter, which means the matter moves back and forth rapidly. The matter shakes, and the shaking motion is called vibration. People with excellent hearing can hear sounds from matter that vibrates between 20 and 20,000 times every second. Bats can hear sounds made by matter that vibrates 100,000 times every second.

Sounds begin, continue, and end with vibrations. Look at Figure 13-1. A sound begins when an object begins to vibrate. Sound continues as the object makes air molecules vibrate. Finally, sound ends when the air vibrations make your eardrum vibrate, sending messages to your brain.

Sounds come from millions of sources. You can recognize thousands of sounds, like the sound of a typewriter or roller skates. The four main categories of sound are vibrating solid objects, surfaces, strings, and air columns. Study the categories in Figure 13-2.

Figure 13-2 Sources of Sound

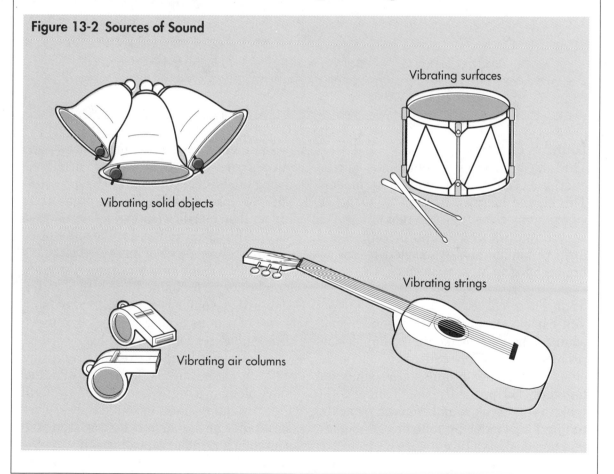

Vibrating solid objects

Vibrating surfaces

Vibrating strings

Vibrating air columns

Predicting

You have learned that sound is caused by vibrations. Study Figure 13-3 and then answer these questions. Try to use the word *vibrations* in each answer.

1. Predict what will occur in each of the tests shown in the figure.
2. Explain your predictions.
3. Predict what would happen if you made the tuning fork and speaker louder.

Figure 13-3

1.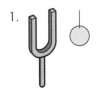
Ball on string touches tuning fork, which emits sound.

2.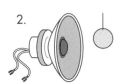
Ball touches audio speaker playing music.

3.
Tuning fork emitting sound is touched to surface of water.

Sound Is Energy

The object making a sound has kinetic energy because of its vibrating motion. This kinetic energy is transferred to the air, where molecules vibrate rapidly over a very tiny distance. Sound, therefore, is a form of kinetic energy created by the vibrating motion of air molecules.

Sound Is a Wave

You can use a tuning fork to understand sound. When you strike the tuning fork, its prongs vibrate. This causes the air molecules next to the prongs to be squeezed together and pushed apart. Tuning forks make a very weak sound. You can make the tuning fork louder by touching its bottom end to a desk or a box.

See Figure 13-4. The tuning fork sends out a series of "pressure puffs" in the air many times every second. The pressure puffs are places where the air pressure is higher than usual. Scientists define air pressure as force per unit area. Thus the pressure puffs are places of increased force.

Scientists diagram sound in several other, simpler ways. Examine Figure 13-5, which shows that sound is a **wave**. A wave is a series of actions or disturbances that move in a regular, repeating way. In hand waving, you move your hand in a regular, repeating way. In water waves, the peaks of water move along in a regular, repeating way until they reach the shore. In sound waves, pressure puffs in the air move away from the source in a regular, repeating way.

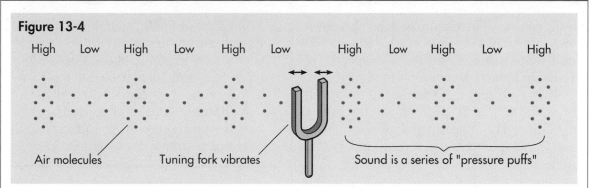

Figure 13-4

High   Low   High   Low   High   Low   High   Low   High   Low   High

Air molecules   Tuning fork vibrates   Sound is a series of "pressure puffs"

Pressure changes caused by a tuning fork

Figure 13-5 Simpler Ways to Represent Sound Waves

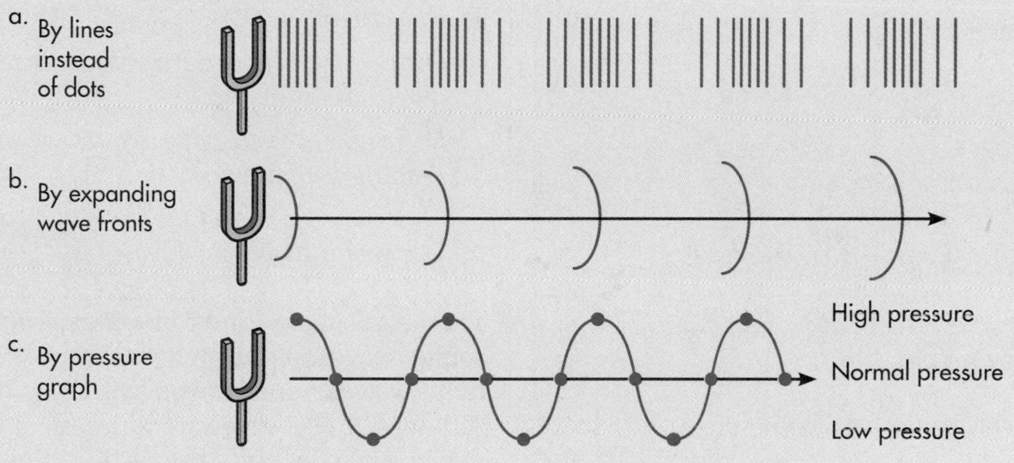

a. By lines instead of dots

b. By expanding wave fronts

c. By pressure graph

High pressure

Normal pressure

Low pressure

How Sound Travels

The material through which sound waves and other waves travel is called the **medium**. Air is the usual medium of sound. Sound can also travel through other *media,* the plural of *medium.* These media include rock, walls, steel, and water. Sound waves must travel through matter. They cannot travel in a vacuum. In a vacuum, there are no air molecules to vibrate. Sound cannot exist.

You can call someone in another room or around the corner, which shows that sound waves do not travel in straight lines. Sound waves spread out and turn around corners.

The Speed of Sound

The speed of sound depends on the type and the temperature of the medium. For example, the speed of sound in air at 0° C is 331 meters per second. You can convert this into 1090 feet per second, or 740 miles per hour. In the language of jet planes, the speed of 740 miles per hour is known as MACH 1. Speeds higher than this are called supersonic. For example, a plane traveling at MACH 4 is going at 4 x 740, or 2960 miles per hour. Compare the speeds of sound in air and other media in Figure 13-6.

These are very fast speeds, but the speed of sound is much slower than the speed of light. The speed of light in air is 300,000,000 meters per second, or 186,000 miles per second.

Figure 13-6	
Speed of sound in air at various temperatures	
At 0° C	331 m/sec
At 10° C	337 m/sec
At 20° C	343 m/sec
At 50° C	361 m/sec
Approximate speed of sound in other media	
Water	1300 m/sec
Steel	3800 m/sec
Rock	6000 m/sec

■■■ Section 13-1 Review ■■■

Write the definitions for the following terms in your own words.

1. **vibrate**   2. **wave**
3. **medium**   4. **supersonic**

Answer these questions.

5. From start to finish, sound is vibrations. Explain this statement.
6. Give a new example of each of the four types of sound source.
7. a. Why should the kind of medium make a difference in sound's speed?
   b. How does speed of sound vary with temperature? Refer to Figure 13-6.

13-2 Characteristics of Sound Waves

■ *Objectives*

☐ *Distinguish between transverse and longitudinal waves.*

☐ *Define the wavelength, frequency, and amplitude of a wave.*

☐ *Match the pitch and loudness of a sound wave with its physical features.*

There are many kinds of waves, such as sound waves, light waves, water waves, radio waves, and waves within a rope. Although these waves are different in their media, speeds, and properties, they do have some common characteristics.

Two Types of Waves

Have you ever watched a boat or a piece of wood floating in water? Next time, notice that while the water waves move horizontally, the objects in the water move vertically, or up and down. This difference reveals something important about all waves. In a moving wave, the disturbances within the medium move and create the wave. The medium itself, however, does not move with the wave.

To imagine this, pretend you are in a room or auditorium where people in one corner start laughing. The laughter spreads toward the people in the center, then to people at the opposite corner of the room. People do not move, but the laughter does. The audience is like a wave's medium. The laughter is a disturbance moving through the audience. Waves are like the laughter.

In water waves, the molecules of water move up and down. They do this in such a way that the disturbances move to the left or right. Look at Figure 13-7. Water waves are **transverse waves**. This means that the medium vibrates perpendicular to, or at 90° across, the direction the wave moves. The water vibrates vertically, but the wave moves horizontally.

In sound waves, the air molecules vibrate a tiny distance left and right. Sound waves do this in such a way that the disturbances also move left and right, as you can see in Figure 13-8. Sound waves are **longitudinal waves** (lan-ju-TOOD-uh-nul). This means that the medium vibrates in the same direction as the wave. The air vibrates horizontally, and the wave also moves horizontally.

You can compare the two types of waves by thinking of two lines of people. In one line, people jump up and down one after another. This is a transverse wave. In the other line, people shove each other in a single direction. This is a longitudinal wave.

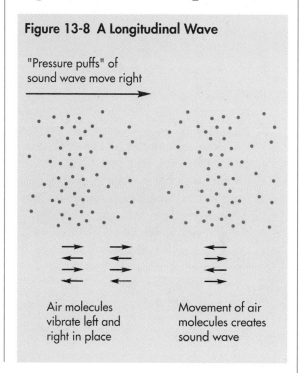

Figure 13-8 A Longitudinal Wave

"Pressure puffs" of sound wave move right →

Air molecules vibrate left and right in place

Movement of air molecules creates sound wave

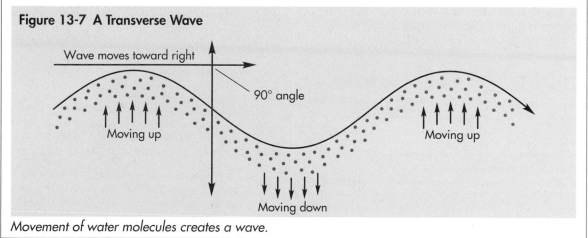

Figure 13-7 A Transverse Wave

Wave moves toward right →

90° angle

Moving up

Moving down

Moving up

Movement of water molecules creates a wave.

What Do Waves Look Like?

Process Skills observing; analyzing

Materials slinky or some other long, stretchable spring; brightly colored pieces of cotton or cloth firmly attached to various spots along the spring

Procedures

1. In procedures 2 and 3, you must note two motions. Observe the motion of the waves themselves. They go back and forth, left and right. Also observe the motion of the medium as the wave travels. Do this by observing how the colored markers move. The waves move quickly, so repeat each experiment several times until you are convinced of what you see.

2. Stretch the spring between two students so that it is in a straight line.

Caution Do not overstretch the spring so that it bends or is permanently stretched. Have one student gather extra coils so that you can see a denser, darker area at one end of the spring. Have the student release this area, and watch its movement.

3. Begin as above, but this time do not gather coils together. Instead, flick the spring so that just one wave, or one bump, moves back and forth.

Conclusions

4. Which experiment created a transverse wave, and which created a longitudinal wave? Explain your answer.

5. Which experiment creates a wave like a sound wave? Explain your answer.

Wave Properties

Figure 13-9 shows a wave moving left to right. It could show air pressure changes in a sound wave. Try to avoid thinking of sound waves as curvy "snakes," however. The curved line just shows how the air pressure changes in a sound wave. It is not a real picture of sound.

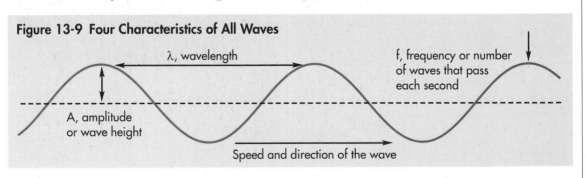

Figure 13-9 Four Characteristics of All Waves

λ, wavelength

f, frequency or number of waves that pass each second

A, amplitude or wave height

Speed and direction of the wave

The distance between the peaks in a wave is its **wavelength**. The symbol for wavelength is a letter from the Greek alphabet λ, which is called "lambda." In sound waves, the wavelength is the distance between the pressure puffs.

The number of wave peaks that pass every second is the **frequency** (FREE-kwun-see) of the wave. The symbol for frequency is f, and frequency is measured in vibrations per second. In sound waves, the frequency is the number of pressure puffs made by the vibrating object every second.

The height of a peak is the **amplitude** (AM-plu-tood) of the wave. The symbol for amplitude is A. In sound, the amplitude represents the strength of a pressure puff.

Hearing and Sound

Your ear instantly analyzes the characteristics of sound waves that hit your eardrum. Your brain then translates this into your feelings about the qualities of the sound. The two key qualities of a sound your brain recognizes are pitch and loudness.

Pitch is how high or low a sound is. A flute, whistle, or soprano voice has a high pitch. A tuba, thunderclap, or bass voice has a low pitch. The pitch of a sound depends on the frequency of its wavelength. Look at Figure 13-10. The greater the frequency of a sound wave, the shorter its wavelength. Frequency and wavelength are described as inversely proportional. This means that as one increases, the other decreases.

The loudness, or volume, of a sound depends on the amplitude of its sound wave. Look at Figure 13-11. Loud sounds are the result of increased pressure on the sound wave. Loudness is measured in

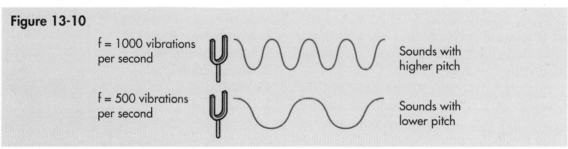

Figure 13-10

f = 1000 vibrations per second — Sounds with higher pitch

f = 500 vibrations per second — Sounds with lower pitch

Frequency determines pitch.

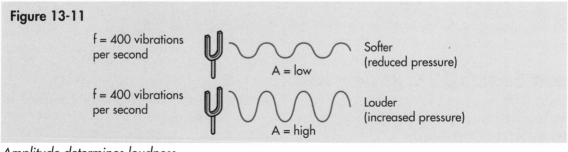

Figure 13-11

f = 400 vibrations per second   A = low — Softer (reduced pressure)

f = 400 vibrations per second   A = high — Louder (increased pressure)

Amplitude determines loudness.

Figure 13-12

Sound	Loudness level, decibels	Description
Threshold of hearing	0	Softest sound humans can hear
Leaves rustling	10	Barely heard
A whisper	20	Quiet
Quiet radio	40	Acceptable
Conversation, air conditioner	60	Intrudes
Heavy traffic, horns	70–90	Annoying, damages ears after 8 hours
Loud stereo, thunder	110	Uncomfortable
Jet plane, loud siren	140	Painful

decibels, abbreviated dB. Figure 13-12 is a table that gives you some ways to compare loudness.

For each increase of ten decibels, the pressure puff is ten times stronger and the sound is twice as loud. So, compared to 0 dB, a 60 dB sound is one million times greater in pressure and 64 times louder. This comes from $10 \times 10 \times 10 \times 10 \times 10 \times 10$ and $2 \times 2 \times 2 \times 2 \times 2 \times 2$.

■ Section 13-2 Review ■

Write the definitions for the following terms in your own words.

1. **transverse wave**
2. **longitudinal wave**
3. **wavelength**
4. **frequency**
5. **amplitude**

Answer these questions.

6. When you talk to your friends, what travels from your mouth to their ears?
7. Give two new examples of a transverse wave and two new examples of a longitudinal wave.
8. What type of wave is sound? Why?
9. How does a 40 db sound compare to a 30 db sound?

13-3 The Behavior of Sound Waves

Have you ever heard your own echo? Sometimes you can hear it near large buildings, cliffs, or hills. When you shout a strong hello, you get back a weak hello echo. Echoing is one special behavior of sound waves.

Absorption and Reflection

When sound waves hit a surface, they are both absorbed and reflected. Study Figure 3-13. Part of the sound energy is taken in by the surface and changes to heat energy. This is **absorption**. Part of the sound energy is bounced back and continues in a new, opposite direction. This is **reflection**. The sound that is reflected has less energy because of absorption. The reflected sound, therefore, is not as loud.

A room with bare walls, bare windows, and wooden furniture tends to be noisy because talking and other sounds reflect from the hard surfaces. A room with a carpet, curtains, couches, and soft ceiling tiles will be less noisy because these materials absorb sound energy well and only a little is reflected.

Echoes

When sound waves are reflected, an echo results. Because sound waves travel fast, the time between the original sound and its echo is very short in most places. Usually, your brain cannot separate the two sounds. You need at least a tenth of a second between a sound and its echo before your brain can tell them apart.

Resonance

Have you ever heard singing that can shatter a wine glass? Shattering can occur if a singer produces a soundwave frequency that exactly matches the natural frequency

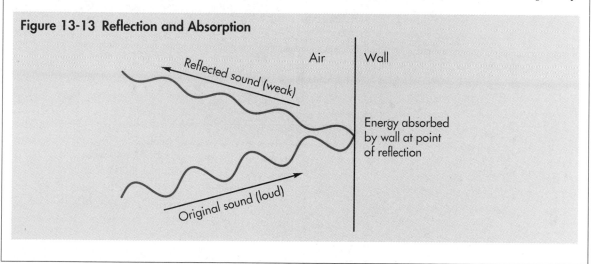

Figure 13-13 Reflection and Absorption

Air | Wall

Reflected sound (weak)

Energy absorbed by wall at point of reflection

Original sound (loud)

the glass would have if you tapped it. The energy in the sound enters the glass and causes it to vibrate so strongly that it shatters. This is an example of **resonance** (REZ-un-uns). Resonance occurs when a sound makes an object vibrate because the frequencies of the sound and the object match.

When you are playing music at home, you may notice that a windowpane or other object may mysteriously vibrate and then stop. In such cases, probably the music hit a frequency that made the windowpane or other object vibrate because of resonance.

Resonance can occur even if sound is not involved. For example, your car may shake strongly at 58 to 60 miles per hour, but not at lower or higher speeds. At that particular speed, the vibrations of the engine match the natural vibration of the car's body and parts. The car's body and parts shake with greater strength.

When soldiers walk over bridges, they are ordered not to march in step because their marching frequency, left-right-left-right, might match the natural vibration of the bridge. If the resonance is very strong, the bridge might shake and loosen.

The Doppler Effect

The next time a car or train goes by, pay attention to its sound. The moment the car or train passes you, you will hear the pitch change from a higher to a lower frequency. This change in pitch is especially noticeable at high-speed car racetracks or near runways at airports. These changes in pitch

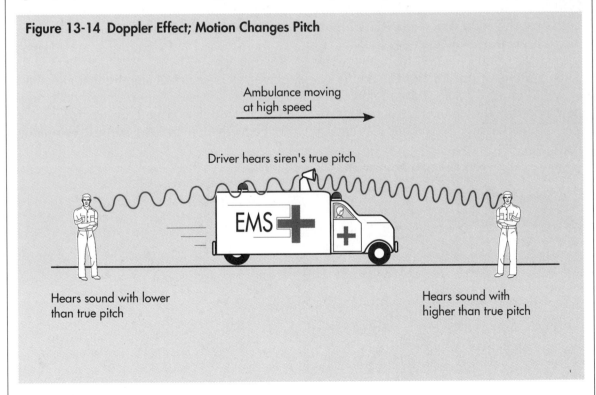

Figure 13-14 Doppler Effect; Motion Changes Pitch

Ambulance moving at high speed

Driver hears siren's true pitch

EMS

Hears sound with lower than true pitch

Hears sound with higher than true pitch

are caused by the **Doppler effect.** The Doppler effect is a change in the pitch of a sound because of the motion of the listener or the source of the sound. Study Figure 13-14. As you can see, the pitch is higher when the listener and source are moving toward each other. The pitch is lower when the listener and source are moving apart. This happens because the number of vibrations that hit your ear each second depends on whether you are stationary, moving toward, or moving away from the sound source.

Music and Noise

Music consists of many sound waves that follow a pattern. Music affects your body, your feelings, and the way you act with others. In contrast, noise consists of unwanted or unpleasant sounds. Noise also affects your body, your feelings, and the way you act with others.

An **oscilloscope** (ah-SIL-u-skohp) is an electronic device that draws pictures of sound waves. A microphone picks up the sound, and a video screen displays the pressure graph of the sound wave. Music tends to have smooth waves with repeated patterns and frequencies. Noise tends to have complicated, sharp waves with few if any repeated patterns and frequencies.

Loud noise can damage your hearing, especially if you are around it for a long time. The damage may be temporary or permanent. You should protect your hearing if you must be near loud sounds. Using earphones to listen to the radio and tapes is very courteous when you are in a crowd. There are, however, some dangers, because the earphones are so close to your eardrums. If played too loudly and for too long, earphones can also damage your hearing.

Noise pollution causes emotional, physical, and social stress. For example, think about how annoying a car alarm is when it stays on for a long time.

Ultrasound

Recall that people with excellent hearing can hear sounds caused by vibrations taking place between 20 and 20,000 times per second. Ultrasound has vibration rates beyond human hearing. Its waves are **ultrasonic.** Such waves are used in medicine. For example, very-high-frequency sound can crumble painful kidney stones and gallstones in people. This kind of ultrasound treatment has been performed instead of surgery.

■ Section 13-3 Review ■

Write the definitions for the following terms in your own words.

1. **absorption**   2. **reflection**
3. **oscilloscope**   4. **ultrasonic**

Answer these questions.

5. Why do some restaurants and buildings seem noisy but others seem quiet, even though they are the same size and contain the same number of people?
6. Describe resonance. Give two examples.
7. Describe the Doppler effect. Give an example.

SCIENCE, TECHNOLOGY, & SOCIETY

Noise Pollution

A protest march is one way for society to influence technology. In Figure 3-15, the protestors are angry about a serious menace to our environment, noise pollution. Research shows that excessive noise has bad short-term and long-term consequences.

An eight-year study compared two similar Los Angeles communities. One community was next to a busy airport, with 650 flights thundering overhead every day. The other neighborhood was quiet. In the noisy community, there were higher death rates from heart attacks, strokes, accidents, violence, and suicide.

Medical studies show that noise causes stress on the body. For example, noise increases heart rate, blood pressure, and levels of stomach acid and hormones. Noise has also been linked to low birth weight and other stress-related problems in newborn babies.

If you feel that a certain technology is harmful, you have a right to try to get it changed. Government has responded to the concerns of citizens regarding noise. Congress passed a Noise Control Act that gave the Environmental Protection Agency (EPA) and the Occupational Health and Safety Administration (OSHA) authority to set noise standards. Some machines have been made quieter and even soundproof. For example, street jack-hammers are not as loud today as they used to be. However, enforcing noise laws is not always a top priority for police or federal agents. This is where public pressure can make an immediate difference.

Follow-up Activity

1. Take five minutes to record the noises you hear around school, while traveling, and at home. For each of these locations, decide which noises seemed unnecessarily loud or annoying.
2. List the actions that concerned citizens can take to change noise problems created by technology.
3. Name two things you believe you could do easily and quickly to reduce noise pollution in your surroundings.

Figure 13-15

KEEPING TRACK

Section 13-1 The Nature of Sound
1. Sound is produced by vibrating matter such as solid objects, surfaces, strings, and air columns.
2. Sound is a form of energy and is a wave.
3. Sound waves travel only through matter.
4. The speed of sound is determined by the type and temperature of its medium.
5. The speed of sound is much slower than the speed of light.

Section 13-2 Characteristics of Sound Waves
1. In waves, the disturbances of the medium move, not the medium itself.
2. Sound is a longitudinal wave.
3. A wave is described by its wavelength, frequency, amplitude, and speed.

Section 13-3 The Behavior of Sound Waves
1. Sound is absorbed and reflected when it hits any surface.
2. Resonance can occur when the frequencies of a sound source and an object match.
3. The Doppler effect is a perceived change in the pitch of a sound caused when either the observer or the sound source is moving.
4. Loud noise, especially if prolonged, can damage hearing temporarily or permanently.

BUILDING VOCABULARY
Write the term from the list that best matches each statement.

amplitude, decibels, Doppler effect, medium, oscilloscope, vibrate, wave

1. shake rapidly
2. a series of disturbances
3. the material that a wave moves through
4. measures of loudness
5. change of pitch explained by motion
6. pictures sound waves
7. height of a wave

Explain the difference between the terms in each pair.

8. transverse wave, longitudinal wave
9. wavelength, frequency
10. absorption, reflection
11. ultrasonic, supersonic

SUMMARIZING
Write the missing word or phrase for each sentence.

1. Sound cannot exist in a ___.
2. The speed of sound depends on ___.
3. Sound is a longitudinal wave because the air molecules vibrate in the ___ as the wave itself.
4. If the wavelength of a sound is decreased, its frequency will ___.
5. If the frequency of a sound is increased, the pitch that is heard will ___.
6. If the amplitude of a sound wave is increased, its ___ will increase.
7. The horn of a car sounds as if it has a higher frequency than normal if the car is ___.

INTERPRETING INFORMATION
Figure 13-16 shows three sound waves. Copy

the waves on your own paper. Then identify the wave with

1. the shortest wavelength,
2. the lowest frequency,
3. the lowest volume, or loudness,
4. the greatest pitch, and
5. the greatest number of decibels. Highlight these features on your paper.

THINK AND DISCUSS

Use the section number in parentheses to help you find each answer. Write your answers in complete sentences.

1. Why are vibrations essential for sound? (13-1)
2. Sound becomes less loud the farther it travels. Why? What happens to its energy? (13-2)
3. Compare the speed of sound in air at 0°C to
   a. its speed in air at 100 °C (13-1)
   b. the speed of light (13-1)

4. Explain why sound is a longitudinal wave. Make a diagram showing the movement of the pressure puffs and the air molecules. (13-2)
5. You can barely hear a violin string when it is plucked if it has been removed from the violin. The materials and shape of the violin increase the loudness of the string's sound. How is this an example of resonance? (13-3)

GOING FURTHER

1. Investigate noise in the workplace. Begin by interviewing people on how they are affected by noise where they work. Take notes and summarize your findings. You might then interview employers in factories and businesses in your area. What precautions do they take to protect their workers from noise?
2. Try to obtain a decibel meter. Record the noise level in classrooms, lunchrooms,

Figure 13-16

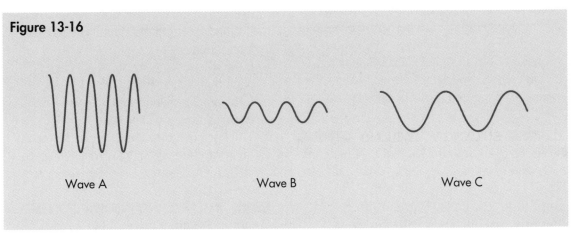

Wave A   Wave B   Wave C

halls, and gyms. How are students and teachers affected by the noise? Were any of the noise levels dangerous, or above 85 dB?

COMPETENCY REVIEW

1. In space, astronauts working outside the space station must talk to each other by radio, even if they are next to each other. The reason they must use radios is that
   a. they wear thick helmets.
   b. there is too much other noise in space.
   c. there is no gravity in space.
   d. there is no air in space.
2. Sound cannot travel in
   a. a vacuum.   b. steel.
   c. dense rock.   d. water.
3. Which statement is true about a fireworks display?
   a. You see and hear the flash and sound at the same time.
   b. You hear the explosion, then see the flash.
   c. You see the flash, then hear the explosion.
   d. The flash can be seen from all directions, but the explosion can only be heard downwind.
4. Human hearing can be damaged by
   a. soft noises that continue a long time.
   b. loud noises.
   c. high-frequency sounds.
   d. low-frequency sounds.
5. Which is the correct order of the things that vibrate when a sound is made?
   a. air, source, eardrum
   b. eardrum, air, source
   c. source, eardrum, air
   d. source, air, eardrum
6. Which statement is *not* true about sound?
   a. It travels as fast as light.
   b. It is a form of kinetic energy.
   c. It can travel around curves and corners.
   d. It only travels through matter.
7. In Figure 13-17, which choice represents the wavelength of the wave?

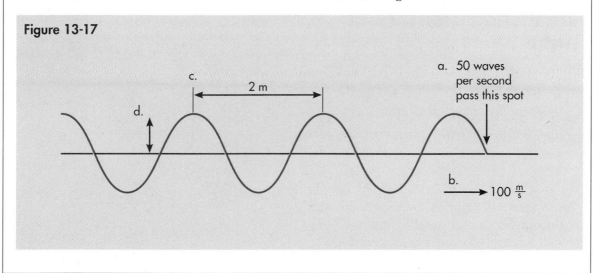

Figure 13-17

c.

2 m

d.

a. 50 waves per second pass this spot

b. 100 $\frac{m}{s}$

CHAPTER 14

ELECTROMAGNETIC ENERGY

You probably do not think much about light, unless your surroundings are too dark or too bright. Light, however, allows you to use your eyes to relate to the world around you. Light can also affect your moods, activities, and health. All living organisms respond to light and depend on light for survival. Plants need light to grow, and plants are the beginning of all food chains that make life possible. Life depends on light.

Did you know that various creatures see light that we cannot see? For example, when birds and bees look at flowers, they see colors and patterns that your eye cannot detect. Dogs, on the other hand, are color blind. They see only various shades of gray. If aliens from another world exist and came here, they might be able to see radio or television signals. Radio and TV signals and visible light are similar. They belong to a large family that also includes X rays and microwaves. When you look in a mirror or notice different colors, certain principles of science are at work. You will be investigating these principles in this chapter.

14-1 Characteristics of Light

■ *Objectives*

☐ *Describe the characteristics of light.*

☐ *Explain how light waves travel.*

☐ *List sources of light.*

What is light? For most of history, people misunderstood light. Many scientists and philosophers thought that light was a very thin fluid that flowed like water. Finally,

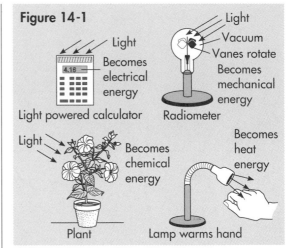

Figure 14-1

Light powered calculator — Light Becomes electrical energy

Radiometer — Light / Vacuum / Vanes rotate — Becomes mechanical energy

Plant — Light Becomes chemical energy

Lamp warms hand — Becomes heat energy

Light can be transformed to other energies.

around 1860, James Maxwell showed that light can be thought of as a wave.

Light as Energy

Light is one form of energy. Other forms of energy can be changed into light energy. For example, chemical energy from burning and electrical energy in light bulbs both produce light. The sun produces light from nuclear energy. Light can also be transformed into other forms of energy, as you can see in Figure 14-1.

Light as Waves and Particles

In many situations, light behaves like a wave. Unlike sound waves, however, light waves can travel in a vacuum. For example, light travels 93 million miles through empty space from the sun to the earth.

Light waves are rapid changes in the electricity and magnetism in space. For this reason, light waves are called **electromagnetic waves**. The electricity and magnetism vibrate at right angles to the direction that the light is traveling. This means that light waves are transverse waves. You should know that, in other situations, light behaves like a stream of small particles. Scientists call these particles **photons**.

Sources of Light

The sun is the major source of light for the earth. Stars, like the sun, generate enormous amounts of light. Some stars give out even more light than the sun. These stars are so far away from the earth that they seem far less bright that the sun, but in fact many are brighter. Lightning and fires produce light. Some insects, fish, and bacteria produce light. Hot objects like melted iron and toaster coils glow with light. Human beings have depended on artificial sources of light such as candles, oil lamps, and TV screens. Of course, light bulbs are the main artificial source of light today.

The Speed of Light

Light travels so fast that scientists once thought its speed could not be measured. In 1927, an American physicist, Michelson Albert, measured the speed of light by using a bright lamp and rotating mirrors set on two mountains. Previously astronomers were also able to estimate the speed of light in a vacuum by observing motions and eclipses in the solar system.

The speed of light is about 300,000 kilometers per second, or 186,000 miles per second. At this speed, light can travel from New York to California in 1/60 of a second. Light can go around the earth seven times in one second. Light travels from the sun to the earth in about eight minutes. Therefore, the light you see outside your window left the sun about eight minutes ago.

In 1905, Albert Einstein stated the hypothesis that light is the fastest thing in the universe. Einstein said that nothing will ever travel faster than light. Starting with this idea, he was able to show that, at very high speeds, objects become heavier and time slows down. Most of Einstein's predictions have been supported by experiments. Scientists, therefore, accept as a law of science the fact that nothing can move faster than the speed of light.

The Path of Light

Light travels in straight lines called **rays**. If you peered through two rubber hoses, one straight and one curved, you would only see light through the straight hose. Think about Figure 14-2. It shows how light travels in a straight line.

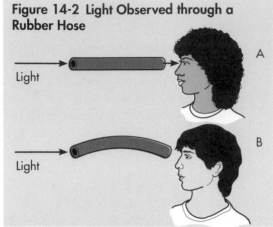

Figure 14-2 Light Observed through a Rubber Hose

Light

A

Light

B

In A, the light is visible. In B, light does not follow the curve and reach the eye.

Light can be represented in several ways because it has many characteristics. Each characteristic is best shown or explained by different pictures. Light can be shown as a wave or as expanding wave fronts, like ripples. It can also be shown as a stream of particles or as **rays**. Compare the diagrams in Figure 14-3.

▬ Section 14-1 Review ▬

Write the definitions for the following terms in your own words.

1. **electromagnetic waves**
2. **photon**
3. **rays**

Answer these questions.

4. What evidence is there that light is a form of energy?
5. What are four sources of light?
6. a. Draw four ways to represent the light coming from a candle.
   b. What does each drawing emphasize about light?
7. Astronomers use a special unit for distance called a light-year. One light-year is the distance that light travels in one year. Calculate how many kilometers and how many miles are in one light-year. Start with the speed of light given in this section, and multiply to find out how many km and miles light would travel in a year. Hint: How many seconds in a minute? How many minutes in an hour?

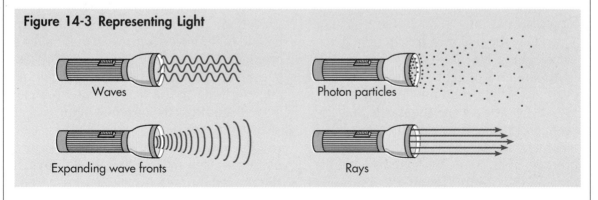

Figure 14-3 Representing Light

Waves

Photon particles

Expanding wave fronts

Rays

14-2 Electromagnetic Energy

■ *Objectives*

☐ *State similarities and differences among kinds of electromagnetic waves.*

☐ *List uses and some dangers of electromagnetic waves.*

Picture yourself lying in the sun during the summer, listening to a radio. Hidden in the picture are electromagnetic waves. They contribute to your listening pleasure, and they can also have some harmful effects. This section examines these topics.

Electromagnetic Waves

The frequency of an electromagnetic wave is measured in waves per second. These waves are vibrations of the electromagnetic field, which is made up of both electricity and magnetism. If the electromagnetic field vibrates one million times every second, then the wave it creates has a frequency of one million waves per second.

The electromagnetic waves you can see are light. Light waves have a frequency of around 500 trillion waves per second. This means the electromagnetic medium is vibrating around 500,000,000,000,000, or

5×10^{14} times every second. It is difficult to imagine such numbers, but your eyes can detect such vibrations. If electromagnetic waves have frequencies much higher or lower than this, your eyes cannot see them.

All electromagnetic waves, whether you can see them or not, have several characteristics:

■ They are created by vibrations of electromagnetic fields.

■ They are forms of energy, and they can move through empty space.

■ They all travel at the speed of light in empty space.

The sun is the major source of electromagnetic waves such as light. It produces waves in a wide range of frequencies. In fact, light waves are only about one millionth of one percent of all the frequencies given out by the sun.

The Electromagnetic Spectrum

Light, radio, and X-ray waves are electromagnetic waves. They differ in frequencies. Certain ranges of frequencies and wavelengths are designated as different kinds of electromagnetic waves. You can see this in Figure 14-4. Notice that, as the frequency decreases, the wavelength increases.

Figure 14-4 Kinds of Waves and Frequencies

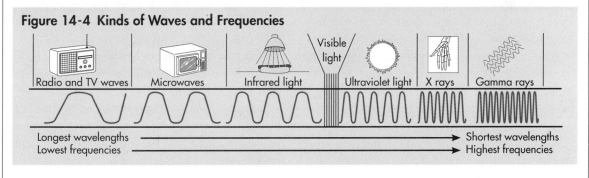

Radio and TV waves | Microwaves | Infrared light | Visible light | Ultraviolet light | X rays | Gamma rays

Longest wavelengths ————————————————→ Shortest wavelengths
Lowest frequencies ————————————————→ Highest frequencies

Figure 14-5 The Electromagnetic Spectrum

Types of waves	Approximate frequency, waves per second	Uses
Radio and TV waves	Millions	Station broadcasting, police-car and airplane communication, ham radios, and so forth.
Microwaves and radar waves	Billions	Telephone communications, cooking food, sighting airplanes and storms from far away
Infrared waves	Trillions	Heat lamps, photography in the dark
Visible light waves	500 trillion	Seeing things
Ultraviolet waves	1,000 trillion	Killing bacteria, skin tanning
X-ray, gamma ray, and cosmic ray waves	Million trillion and higher	X-rays and gamma rays are used in medicine and industry

When electromagnetic waves are arranged in order of their frequencies and wavelengths, the result is the **electromagnetic spectrum**. The spectrum can be presented in table form, as in Figure 14-5. This table also describes some uses of each kind of wave.

Harmful Effects

Figure 14-5 listed some benefits of electromagnetic waves. Overexposure to electromagnetic waves, however, may be harmful to humans. The most dangerous electromagnetic waves are those with very high frequencies and short wavelengths, such as X rays and gamma rays.

X rays, gamma rays, and cosmic rays can cause illness or death. Gamma rays are given off by certain radioactive elements, which you should not handle. Radioactive elements should be kept in lead containers. Lead can stop gamma waves. When you have an X ray taken, your doctor or dentist should cover the rest of your body with a lead apron.

Ultraviolet waves can cause skin cancer and burn skin tissue. They cause sunburn. Because of ultraviolet waves, you should stay out of strong sunlight or stay in it only briefly. At the beach, protect yourself by wearing sunscreen lotions and sunglasses that filter out ultraviolet light.

Infrared waves can also burn your skin. People use infrared heat lamps to relax muscles and to keep food warm in some restaurants. Do not let your skin get too close to such lamps. Do not stare at such lamps. The infrared waves can produce heat inside your eyes. Very bright light can also cause permanent eye damage. Do not stare at the sun or at extremely bright lights.

Microwaves can be harmful. Most microwaves ovens keep this energy inside. Still, do not stand next to microwave ovens for long periods of time. You want them to cook your food, not you!

Very-low-frequency electromagnetic waves exist around electrical wires. Recent research suggests that people who live near high-voltage electricity lines have higher rates of cancer. This may be due to their long-term exposure to the electromagnetic waves the power lines give out. Recent research also suggests that electric blankets pose a smaller but similar risk because your body is close to the wires. The wires give off weak, low-frequency electromagnetic waves.

■ Section 14-2 Review ■

Write the definition for the following term in your own words.

1. **electromagnetic spectrum**

Answer these questions.

2. Memorize the kinds of electromagnetic waves, from lowest frequency to highest frequency. It is not necessary to memorize their frequency numbers or their uses. After a five-minute wait, list the kinds of electromagnetic waves in order, from memory.

3. In what ways are all kinds of electromagnetic waves similar? In what ways do they differ?
4. For each kind of electromagnetic wave, write one use and one danger.
5. Which waves are most dangerous? What precautions should you take to avoid their harmful effects?

14-3 Properties of Light

■ *Objectives*

☐ *Explain how light waves and objects get their colors.*

☐ *Sketch what occurs when light strikes a surface.*

☐ *Describe what lasers are and what they can do.*

☐ *Distinguish between intensity and illumination.*

If you carefully examine a diamond or crystal glass, you will see the light broken into beautiful colors. An oil slick on the road often produces a rainbow of colors. In water sprayed from a sprinkler, you may see a rainbow. The rainbow forms because drops of water can also break sunlight into many colors.

The Color of Light

Isaac Newton was the first to suggest how rainbow colors are formed. White light, he said, was really a mixture of many colors. Diamonds, raindrops, and oil slicks can unmix the colors and show them separately. If the colors are remixed, they will again form white light. These facts are best demonstrated using glass prisms, as you can see in Figure 14-6.

You can memorize the colors of the rainbow as "ROY-G-BIV" for red, orange, yellow, green, blue, indigo, and violet. This does not mean there are exactly seven colors. Most people do not see indigo and violet as separate colors, but as two shades of purple. In some cultures, people see orange and red as different shades of one color. Painters and interior decorators can distinguish and name hundreds of colors.

Colors have the same relation to light as musical notes have to sound. Every note of music is a sound wave of a particular frequency Every shade of color is a light wave of a particular frequency. For example, regular green has a frequency of 580 trillion waves per second. This is 580,000,000,000,000, or 5.8×10^{14} waves per second. Figure 14-7 shows the frequencies of the colors.

You can observe several things in this table. For example, a particular shade of a color will have a frequency within the ranges given for that color. Aqua is a blue

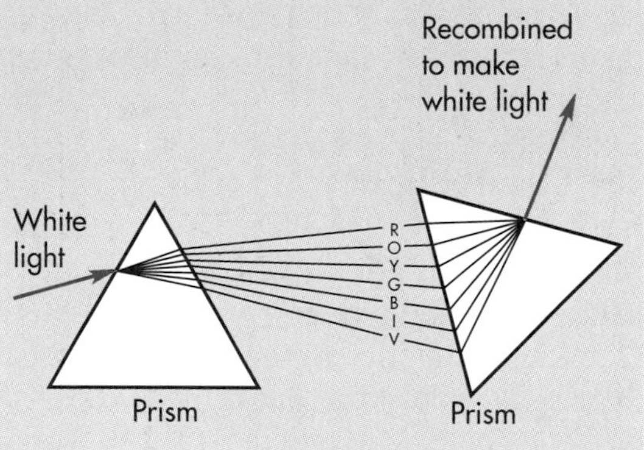

Figure 14-7

Colors	Frequency, waves per second
Reds	400 to 450 trillion
Oranges	450 to 510 trillion
Yellows	510 to 550 trillion
Greens	550 to 610 trillion
Blues	610 to 700 trillion
Purples	700 to 750 trillion

shade that borders on being green. It could have a frequency of 620 trillion waves per second. This is within the blue range but near green.

The table shows the small range of electromagnetic-wave frequencies humans can see. The range is from 400 trillion to 750 trillion waves per second. Note, also, that white and black are not listed. They are not true colors; they do not have a particular frequency. Pure white is a mixture of all the colors, entering your eye together. Pure black is the absence of any visible light.

Reflection

Look at a friend through a glass door or window. You can see each other, which means light waves pass through the glass. You may also see a weak reflection of yourself in the glass, as in a mirror. Three things are happening to the light from your faces in these situations.

Figure 14-6 Experiment with Prisms

White light → Prism → R O Y G B I V → Prism → Recombined to make white light

When light waves strike a surface, some will be reflected, some will be absorbed, and some may be transmitted. Look at Figure 14-8. **Reflected light** is made up of waves that bounce off the surface. **Absorbed light** consists of the waves that "die out" at the surface and within the material below the surface. The energy of absorbed light becomes heat. **Transmitted light** is made up of the waves that pass through a surface and the material underneath.

Various materials reflect, absorb, and transmit light in different amounts. For example, mirrors reflect a high percentage of the light that hits them. Sunglasses absorb much of the light that hits them, but reading glasses permit transmission of most of the light.

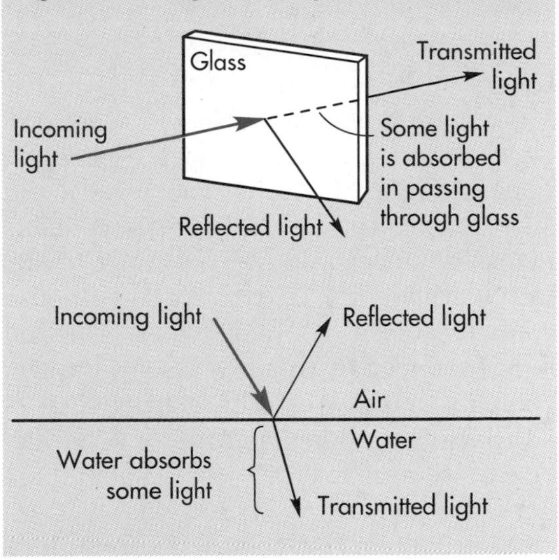

The Color of an Object

Why is a plant green? When white light shines on a plant, chemicals in the plant absorb most of the colors. Plants do not absorb, however, the green waves. Plants

Figure 14-8 Light Striking Glass and Water

reflect green light, so your eye sees only green light coming from the plant. Similarly, an apple is red because it absorbs all the frequencies that shine on it, except for the red frequencies, which it reflects. Look at Figure 14-9. An object is white because it reflects all colors to your eye. An object

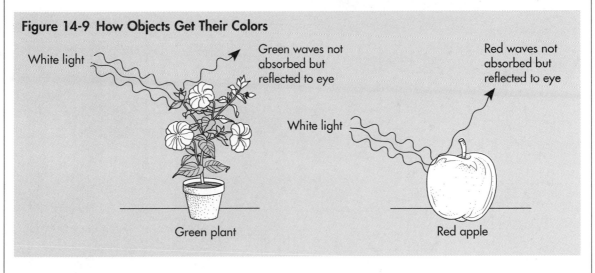

Figure 14-9 How Objects Get Their Colors

is black because it absorbs all colors and sends no light to your eye.

Lasers

Lasers are a new technology. Lasers produce an exact, single frequency of light. Look at Figure 14-10. Ordinary red light, such as that in car or traffic lights, contains many frequencies in the red range of 400 to 450 trillion waves per second. It also contains traces of other colors. The red light produced by a "ruby" laser, however, contains only waves with a frequency of 431.8 trillion waves per second. So far, lasers are available only for a small number of frequencies.

Laser light is also special because all its waves are in step, as you can see in the figure. For this reason, laser light is called coherent light. Because laser light is of a single frequency and because all its waves are in step, it can transfer lots of energy in a narrow beam. The beam will not spread out over long distances as much as ordinary light beams. Lasers can cut metal. They are being used today in medical operations, in light-wave telephone communication, and in research labs.

Intensity and Illumination

When you read or study, do you make sure the light is bright enough? Dim light can cause eyestrain. Recent research shows that employees are happier, healthier, and more productive when their work area is kept bright.

Brightness is a broad term that covers two different features of light, **intensity** and **illumination**. Look at Figure 14-11. Intensity is the brightness of the light source itself. A 150-watt bulb has a greater light intensity than a 50-watt bulb. Illumination is the brightness of the light as seen by your eyes. It is the amount of light falling on a surface or an object. If you

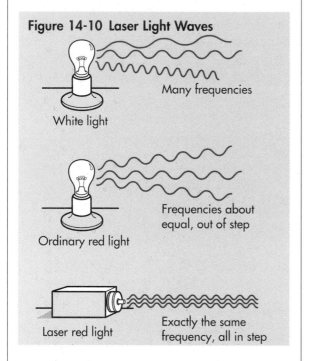

Figure 14-10 Laser Light Waves

White light — Many frequencies

Ordinary red light — Frequencies about equal, out of step

Laser red light — Exactly the same frequency, all in step

Figure 14-11 Aspects of Brightness

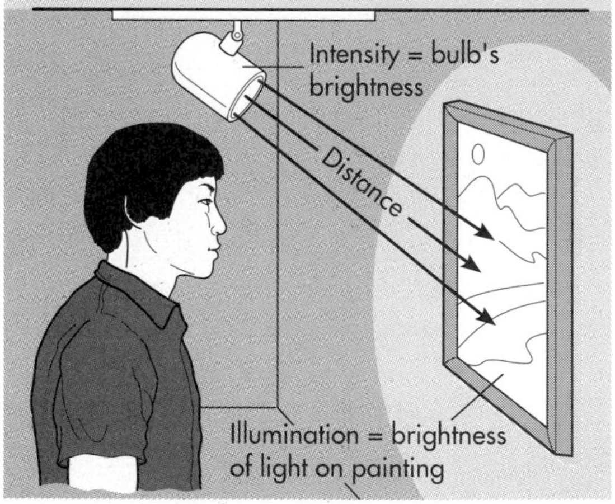

Intensity = bulb's brightness

Distance

Illumination = brightness of light on painting

have trouble reading a newspaper, maybe the illumination on the newspaper is too low. If the illumination is too low, you can get a brighter bulb or move the lamp closer. Illumination on a surface increases as the intensity of the source increases, and as the distance from the source decreases.

▰▰ Section 14-3 Review ▰▰

Write the definitions for the following terms in your own words.

1. **reflected light**
2. **transmitted light**
3. **laser**
4. **intensity**
5. **illumination**

Answer these questions.

6. a. What makes yellow light yellow?
   b. What makes a yellow object yellow?
   c. What makes a white object white?
   d. What makes a black object black?
7. How is the illumination on a surface related to the intensity and distance of the light source?
8. Sketch three events that occur as light rays strike a surface such as a piece of glass. Label your sketch to explain what is happening.
9. What is the color of light waves at each of the following frequencies?
   a. f = 490 trillion
   b. f = 634,789,361,739,202
   c. f = 7.1×10^{14}
   d. f = 800 trillion

14-4 Mirrors and Lenses

■ *Objectives*

☐ *Draw sketches that show the laws of reflection and refraction.*

☐ *Describe the images formed in plane, convex, and concave mirrors.*

☐ *List various uses for lenses.*

Are you wearing glasses? If so, you are using lenses to adjust the light rays entering your eyes. You probably saw yourself in a mirror today. A mirror is a device that reflects light waves according to certain laws.

Law of Reflection and Mirrors

When light is reflected by a surface, it does not bounce off in any direction it pleases. The **law of reflection** says that the angle at which a light ray strikes a flat surface is equal to the angle the reflected ray makes with the same flat surface. Study Figure 14-12. This is an example of light behaving

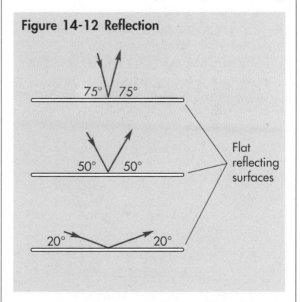

Figure 14-12 Reflection

75° 75°

50° 50°

Flat reflecting surfaces

20° 20°

as if it were made of hard particles. This same law of reflection applies when hard objects, such as balls, bounce off surfaces like the edge of a pool table.

The law of reflection helps to explain how plane mirrors work. Plane mirrors are flat. You use plane mirrors at home, at school, and in stores. The inside rearview mirrors in cars are also plane mirrors. When you look in a plane mirror, there are several things to note about your image.

- The images are right side up. Your eyes are above your mouth.

- The images are reversed, left to right. Your right hand becomes the left hand of your image in a mirror. If you wear a bracelet on your left arm, it will be on the right arm of your image.

- The images appear to be as far behind the mirror as the object is in front of the mirror. If your mother is behind you as you look into a mirror, then her image will be behind your image in the mirror.

- The images are illusions, created because light waves reflect and enter your eye.

S K I L L   B U I L D E R

Evaluating

Remember that a scientist conducts an experiment to test a hypothesis. A laboratory report on the experiment should indicate whether or not the data support the hypothesis. Read this description of an experiment and then answer the questions.

Julia performed an experiment to see how light is reflected from a plane mirror. Look at Figure 14-13, which shows the angle of incidence and the angle of reflection. Julia thought that the angle of reflection would increase if the angle of incidence increased. With a flashlight, she shone a narrow beam of light on a mirror. She measured the angle of incidence and the angle of reflection, and recorded her measurements in a table. She repeated the procedure four times. Each time, she made the angle of incidence a bit larger, and observed the effect on the angle of reflection. Finally, Julia plotted her results on a line graph. The data showed that the angle of reflection increased as the angle of incidence increased.

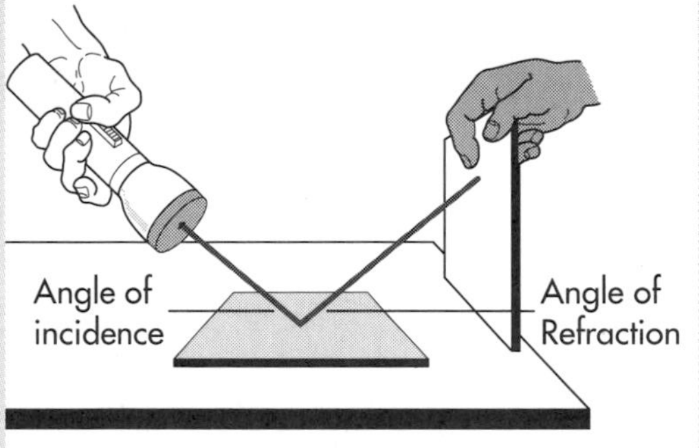

1. What was the purpose of this experiment?
2. What was the hypothesis?
3. What data did Julia collect?
4. How did Julia organize the data?
5. How did Julia analyze her data?
6. Did the data support the hypothesis?

Figure 14-13

Angle of incidence

Angle of Refraction

Your brain is tricked into thinking that these reflected rays came from some object behind the mirror. That is why your brain "sees" an image behind the mirror.

People also use curved mirrors. See Figure 14-14. A **convex** mirror bulges outward and lets you see a wider view than a plane mirror can show, but the images are much smaller. Convex mirrors are used on the sides of cars and trucks and in hallways. A **concave** mirror bulges inward. It produces a larger image when held close, so it is used in magnifying mirrors. Concave mirrors can also gather light and focus it on one spot. Because of this, they are used in microscopes and light-energy collectors.

Refraction

As light rays pass from one material to another, they bend. This occurrence is known as **refraction**. Refraction happens because light moves at slightly different speeds in different materials. This can cause odd illusions. For example, look at a pencil through a glass of water. Does the pencil look like it usually does? If it does not, that is because light has been refracted by the water in the glass, changing the image you see.

You can understand refraction by studying Figure 14-15. The line that is 90 degrees to the surface that a light ray strikes is called the normal line. The **law of refraction** says that when a light ray

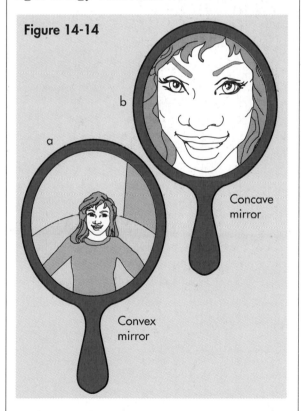

Figure 14-14

a

b

Concave mirror

Convex mirror

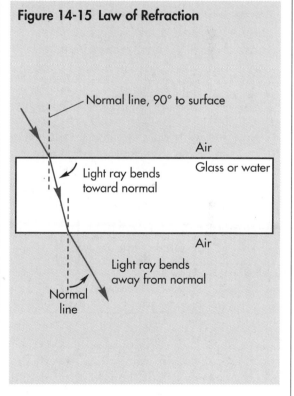

Figure 14-15 Law of Refraction

Normal line, 90° to surface

Air

Glass or water

Light ray bends toward normal

Air

Light ray bends away from normal

Normal line

How Are Lenses Used?

Process Skill observing

Materials piece of white cardboard; small convex magnifying glass or other available convex lenses; concave lenses

Procedures

1. Hold the magnifying glass or convex lens an arm's length away. Look through it at objects outside a window. What is the image like?
2. Stand with your back to the window and hold the white cardboard in front of you. Hold the convex lens in front of the cardboard and adjust its position until you see a sharp image on the cardboard of the buildings, objects, or people you saw through the lens. Study the image carefully and record what you observe about it.
3. Try to repeat procedure 2 with a concave lens. Can you get an image on the cardboard? Look through the concave lens at objects outside the window. What is the image like?

Conclusions

4. How do lenses illustrate the law of refraction?

enters from the air into another material, such as glass or water, its speed decreases and it bends toward the normal line. When a light ray leaves a material and reenters the air, its speed increases and it bends away from the normal line.

Lenses

With lenses, people use refraction for practical purposes. Lenses bend the light to form images that may be larger, the same size as, or smaller than the object being viewed. Study the chief actions of convex and concave lenses in Figure 14-16.

Figure 14-16 Lenses Bend Light Rays

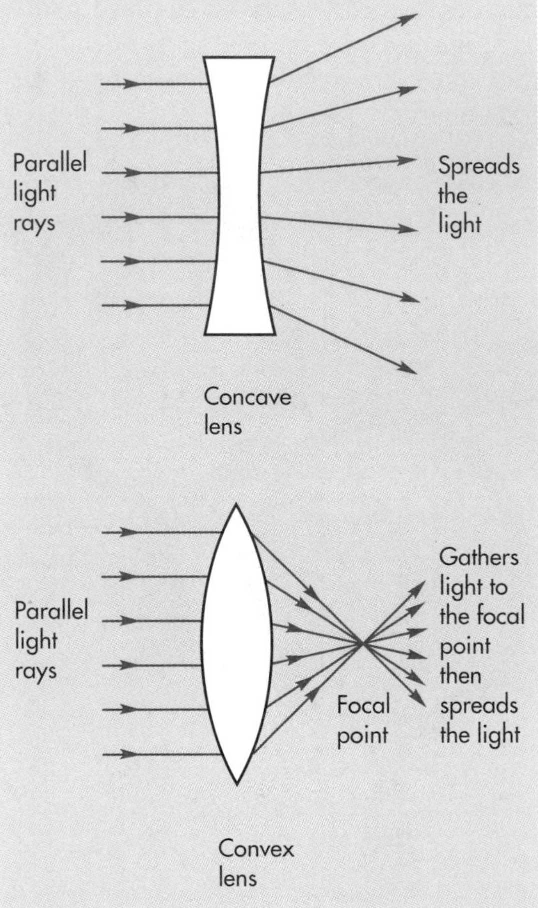

Parallel light rays

Spreads the light

Concave lens

Parallel light rays

Gathers light to the focal point then spreads the light

Focal point

Convex lens

Figure 14-17

Example	Type lens	Image size
Camera	Convex	Smaller than object being photographed
Photocopier	Convex	Same size as object being copied
Movie projector	Convex	Larger than film or object being viewed
Microscrope, magnifying glass	Convex	Much larger than objects being viewed
Peephole in a door to see hallway	Concave	Tiny images, but wide view of hallway

Convex lenses in cameras make things look smaller than they really are. The convex lens in a telescope makes objects appear larger than without the telescope. If you have ever used a peephole in a door to see who is knocking, you have used a concave lens. It creates tiny images, but gives a widened view of the hall or steps. The chief actions of convex and concave lenses are summarized in Figure 14-17.

■■■ Section 14-4 Review ■■■

Write the definitions for the following terms in your own words.

1. **law of reflection**
2. **refraction**
3. **law of refraction**

Answer these questions.

4. Figure 14-18 shows a ray of light striking calm water. Some of the light will be reflected, and some will be transmitted into the water.

a. Copy the figure on your paper and show exactly the paths taken by the reflected and transmitted rays.
b. Within the diagram, note where the law of reflection and the law of refraction are obeyed.
5. a. What would the following symbols and letters look like in a mirror?
   V P L T K Δ < Σ
b. What two characteristics of an image in a plane mirror are demonstrated?
6. Where are convex and concave mirrors and lenses used?

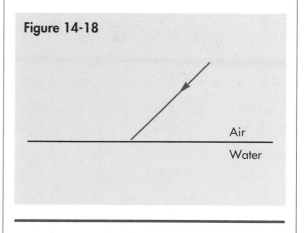

Figure 14-18

Air

Water

SCIENCE, TECHNOLOGY, & SOCIETY

The Federal Communications Commission

Imagine turning on your radio and hearing a station interfering with the one you have chosen. This can happen if both stations are broadcasting at the same frequency. The Federal Communications Commission, or FCC, assigns a broadcasting frequency to each radio and television station in the country. Figure 14-19 shows AM radio stations that the FCC has assigned the frequency 890 kiloHertz, which is 890,000 waves per second. The FCC also determines the signal power and hours of operation of each station. This is so that stations in different cities with identical frequencies will not interfere with each other. The FCC regulates any device that gives off electromagnetic waves for the purpose of communication.

Some countries jam radio broadcasts from other nations so their people cannot hear the broadcast. Jamming is done by producing a strong, noisy signal at the same frequency as the unwanted broadcast. The result is that people hear only a loud buzz on the radio. Fortunately, jamming has been reduced in recent years and may soon end.

Follow-up Activities
1. What problems might arise if broadcasters could use any frequency at any strength?
2. What is jamming? Do you think this practice is ever necessary?

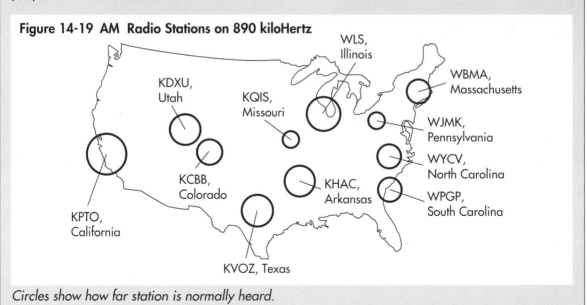

Figure 14-19 AM Radio Stations on 890 kiloHertz

WLS, Illinois

KDXU, Utah

WBMA, Massachusetts

KQIS, Missouri

WJMK, Pennsylvania

KCBB, Colorado

WYCV, North Carolina

KHAC, Arkansas

WPGP, South Carolina

KPTO, California

KVOZ, Texas

Circles show how far station is normally heard.

KEEPING TRACK

Section 14-1 Characteristics of Light
1. Light is a form of energy.
2. Light waves travel away from a source in straight paths.
3. Nothing can move faster than the speed of light.

Section 14-2 Electromagnetic Energy
1. Visible light is a small part of the family of electromagnetic waves.
2. The kinds of waves within the electromagnetic spectrum differ only in their frequencies and wavelengths. They all travel at the speed of light and through vacuums.
3. Electromagnetic waves have many uses in modern society. Radio waves are used in communication; microwaves are used in communication and in cooking food; X rays are used in diagnosing diseases.
4. Overexposure to electromagnetic waves may be harmful to humans.

Section 14-3 Properties of Light
1. The color of light comes from its frequency. Humans can see frequencies between 400 trillion and 750 trillion waves per second.
2. When light strikes a surface, some light is reflected, some is absorbed, and some may be transmitted.
3. The color of an object is determined by the light waves it reflects.

Section 14-4 Mirrors and Lenses
1. The law of reflection determines the angle at which a light ray bounces off a surface.
2. The law of refraction determines how a light ray bends as it enters a new material.
3. Images in a plane mirror are reversed, left and right.
4. Lenses can be used to form images that may be larger, the same size as, or smaller than the object.

BUILDING VOCABULARY
Write the term from the list that best completes each sentence.

absorbed light, illumination, intensity, laser, photons, rays

Light waves move along straight lines called ___1___. Light can also be considered to be a stream of particles known as ___2___. The light energy that turns to heat when it hits a surface is called ___3___. Light waves from a ___4___ are of one frequency and color. The brightness of a source of light is the ___5___ but the brightness of light shining on a surface is the ___6___.

SUMMARIZING
If the statement is true, write *true* . If the statement is false, change the *italicized* term to make the statement true.

1. Heat and *ultraviolet* light waves can kill germs and bacteria.
2. A purple coat *absorbs* purple light waves.
3. *Transmitted light* immediately dies away and produces heat.
4. A *convex* mirror can produce a larger-than-life-size image of a person's face.

5. The convex lens in a *microscope* forms an image that is larger than the object being viewed.
6. Radio waves and light waves are electro-magnetic waves that differ in their *speeds*.

INTERPRETING INFORMATION

The two graphs in Figure 14-20 show how illumination changes with the intensity of the source and the distance to the source.

1. What obvious relationships are shown by these graphs? Relate distance and intensity to the total illumination.
2. By how much was the illumination multi-plied when the intensity was
   a. doubled, from 1 to 2
   b. multiplied by 5, from 2 to 10
3. By how much was the illumination divided when the distance was
   a. doubled, from 1 to 2
   b. multiplied by 5, from 2 to 10

THINK AND DISCUSS

Use the section numbers in parentheses to help you find each answer. Write your answers in complete sentences.

1. Astronauts placed special mirrors on the moon. Scientists can send a tiny burst of laser light to the moon and detect its reflection about 2.4 seconds later. From this fact, find how many kilometers it is to the moon. (14-1, 14-3)
2. For many years, people thought that enjoy-ing a day in the sun and getting a tan was a healthful thing to do. Now doctors and scientists warn of the dangers of doing this. What is the problem? (14-2)
3. White and black are not true colors in science. Why not? What is a true color? (14-3)
4. Why must reflected light always be less bright than the original light before it was reflected? (14-3)

Figure 14-20

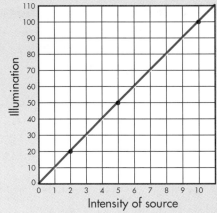

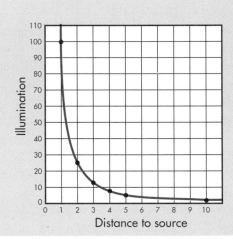

GOING FURTHER

1. Do research about how human eyes perceive color and how colors can be "added" and "subtracted." You may want to look for this information in biology, psychology, and physical science books.
2. Make a report on a topic related to light. Possible topics include
   - Color blindness
   - How lasers work
   - How light bulbs work
   - Types of telescopes
   - Lenses for correcting vision problems
   - Photons and light energy

COMPETENCY REVIEW

1. Which statement is not true about light?
   a. Its speed cannot be measured.
   b. It travels in a vacuum.
   c. It travels in straight lines.
   d. It travels in some liquids and solids.
2. The greatest source of light and electromagnetic energy is
   a. atomic reactors and atomic bombs.
   b. light bulbs and TVs.
   c. lightning and fires.
   d. the sun.
3. Figure 14-21 shows a light ray striking a glass block. Which ray is the reflected light?
   a. 1   b. 2
   c. 3   d. 4
4. Doctors can detect and treat some diseases by using
   a. radio waves.   b. X-ray waves.
   c. infrared waves.   d. microwaves.

Figure 14-21

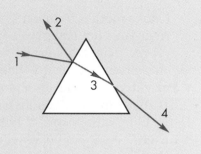

5. Which are used for both communication and cooking?
   a. radio waves
   b. X-ray waves
   c. ultraviolet waves
   d. microwaves
6. The damage done to living cells by electromagnetic waves increases as the frequency increases. Therefore, which kind of wave is most dangerous to humans?
   a. X rays
   b. visible light
   c. microwaves
   d. radio and TV
7. In direct sunlight, black clothing will make you warmer than white clothing because black clothing
   a. is always thicker.
   b. absorbs more light.
   c. reflects more light.
   d. transmits more light.
8. In a plane mirror, the letter **b** will appear as
   a. **p**   b. **b**
   c. **q**   d. **d**

HEAT ENERGY

"That is too hot to touch!"

"Wear your hat and gloves in the cold!"

How many hundreds of times were you told these things as you were growing up? Still, at some point in your life, you may have burned your hand on a stove or felt cold when you did not wear your hat and gloves. A child must experience hot and cold to understand them. In other words, hot and cold cannot be explained, they must be felt.

Later you learned that how hot or how cold something is can be measured. Three examples follow: The weather forecaster said it will be 10°F below zero today. The turkey cooks at 325°F in the oven. The room is comfortable between 65°F and 70°F.

Now you are ready to learn what heat really is and what its effects are. Heat is a form of energy, and it is measured by a special unit. Heat energy can move from one spot to another, and it can be transformed into other forms of energy as shown in Figure 15-1. You will learn about some uses of extreme cold and also why even small increases in the average temperature of Earth's atmosphere may be a danger to our environment.

15-1 Heat and Temperature

■ *Objectives*

☐ *Give evidence that heat is a form of energy.*

☐ *Explain the effects of heat energy in expansion, contraction, and phase changes.*

☐ *Distinguish between heat and temperature.*

Before 1800, scientists regarded heat as a fluid that could flow between objects.

Figure 15-1

Hot-air balloon

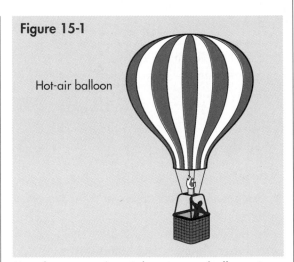

Heat becomes potential energy as balloon rises.

They called the fluid "caloric." An object became hotter when it contained more of the "caloric." This view was abandoned between 1800 and 1850 because of the separate work of two British physicists, Count Rumford and James Joule. Rumford showed that an object did not weigh more when he made it hotter. If heat were a material put into an object, then the mass of the object should be greater when hotter. Joule showed that the amount of heat created in an object and the amount of work done to the object were equal. This opened the way to the view that heat is a form of energy.

Heat as Energy

One sign that heat is energy is that work can be changed into heat. Touch a thin rubber band to your cheek. Then stretch it quickly and touch it to your cheek while it is stretched. Can you feel its warmth? By stretching the rubber band, you did work. Work now appears in the rubber band as energy. Figure 15-1 shows another example of heat energy being transformed.

Heat and Matter

You learned that all matter consists of particles, either atoms or molecules. These particles are in constant motion. They move from spot to spot and rotate in liquids and gases. Even if they do not move much, as in solids, they vibrate in place. Because they are in motion, the particles in matter must contain kinetic energy. This is the real nature of heat. Heat is the sum of the kinetic and potential energies of the particles within a piece of matter.

Heat and Expansion

When you make an object or material hotter, it gains heat energy. This means its particles gain kinetic and potential energy. They move faster, and they also move farther apart. This is why objects and materials expand when heated and contract when cooled. Expansion means an increase in size, and contraction means a decrease in size.

Engineers must always plan for expansion or contraction when they build large structures. For example, look at Figure 15-2. Sections of concrete sidewalks have gaps, or "cracks," so that the concrete can expand in hot weather without buckling and cracking. The gratings between a metal bridge and the road on either side permit the bridge to expand and contract as the temperature changes.

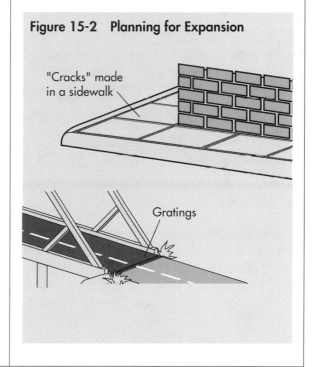

Figure 15-2 Planning for Expansion

"Cracks" made in a sidewalk

Gratings

Heat and Temperature

Imagine a bathtub filled with hot water and a pot full of water at the same temperature. It took more energy to heat the large amount of water in the tub than it took to heat the water in the pot. The water in the tub contains more heat energy than the water in the pot, even though they are at the same temperature. This is the first difference between heat and temperature. The total heat energy in an object depends on its mass, but temperature does not depend on mass.

As you add heat energy to a material, its temperature increases. Different materials heat up at different rates. Figure 15-3 compares how long it takes for identical amounts of water, olive oil, and iron to go from the same starting temperature to the same final temperature over the same burner.

Figure 15-3	
Material	**Time needed**
Water	5 min
Olive oil	2 min
Iron	0.5 min

This shows that the heat energy needed to raise the temperature of an object also depends on the object's composition. When a pot of water and a pot of oil of identical mass have their temperatures raised by the same amount, the oil needs less heat than the water.

You can see that the concepts of temperature and heat are related but different. Temperature depends only on the speed of the particles in a material. In scientific terms, temperature depends on the average kinetic energy of the particles in a material. When a material is made hotter, its particles move faster and have a higher average kinetic energy. When a material is made colder, its particles move more slowly and have a lower average kinetic energy.

In contrast, the heat in an object depends on the total amount of energy in its particles. Heat depends on kinetic energy plus potential energy. This means that the heat energy in an object depends on its temperature, mass, and material.

Thermometers measure temperature. They work by the expansion of a liquid or gas inside a tube. Most thermometers use red-colored alcohol or natural mercury, which has a silvery color. There are two widely used temperature scales, degrees Fahrenheit, or °F, and degrees Celsius, or °C. Many thermometers have both scales. Scientists use the Celsius scale.

Heat and Change of Phase

Heat energy is transferred during a change of phase. When a solid melts or a liquid boils, heat energy must be added. When a gas condenses or a liquid freezes, heat must be removed.

It may seem odd that during melting and boiling, the temperature stays the same even though heat is added. The heat energy separates the particles, giving them increased potential energy. This allows them to escape the attractions that kept them bound together as a solid or a liquid. Study Figure 15-4. The heat energy does not increase the kinetic energy of the particles during phase changes. That is why the temperature does not change, for temperature depends only on the kinetic energy of the particles. The table in Figure 15-5 summarizes these facts.

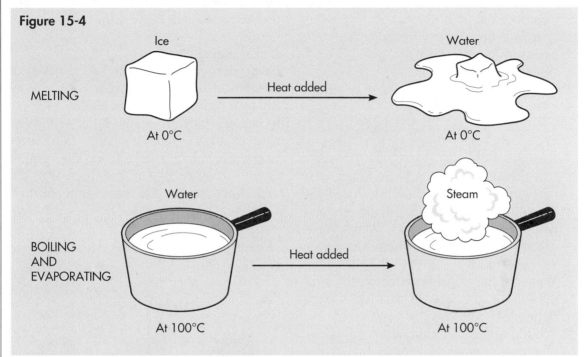

Figure 15-4

MELTING

Ice — At 0°C

Heat added

Water — At 0°C

BOILING AND EVAPORATING

Water — At 100°C

Heat added

Steam — At 100°C

During these phase changes, heat is added.

Figure 15-5 Energy Changes

	Temperature	Kinetic energy of particles	Potential energy of particles
Warming	Increases	Increases	Increases
Cooling	Decreases	Decreases	Decreases
Melting and boiling	Stays same	Stays same	Increases
Freezing and condensing	Stays same	Stays same	Decreases

Write the definitions for the following terms in your own words.

1. **expansion**
2. **contraction**
3. **average kinetic energy**
4. **degrees Celsius**

Answer these questions.

5. What evidence suggests that heat is energy?
6. If a metal cover on a jar is stuck, you can run hot water over the cover for about 15 seconds. This usually lets you twist the cover off easily. Why? What occurs to the atoms within the cover as it gets hotter?
7. What is the difference between temperature and heat? On what does each depend?
8. Why does the temperature of water change as you add heat to it, but not after it starts to boil?

15-2 Heat Flow

■ *Objectives*

☐ *Describe the circumstances under which heat energy transfers between regions and objects.*

☐ *Calculate the heat required, in calories, to change the temperature of water.*

☐ *Contrast conduction, convection, and radiation as methods of heat flow.*

Heat energy can move from one spot to another, within an object or between two objects. For example, in a gas oven, the heat energy of the flames moves to the food being cooked. You will see that this transfer occurs in three different ways.

Transferring Heat

Heat energy is transferred as a result of differences in temperatures. Heat energy can only flow from warmer objects or spots to cooler object or spots. Look at the left side of Figure 15-6. In this and other cases, the warmer object becomes cooler and the cooler object becomes warmer, until both are at the same temperature. Then heat transfer stops.

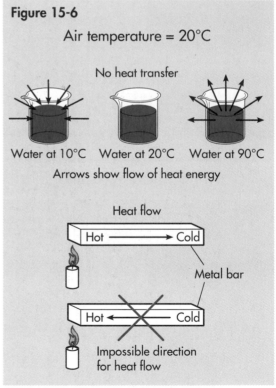

Figure 15-6

Air temperature = 20°C

No heat transfer

Water at 10°C   Water at 20°C   Water at 90°C

Arrows show flow of heat energy

Heat flow

Hot ——→ Cold

Metal bar

Hot ←——✕—— Cold

Impossible direction for heat flow

Heat always flows from hot to cold spots.

Sometimes people speak about "the cold" coming through a window. They speak as if cold is a substance or an energy that flows into warmer places. This is incorrect. Cold just means a lower temperature. Only heat energy exists and moves from a warmer place to a colder place. If you put your hands into cold water, some heat energy from your hands flows into the water. You feel the cooling sensation. Cold did not transfer into you; heat transferred out of you.

The amount of heat transferred between two objects increases as the temperature difference between them increases or the mass of either object increases. For example, a large potato will transfer more heat energy than a small potato. Also, a 200-gram potato at 100° C will transfer more heat energy to the air than a 200-gram potato at 50° C.

Measuring Heat

The metric unit of heat energy is the calorie (cal). A kilocalorie (kcal) equals 1000 calories. The American unit for heat energy is the British Thermal Unit (BTU). Calories and kilocalories are used in science. BTUs are used in engineering.

Examine Figure 15-7. One calorie is the amount of heat energy needed to raise the temperature of one gram of water by 1° C. Therefore, you could use two calories to raise the temperature of two grams of water by 1° C, or one gram of water by 2° C.

Calculate heat by using this formula.

$$\begin{array}{c}\textbf{Heat} = \textbf{mass of} \times \textbf{temperature} \\ \textbf{water} \qquad \textbf{change} \\ \text{(cal)} \qquad \text{(g)} \qquad (°C)\end{array}$$

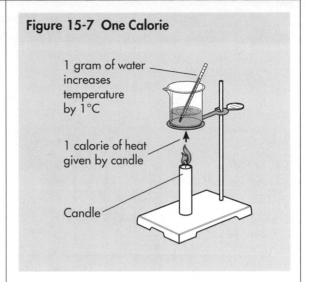

Figure 15-7 One Calorie

1 gram of water increases temperature by 1°C

1 calorie of heat given by candle

Candle

Example How much heat is needed to warm 300 grams of water from 20° C to 30° C?

Heat = mass of water × temperature change
Heat = 300 g of water × (30° C − 20° C)
Heat = 300 g × 10° C
Heat = 3000 cal, or 25 kcal

Example How many kilocalories (kcal) will change the temperature of 500 grams of water at 46° C
a. to 50° C?
b. by 50° C?

As you work through this example, note that in part a, the temperature must be raised four degrees to 50° C. In part b, the temperature must change by 50° C; that is, the final temperature must be 96° C.
a. Heat = 500 g × (50° C − 46° C) = 500 × 4
   Heat = 2000 cal, or 2 kcal
b. Heat = 500 g × 50° C
   Heat = 25,000 cal, or 25 kcal

Calculating

Sharon and Harvey wanted to compare the heat given out by a Bunsen burner and an electric heater. Each student heated 200 grams of water in a beaker, using each heater. They stirred and recorded the temperature of the water every minute for three minutes. Their data is shown in Figure 15-8.

Figure 15-8

For Bunsen burner:

Time of heating, min	0	1	2	3
Temperature of water, °C	18	35	52	69

For electric heater:

Time of heating, min	0	1	2	3
Temperature of water, °C	18	27	36	45

1. Calculate the heat, and the heat per minute, given out by each heater after one, two, and three minutes of heating. For an example, study these calculations for the Bunsen burner after two minutes:

$$\text{Heat} = \text{mass of water} \times \text{temperature change}$$

$$\text{Heat} = 200 \text{ g} \times (52°C - 18°C)$$
$$\text{Heat} = 200 \text{ g} \times 34°C$$
$$\text{Heat} = 6800 \text{ cal}$$

$$\textbf{Heat per minute} = \textbf{heat / time}$$
$$\text{Heat per minute} = 6800 \text{ cal} / 2 \text{ min}$$
$$\text{Heat per minute} = 3400 \text{ cal}$$

2. Place the results of your calculations into new data tables, like those shown in Figure 15-9.
3. What do you observe about the calories per minute given out by each heater? Which heater gave out more heat over the same amount of time?

Figure 15-9

For Bunsen burner

Time of heating, min	1	2	3
Heat given, cal		6800	
Heat per minute, cal/min		3400	

For electric heater

Time of heating, min	1	2	3
Heat given, cal			
Heat per minute, cal/min			

Methods of Heat Flow

Heat energy flows from warmer to cooler areas by three different methods. These methods are conduction, convection, and radiation.

Conduction When heat energy transfers from particle to particle in a material, conduction occurs. Look at Figure 15-10. The heat passes through the material, with faster particles bumping slower ones. If you place one end of a metal spoon into hot water, the water transfers heat directly into the spoon by contact, and heat slowly moves to the other end of the spoon. Both are examples of conduction.

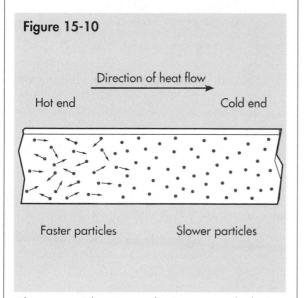

Figure 15-10

Direction of heat flow

Hot end   Cold end

Faster particles   Slower particles

Slower particles are made to move and vibrate faster by faster neighbors.

Some materials conduct heat well, and they are called conductors. Other materials, called insulators, do not conduct heat well. Metals are good heat conductors; this is why pots and pans, as well as car radiators, are made of metal. Pots and pans must conduct heat from the flame into the pot. Car radiators must conduct heat from the engine into the air, or else the engine will overheat. Wood, cloth, and styrofoam are good insulators. This is why they are used for pot handles, pot holders, coffee cups, and insulation around hot pipes. They keep heat in, and they allow people to handle hot objects. Poor conductors are excellent insulators.

Air and water are good insulators when they cannot swirl around. Thermal windows have a layer of air sandwiched between two panes of glass. The glass holds in the air, and this causes insulation. Wool sweaters and blankets keep heat from escaping mainly because of air trapped in their fibers. Divers use a rubber "wet suit" that keeps a layer of water around their bodies for insulation.

Convection When heat energy transfers because warmer material swirls and mixes with cooler material, convection occurs. Convection takes place when you stir a pot of soup or oatmeal. The warm soup nearest the flame is forced to mix with cooler soup near the top.

Convection occurs naturally, without stirring or fans, in liquids and gases that are unevenly heated. The molecules in the heated part of the liquid or gas move faster. As they move faster, they spread out. The warmer part becomes less dense, and starts to rise. The cooler and denser material sinks. This sets up a convection current that mixes hot and cool material. Most rooms are warmed by convection, as warm air from hot radiators rises, swirls, and carries heat throughout the room. Study Figure 15-11.

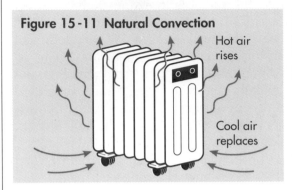

Figure 15-11 Natural Convection

Hot air rises

Cool air replaces

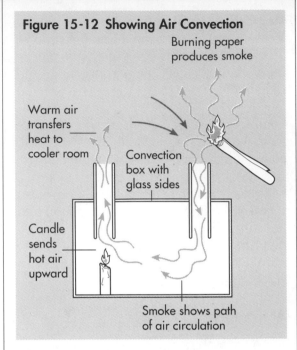

Figure 15-12 Showing Air Convection

Burning paper produces smoke

Warm air transfers heat to cooler room

Convection box with glass sides

Candle sends hot air upward

Smoke shows path of air circulation

Figure 15-12 shows how convection currents work. The hot air around the candle rises up its chimney, causing cooler air to flow down into the other chimney. The air movement can be seen by watching what happens to the smoke.

Radiation   The transfer of heat energy by infrared waves is called radiation. Infrared waves can pass through a vacuum. Hot and warm objects glow with infrared waves, but your eyes cannot see such waves. Since this glowing uses up energy, the object cools off. The infrared waves given out are absorbed by the air, walls, and other objects. This absorption causes them to warm up. This is heat transfer by radiation.

Radiation is the method by which heat energy gets from the sun to Earth. It is the only method by which heat energy can be transferred through a vacuum. You can also feel radiation in the air. When you stand in front of a wood or coal fire you can feel the strong infrared

radiation it gives off. After the flames on a stove's burners are shut off, you can still feel the heat coming from the hot burners. The heat is caused by radiation.

Like light, infrared waves are part of the electromagnetic spectrum. They have frequencies just below visible light, so you cannot see them. However, you can feel infrared rays. This is because infrared waves are easily absorbed by skin and most materials. Instead of being reflected, infrared waves readily turn into heat. They are, therefore, also called heat waves. As with visible light, white and silvery surfaces reflect infrared heat waves. Dark material absorbs heat waves best.

Applications

There are many situations where one, two, or all three methods of heat flow occur, as in a lamp. Examine Figure 15-13. Look for the three kinds of heat flow.

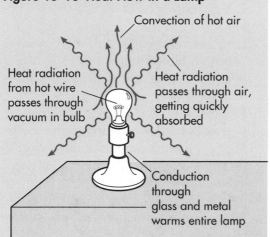

Figure 15-13 Heat Flow in a Lamp

Convection of hot air

Heat radiation from hot wire passes through vacuum in bulb

Heat radiation passes through air, getting quickly absorbed

Conduction through glass and metal warms entire lamp

If you touch the bulb and the lamp base, they will be hot because of conduction. Heat has traveled through the glass and metal base. If you place your hands above the bulb, you can feel the warm air rising up from it, illustrating convection. If you hold your hands to the side or below the bulb, you can still feel heat. Since your hands are to the side, the heat cannot be transferred by convection, which makes the hot air rise. It cannot transfer by conduction, because air is not a good conductor. It must transfer by radiation.

Next time you use a toaster or an oven, look for examples of conduction, convection, and radiation. For example, the toast is heated by radiation, while air around the toaster is warmed by convection.

ACTIVITY

What Is the Difference between Conduction and Convection?

Process Skill *observing*

Materials candle, firmly supported on a stand; long iron nail, about three inches; metric ruler; watch

Procedures
1. Light the candle. Hold the nail horizontally near its head and place just the point in the flame.
2. Remove the nail from the flame the moment you begin to feel a slight warmth in the nail's head.
   Caution Do not let the nail get hot, or it may burn you. Be careful not to touch the nail's point.
3. With the nail removed from the flame, continue to hold it. What do you observe during the minute after it is removed from the flame?
4. Place the nail on a dish or pad, or in water, to cool off.
5. Light a match and let it burn for a few seconds. Gently blow it out. It will release some smoke for a few seconds. While it is smoking, hold the matchstick near the candle flame. What happens to the smoke? Why does this occur?

Conclusion
6. Explain how procedures 1-3 show conduction, and how procedure 5 shows convection. What are the similarities and differences?

■ Section 15-2 Review ■

Write the definitions of the following terms in your own words.

1. **conduction**
2. **insulator**
3. **convection**
4. **radiation**

Answer these questions.

5. Many people think that a refrigerator puts cold into the food placed inside. What is a more accurate way to state what occurs?

6. A beaker contains 400 grams of water at 25° C. How many calories are needed to increase its temperature
   a. to 30° C?
   b. by 30° C?

7. Describe the main method(s) of heat flow in each device.
   a. a clothes dryer
   b. an electric iron for pressing clothes
   c. an electric heater with red-hot wires but no blower fan
   d. an electric heater with red-hot wires and a blower fan

8. Figure 15-14 is a cut-away diagram of a thermos bottle containing hot coffee. Name at least five reasons why the thermos bottle keeps the coffee hot for a long time. How does it prevent the heat from leaking out from the coffee? State your answer in terms of the three methods of heat flow.

Figure 15-14

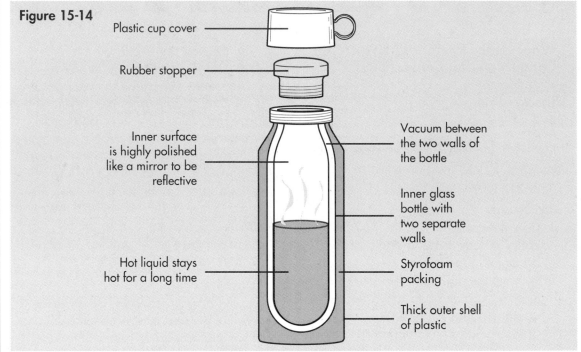

Plastic cup cover

Rubber stopper

Inner surface is highly polished like a mirror to be reflective

Hot liquid stays hot for a long time

Vacuum between the two walls of the bottle

Inner glass bottle with two separate walls

Styrofoam packing

Thick outer shell of plastic

How does the thermos bottle prevent the heat from leaking out from the coffee?

15-3 Special Topics on Heat

■ *Objective*

☐ *Describe some special effects of heat energy and temperature, and their applications.*

The Bimetallic Strip

A **bimetallic strip** is a thin piece of two different metals that have been welded together. When a bimetallic strip is heated, it bends. When it is cooled, it unbends. This property makes it a key part of a thermostat, an electrical device that controls the temperature in rooms, ovens, refrigerators, and car engines. Look at Figure 15-15. When a room gets too cool, a bimetallic spring inside the thermostat begins to unbend and switches on the electricity for the furnace. As the room warms up, the bimetallic spring bends in and switches off the electricity and furnace.

The bimetallic strip bends because the two metals expand by different amounts. One metal expands more than the other metal when they are warmed. Because they are welded together, the only way the metals can expand different lengths is by bending. The metal that expands more is on the outside of the curve that is made.

The Odd Behavior of Water

When most liquids are cooled, they shrink in volume. This is contraction. When the temperature goes down to a liquid's freezing point, the liquid continues to shrink as it changes into a solid. In the solid phase, the particles of the material are squeezed into a smaller volume than in

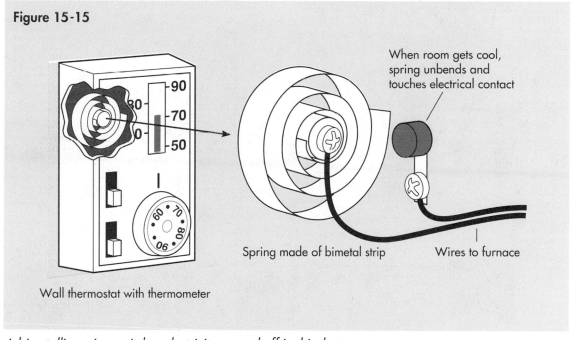

Figure 15-15

When room gets cool, spring unbends and touches electrical contact

Spring made of bimetal strip

Wires to furnace

Wall thermostat with thermometer

A bimetallic spring switches electricity on and off in this thermostat.

the liquid phase. The density of the solid material is greater than that of the unfrozen liquid and, so, the solid sinks.

This is true for all liquids except water. Water expands as it freezes into ice, as shown in Figure 15-16. This means that ice occupies more volume than the liquid it came from. The density of ice is less than the density of water and, therefore, ice floats in water.

This odd behavior of water has some important consequences. If the water in a metal pipe freezes because of cold temperatures, the water will expand and burst the pipe. Water that gets into cracks in rocks, sidewalks, and streets expands during freezing weather. This process causes the gradual erosion of rocks over millions of years. Entire mountains have been slowly cracked apart by water.

Freezing water also causes the potholes and cracks you see in the streets and sidewalks.

Some biologists believe that this unusual property of water permitted life to evolve as it has. During winter, lakes and ponds begin to freeze. If water behaved like other materials, the ice would sink. The water above would continue to freeze until the whole lake turned into solid ice. No fish could survive. Because ice expands, however, it floats on water. The ice layer that forms on top of a lake actually protects the water underneath from the freezing air. Thus, life goes on in the cold water under the ice.

Heat Energy in Joules

All energy and work is measured in joules. For convenience and historical reasons,

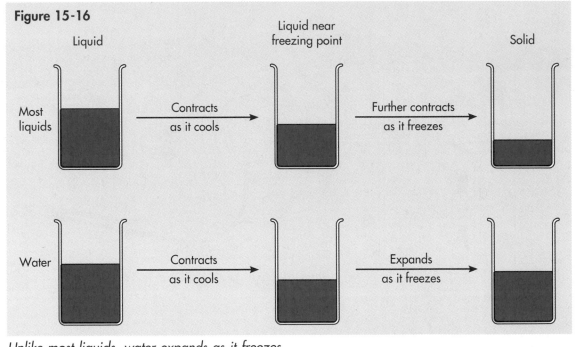

Figure 15-16

Liquid | Liquid near freezing point | Solid

Most liquids — Contracts as it cools → — Further contracts as it freezes →

Water — Contracts as it cools → — Expands as it freezes →

Unlike most liquids, water expands as it freezes.

heat energy is also measured in calories. However, calories can be converted into joules. This is done, for example, when scientists calculate how much heat energy, in calories, will come from an amount of kinetic, potential, or electrical energy, in joules. Experiments show that

1 calorie = 4.2 joules

Consider the question, how much heat energy, in joules, is released by 1000 grams of hot water at 100° C as it cools to 20° C?

Heat = mass of × temperature
water change

Heat = 1000 g × (100° C − 20° C)
Heat = 1000 × 80 = 80,000 cal
Since there are 4.2 J in every 1 cal,
Heat = 80,000 cal × 4.2 J per cal
= 336,000 J

Second Law of Thermodynamics

Recall from Chapter 11 that heat is always produced during any transformation of energy. This is the second law of thermodynamics. For example, in a motor, electrical energy is changed mainly into kinetic energy. This is the purpose of a motor. However, heat within the wires, friction between the parts that turn, and even some air resistance cause some electrical energy to become heat. The motor gets hot.

By means of better design and lubrication, engineers try to reduce the amount of energy that gets wasted as heat. But there is a limit to how much they can improve machinery and devices of technology.

The Greenhouse Effect

The greenhouse effect is discussed in some detail within the Science, Technology, and Society feature that follows this section. Briefly, the greenhouse effect means that air pollution is making the air increasingly warmer. Polluted air absorbs the sun's heat rays and keeps them from going back into space. The result may be that temperatures will increase worldwide. Many agricultural, climate, and international changes would result from such a warming effect.

Infrared Photography

Special film and cameras can take pictures in total darkness. They "see" the infrared waves that emerge from any hot or warm object. The military uses infrared photography to locate targets at night or under cover. Hospitals use infrared equipment to detect warm areas of the body or brain that may indicate disease or some other medical problem. Many people have used infrared experts to locate where heat is being wasted in their homes or apartments. Special equipment can "see" the heat escaping from windows, chimneys, or walls that have no insulation material. Such information can be helpful in reducing energy use and energy bills.

Thermal Pollution

As machines convert energy, they waste heat. When large amounts of waste heat enter the atmosphere or bodies of water, the environment can suffer. For example, electricity is made by electrical power plants. These power plants may use coal, oil, gas, dams, or uranium as their source of energy. The plants produce plenty of heat in the boilers and machinery. This heat must be removed, or the equipment will overheat and stop working. Usually, cold water from a river or a pond is pumped through the equipment to keep it cool. Naturally, the water becomes warm. When warm water is sent back to the river or pond, it may do damage. The fish and plants of the river may not be able to survive in the warmer water. The warming can affect the environment near the power plant. Geese that fly thousands of miles and usually land near a river or pond may now find it unusable. This is one type of thermal pollution, or heat pollution. Look at Figure 15-17.

Air conditioners in large buildings and on buses and trains must get rid of their heat somewhere. That somewhere is the air. This hot air is released into the city and train stations and certainly makes the environment worse for people, especially on a hot summer day.

Cryogenics

Cryogenics (kry-uh-JEN-iks) is the study of the special properties that materials have at very low temperatures. Cryogenics is an important part of science today. Especially important is the freezing of living tissue and organisms so that they can be thawed after a long time and still be healthy. Freezing of blood, human organs, sperm and eggs, and other body parts is part of new procedures made available to hospitals by cryogenic technology.

There is another important side of cryogenics. When electric wires are made extremely cold, they appear to offer no resistance, or friction, to the flow of large amounts of electricity. Heat is no longer

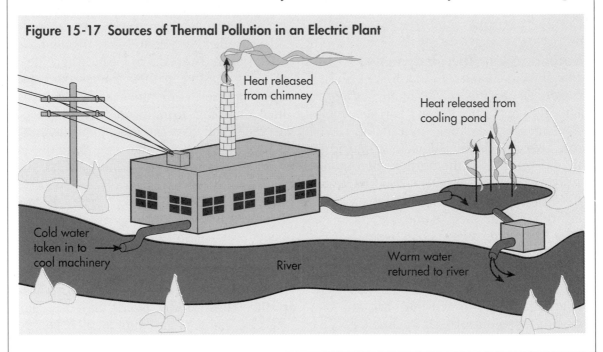

Figure 15-17 Sources of Thermal Pollution in an Electric Plant

Heat released from chimney

Heat released from cooling pond

Cold water taken in to cool machinery

River

Warm water returned to river

made as the electricity flows through the wires. These effects are part of superconductivity, which will have important applications in making extremely strong magnets that can be used instead of wheels on trains and cars of the future.

Absolute Zero

Did you know there is such a thing as the coldest temperature? It is called absolute zero, and it is $-273°C$. This is the temperature at which all particle motions and vibrations stop within a material. The average kinetic energy of particles is zero. This is the "true" zero temperature, and there is no colder temperature. Absolute zero is an important temperature in science. Many mathematical formulas in science are based on the temperature as it is measured from absolute zero, not as measured by the Celsius scale.

There is no such thing as the highest temperature. Particles can always be given greater kinetic energy. Therefore, there is no upper limit to the temperature matter can have. Temperatures within the sun's core are millions of degrees Celsius.

Fans and Air Conditioners

Did you know that it can be useless to keep a fan going in an empty room? Unless a fan is blowing cooler air into a room, a fan is only useful when the air blows over your skin. Its breeze causes more water to evaporate from your skin, where pores contain tiny water droplets. This evaporation removes heat energy from your body. Fans have a cooling effect because of water evaporation.

An air conditioner works differently. It removes heat energy from the air in the room and pumps it outside. The room air is cooled, whether people are in the room or not. A refrigerator works in the same way. It removes heat energy from the food inside and passes it to the back of the refrigerator. With adult supervision, carefully feel the back of a refrigerator or the outside part of an air conditioner. Do you feel the warmth? That is the heat energy removed from inside the refrigerator or room.

■ Section 15-3 Review ■

Write the definitions for the following terms in your own words.

1. **thermostat** 2. **thermal pollution**
3. **cryogenics** 4. **greenhouse effect**
5. **absolute zero**

Answer these questions.

6. What is the purpose of a bimetallic strip within a thermostat?
7. a. Why is it odd that ice cubes float in a glass of water or soda?
   b. What does this fact have to do with the breaking of rocks, roads, and water pipes in winter?
8. What are some sources and effects of thermal pollution?
9. Describe two areas in which cryogenics has practical applications.
10. The finest machines and other technological devices must always "waste" some energy. Explain.

SCIENCE, TECHNOLOGY, & SOCIETY

The Greenhouse Effect

Have you ever visited a greenhouse? It is made of glass, and the air inside is usually warm and moist, just right for growing plants. The sunlight passes through the glass and warms the air and the ground. In nature, most of such heat energy is usually reflected from the ground back into space as infrared waves. Glass, however, does not allow many infrared waves to pass through, so heat is trapped inside the greenhouse. As a result, the air inside the greenhouse is warmer than the air outside.

Certain gases have an effect like the glass of the greenhouse. For example, carbon dioxide (CO_2) and methane (CH_4) are heat-absorbing gases that make the atmosphere behave like a blanket. This warming is called the greenhouse effect. New technologies have put more carbon dioxide into the air than was present naturally. Methane comes from farm animals, swamps, rice paddies, and garbage dumps.

Figure 15-19 shows carbon dioxide trends. The trends are projected to the year 2050.

If the amount of CO_2 in the atmosphere were to double from present levels, some scientists estimate that the average temperature of the earth would rise by 3°C to 8°C. Some growing areas, for example, the central United States, might become much less productive. Energy demand for air conditioning would increase tremendously.

If the earth warms up too much, the ice caps at the North and South Poles will melt, raising the ocean levels by several meters. This, in turn, would flood many of the major coastal cities of the world, affecting possibly one fourth of the world's population.

Follow-up Activity

Gather recent reports on how experts plan to attack the greenhouse problem. Present your information and discuss the issue in class.

Figure 15-18

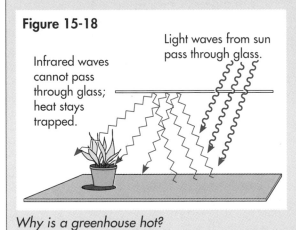

Light waves from sun pass through glass.

Infrared waves cannot pass through glass; heat stays trapped.

Why is a greenhouse hot?

Figure 15-19

Year	CO$_2$ in air, in parts per million
1860	280
1980	400
2050	600

KEEPING TRACK

Section 15-1 Heat and Temperature
1. Heat is a form of energy created by the motion of particles within a material.
2. Objects expand when heated and contract when cooled.

Section 15-2 Heat Flow
1. Heat flows from warmer objects or regions to cooler objects and regions. Insulation reduces the flow of heat.
2. Heat is measured in calories. It equals the mass of water in grams times the change in temperature in degrees Celsius.
3. Heat energy flows by conduction, convection, and radiation.

Section 15-3 Special Topics on Heat
1. Unlike all other materials, water expands when it freezes.
2. Thermal pollution and the greenhouse effect are problems society must handle.
3. Absolute zero is $-273°$ C and is the coldest possible temperature. All particle motion stops at absolute zero.
4. A fan cools by increasing water evaporation from skin. Air conditioners and refrigerators cool by moving heat from the inside to the outside.

BUILDING VOCABULARY
Write the term from the list that best matches each statement.

absolute zero, bimetallic strip, calorie, insulator, radiation

1. measurement of heat energy
2. does not conduct heat well
3. transfer of heat through infrared "heat waves"
4. bends when heated
5. no atomic or molecular movement

Explain the difference between the words in each pair.

6. expansion, contraction
7. conduction, convection
8. thermal pollution, greenhouse effect

SUMMARIZING
Fill in or complete each statement with the best word or phrase.

1. Most objects ___ in length and volume as the temperature increases.
2. The heat energy in an object depends on its temperature, its material, and its ___.
3. One calorie will change the temperature of one gram of water from $26°$ C to ___.
4. Heat energy can only flow from a ___ spot to a ___ spot.
5. The sun's rays heat the ground by ___. The warm air near the ground rises, bringing heat to the atmosphere by the method of ___.
6. A fan or a breeze are cooling because they increase ___ from your skin.

INTERPRETING INFORMATION
Helen and Earl place 100 grams of ice into a beaker. They gently heat the beaker for 12 minutes with a Bunsen burner, stirring

continuously. Every minute they record the temperature of the water in the beaker. Their results are graphed in Figure 15-20.

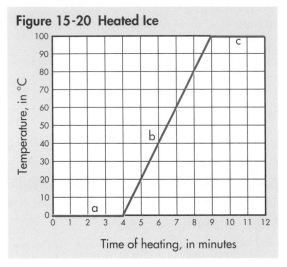

Figure 15-20 Heated Ice

1. What was the temperature of the water after three, six, nine, and 12 minutes?
2. After how many minutes was the water at 50°C?
3. If Helen and Earl kept heating beyond 12 minutes, what would eventually happen?
4. What does this graph tell you about cooking food in boiling water? Once water has started to boil, is there advantage to keeping a large flame under the pot instead of a small flame? Why is vigorous boiling with a large flame a waste of energy?

THINK AND DISCUSS

Use the section numbers in parentheses to help you find each answer. Write your answers in complete sentences.

1. How is it possible for object A to have a higher temperature than object B, yet have less heat energy than object B? (15-1)
2. How many calories are needed to raise the temperature of 1 gram of water at 12°C
   a. to 20°C?
   b. by 20°C? (15-2)
3. A red-hot horseshoe is laid upon the ground. Describe the three ways by which it cools off. (15-2)
4. a. A jar is filled to the top with water and capped tightly. When placed into a freezer, the jar cracks as the water freezes. Why? (15-3)
   b. If this is done with cooking oil, paint remover, or pure alcohol, the jar will not crack as the liquid freezes. Why? (15-3)
5. Imagine a large refrigerator with its door left wide open in a small room. If the refrigerator is turned on for a few hours, will the room be made cooler, warmer, or remain unchanged in temperature? Explain your answer. (15-3)

GOING FURTHER

1. Make a listing of the calories contained in your favorite foods. For example, one teaspoon of sugar contains 16 food-calories. **Important:** One food-calorie is 1,000 times larger than the standard calorie discussed in this chapter. In other words, 16 food-calories equal 16,000 regular calories. Give your answers as regular calories, not as the food-calories listed in books and labels. This means multiplying your answers by 1,000.

2. Discuss one of the topics about heat presented in this chapter. Topics include
   - how thermometers are made
   - how materials conduct heat
   - how absolute zero was discovered and measured

COMPETENCY REVIEW

1. The average kinetic energy of particles within an object determines the object's
   a. motion.
   b. heat energy.
   c. temperature.
   d. mass.
2. What occurs at absolute zero?
   a. Molecules stop vibrating and moving.
   b. Water freezes.
   c. Ice melts.
   d. Objects shrink to zero volume.
3. Figure 15-21 shows telephone lines on a cold winter day and on a hot summer day.

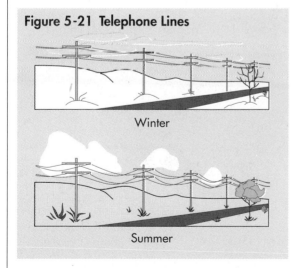

Figure 5-21 Telephone Lines

Winter

Summer

Why do the wires sag more during the summer?
   a. They increase in length because of heat.
   b. They partly melt because of heat.
   c. They support many summer birds.
   d. They carry extra telephone calls and electricity in the summer.
4. Insulators like coats and blankets keep you warm on cold days because they
   a. produce heat energy.
   b. reduce the flow of heat out of your body.
   c. reduce the flow of heat into your body.
   d. reduce the flow of cold into your body.
5. Which is *not* a unit for measuring heat energy?
   a. calories
   b. BTU
   c. joules
   d. degrees Celsius
6. Heat passes through solid rock by
   a. conduction.
   b. convection.
   c. radiation.
   d. friction.
7. Which pair lists an excellent heat conductor first, followed by an excellent heat insulator?
   a. styrofoam, wood
   b. copper, iron
   c. glass, metal
   d. aluminum, wool
8. Which is *not* an application of cryogenics?
   a. freezing body parts
   b. photography in the dark
   c. superconductivity of electricity
   d. perserving living things for a long time

CHAPTER 16

MAGNETIC ENERGY

Magnets attract certain materials and utterly fascinate many people. Perhaps you have played with a magnetic toy or used a compass, which is a magnet, to find directions on a hike in the woods. Magnets can be fun. Magnets, however, are also used for many purposes in your home and in industry. Most likely, while you are reading this, you are not far from a magnet. You may not even be aware of it, however. Your closet door may have a magnetic lock, and your phone has a magnet that enables it to ring and produce sounds. As in telephones, magnets in televisions and radios help turn electrical impulses into sounds. Very small magnets store images and sounds on tapes. Similar magnets store games and programs on computer disks.

Magnets are influenced by electricity. Scientists study magnets to learn about the relation between electricity and force. They examine how magnets affect other objects. Magnets have properties that help explain the behavior of electrical charges and, therefore, the behavior of atoms. Researchers use magnets in very helpful experiments. In this chapter, you will begin to understand what the big attraction is by exploring some properties and uses of magnets.

16-1 Properties of Magnets

■ *Objectives*

☐ *List types of magnets and magnetic materials.*

☐ *Compare and contrast the two different poles of a magnet.*

☐ *Explain why a compass works.*

Magnets were discovered thousands of years ago. People noticed that some rocks behaved in an unusual way. These rocks

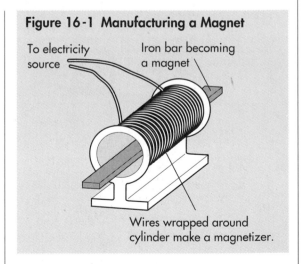

Figure 16-1 Manufacturing a Magnet

To electricity source

Iron bar becoming a magnet

Wires wrapped around cylinder make a magnetizer.

attracted small pieces of iron. These rocks could also attract and repel each other. Rocks with the ability to attract things through an unseen force were said to have **magnetism**.

People also discovered that they could magnetize a needle. They did this by stroking an iron needle in the same direction, many times, against magnetic rocks. If such a needle is allowed to swing freely, by hanging on a string, rotating on a pin, or floating in water, it points toward the north. Although the reasons for this behavior were not clear, this is how early compasses were made. Compasses played a very important role in the exploration of unknown lands and in sea travel. A compass works because of properties found in all magnets.

Types of Magnets

The most common type of magnetic rock found in the earth's crust is magnetite, also known as **lodestone** (LOHD-stohn). Lodestone is an example of a natural magnet. It attracts and repels other magnets.

Today, most magnets are manufactured. This is done by placing a piece of metal in a magnetizer, which is a cylinder wrapped with many turns of wire. Examine Figure 16-1. When electricity flows through the wires for a length of time, the metal becomes magnetic. Magnets can be made of the metals iron, cobalt, and nickel. They can also be made of steel, an iron alloy. The alloy usually used for magnets is alnico, which contains *al*uminum, *ni*ckel, and *co*balt.

Some magnets keep their strength for a long time, even for many years. These are called **permanent magnets**. Other magnets lose their magnetic strength quickly. These are temporary magnets. A common temporary magnet is the **electromagnet**, which is magnetic only when someone supplies it with electricity. Look at the electromagnet in Figure 16-2.

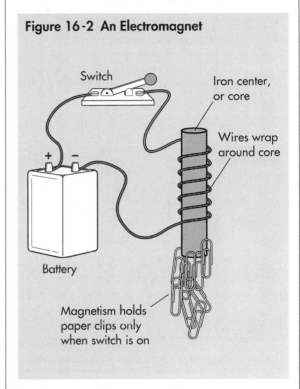

Figure 16-2 An Electromagnet

Switch

Iron center, or core

Wires wrap around core

+ −

Battery

Magnetism holds paper clips only when switch is on

Magnetic Poles

If you hang a plain bar magnet from a string at its center, away from other metal objects, the magnet will line up in a north-south direction. One end always seeks north; and the other end always seeks south. Each end of the magnet contains a different type of magnetism, called north and south. These are the **magnetic poles**, or ends, of a magnet.

The north pole of a magnet will point northward if the magnet is permitted to swing freely. The south pole of a magnet will point southward. If a single magnet is broken into two or more parts, each part forms a new magnet with north and south poles. You can see this in Figure 16-3.

Figure 16-3

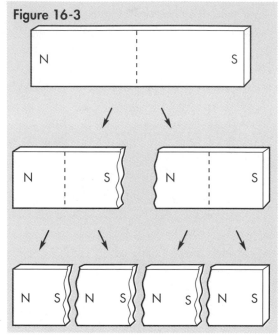

Cutting a magnet only creates smaller magnets.

Attraction and Repulsion

If the north pole of one magnet is brought near the south pole of another magnet, a force pulls them together. This is magnetic attraction. If the south pole of one magnet is brought near the south pole of another magnet, a force pushes them apart. The same occurs if a north pole is brought near another north pole. This is magnetic repulsion. Compare these by studying Figure 16-4.

Figure 16-4 Magnetic Attraction and Repulsion

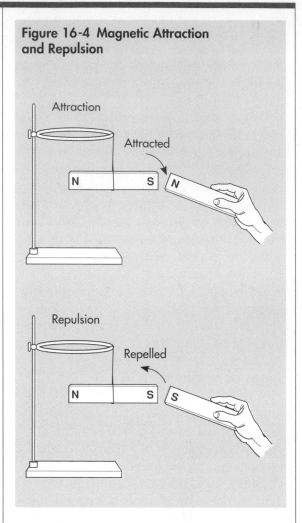

These results are often stated as a law of magnetism. This law says that unlike poles attract, and like poles repel. You may also hear this said as "Opposite poles attract, similar poles repel." Many magnetic toys and games operate on this principle of attraction and repulsion.

Compasses and Earth

A compass works because the earth is like a gigantic magnet, and the needle of the

How Can You Measure the Strength of a Magnet?

Process Skills observing; gathering and recording data; stating conclusions

Materials several bar magnets; a box of paper clips or brads, which are small iron nails; a ruler, string, and tape

Procedures

1. Completely cover a magnet with many paper clips or brads. Carefully lift the magnet away from the pile. Gently shake off any paper clips or brads that are loosely attracted.

2. Sketch what you observe. Follow the style of Figure 16-5, but record your own observations. Where does the magnet appear to have the greatest strength?

Figure 16-6

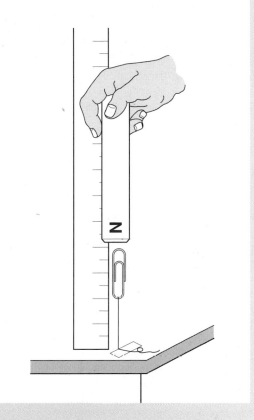

Figure 16-5

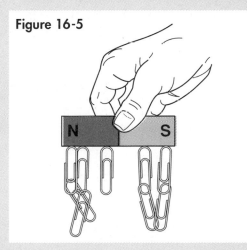

3. Tie a string to a paper clip and attach it to your desk with tape, as shown in Figure 16-6.

4. Lift up the paper clip with a magnet until the string is taut. Keep lifting the magnet until the paper clip drops. Record the distance between the end of the paper clip and the magnet when the drop occurs.

Conclusions

5. Repeat procedures 3 and 4 with other magnets. Which magnet is weakest? Which is strongest? How do you know?

compass is a thin magnet. Look at Figure 16-7. The north pole of the needle points northward because of the earth's magnetism. If a magnet is brought near a compass, the north pole of the compass will point toward the south pole of the magnet. The magnetism of the nearby magnet is far stronger than the magnetism of the earth. Thus, the nearby magnet attracts the compass needle.

Figure 16-7

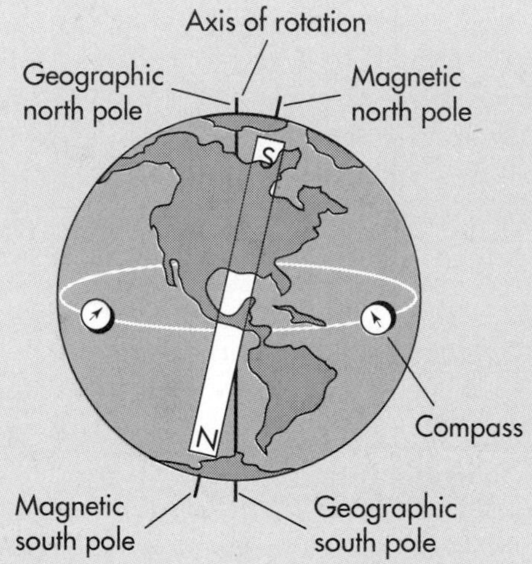

Axis of rotation

Geographic north pole

Magnetic north pole

Compass

Magnetic south pole

Geographic south pole

Earth acts like a large magnet.

Magnetic Fields

The strength and direction of the forces around a magnet can be pictured as an invisible **magnetic field**. One way to see the invisible magnetic field is to shake iron filings on paper or glass that has been placed over a magnet. Examine Figure 16-8.

Figure 16-8 Magnetic Field around a Bar Magnet Shown by Iron Fillings and Compasses

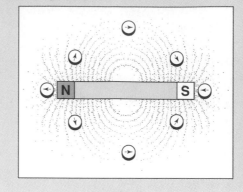

Another way to detect a magnetic field is to place many small compasses around a magnet. The compasses will show the direction of the magnetic field because their north ends are attracted to the south pole and repelled by the north pole of the magnet. The shape of a magnetic field depends on the shape of the magnet and on what poles may be brought near one another. For example, compare the two magnetic fields in Figure 16-9.

Figure 16-9 Magnetic Field Shapes

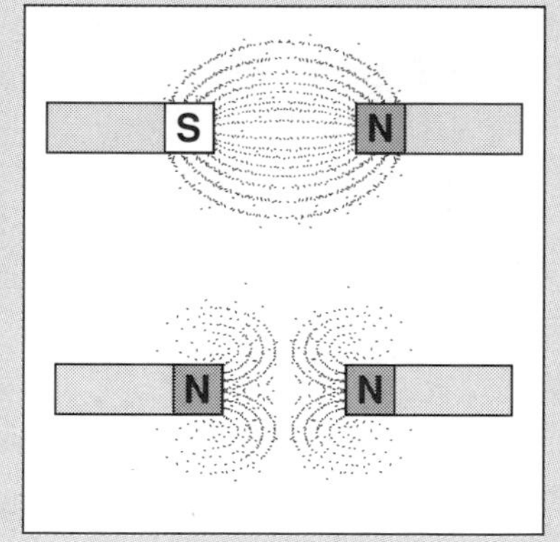

How Can You Observe Magnetic Fields?

Process Skills observing; analyzing; evaluating

Materials bar and horseshoe magnets; iron filings; a piece of glass, clear stiff plastic, or thin cardboard

Procedures

1. Place a bar magnet on your desk. Cover it with the glass, plastic, or cardboard.
2. Shake iron filings around the entire outline of the magnet. Do not pour too much of the iron at any one spot, or you will not see the lines of the magnetic field clearly.
3. Gently tap the glass with your finger so that the filings shake into positions clearly showing the lines of magnetism around the magnet.
4. Sketch the result.
5. Repeat procedures 1-4, using a horseshoe magnet.
6. Repeat this procedure using two magnets, first with their opposite poles near each other, then with their similar poles near each other. Refer to Figure 16-10.

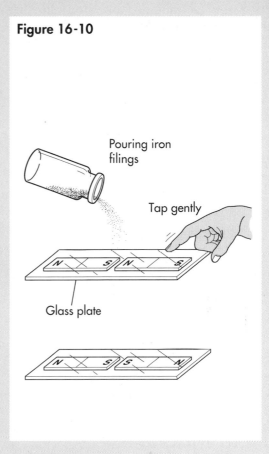

Figure 16-10

Pouring iron filings

Tap gently

Glass plate

Conclusions

7. Compare all of your sketches. What do they suggest about the magnetic fields?

Magnetic Materials

Magnets only attract objects that are made of iron, cobalt, nickel, or alloys of these three elements. Magnets will not pick up or attract objects made of pure aluminum or other metals, nor paper, wood, or plastic. The atoms of iron, cobalt, and nickel have special arrangements of their electrons. That is why magnets only attract objects made of those elements.

■ Section 16-1 Review ■

Write the definitions for the following terms in your own words.

1. **lodestone**
2. **permanent magnet**
3. **electromagnet**
4. **magnetic poles**
5. **magnetic field**

Answer these questions.

6. a. What materials are used to make magnets?
   b. What materials are attracted to magnets?
7. The north pole of a compass needle points toward the north geographic pole of the earth, which is located in northern Canada. What type of magnetism, north or south, must exist at that spot? Explain your answer.

16-2 Understanding and Using Magnets

■ *Objectives*

☐ *Explain the model of magnetism.*

☐ *Illustrate magnetic energy.*

☐ *List some uses of magnets.*

The reason magnets are so fascinating is that they exert a force on an object in any direction, even upward, but without actually touching the object. How magnets work is an advanced topic in physics. Some key ideas are presented here.

A Model for Magnetism

You will recall that a model is a description or picture that helps scientists explain why things behave as they do. Scientists believe that the individual atoms of iron, cobalt, or nickel act as tiny magnets. The electron patterns in these atoms are special. The patterns can explain why they are magnets. One end of each atom is a north pole, and the other end is a south pole. Study Figure 16-11.

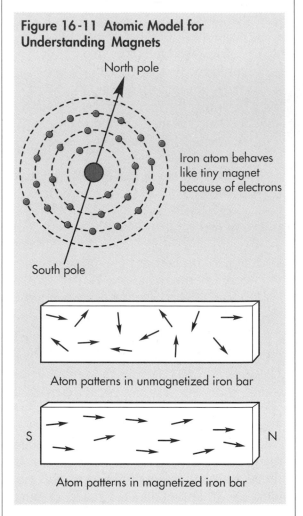

Figure 16-11 Atomic Model for Understanding Magnets

North pole

Iron atom behaves like tiny magnet because of electrons

South pole

Atom patterns in unmagnetized iron bar

S   N

Atom patterns in magnetized iron bar

Normally, the trillions and trillions of atoms in a piece of metal are arranged randomly. This means that their magnetic poles point in all possible directions. The magnetic strength of each atom is cancelled out by the atom's neighbors. However, if some of the atoms have their magnetic poles arranged in one direction, then the whole piece of metal will show magnetic properties. As the number of atoms pointing in one direction increases, the magnetic strength of the metal will increase too.

As you can see, magnetic materials can be thought of as made up of many small magnets. This model explains what happens when a piece of metal is placed inside a magnetizer or when an electromagnet is turned on. The atoms are forced to line up. This model also helps explain other behavior of magnets. For example, when a magnet is cut, you will recall that the pieces are also magnets, with two poles. Like the larger magnet, each piece has its atoms pointing in one direction. You can see this in Figure 16-12.

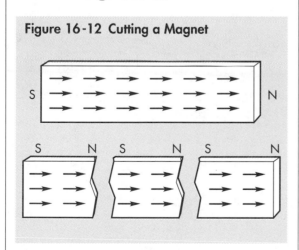

Figure 16-12 Cutting a Magnet

The model helps explain why a magnet loses its magnetic strength when it is beaten or heated. As atoms are knocked around and start pointing in all different directions, the overall strength of the magnet decreases.

Magnetic Energy

A magnet can pull on a piece of iron. It can also pull or push another magnet through a distance. This means a magnet has the ability to do work on another object. This ability explains why magnets are said to have **magnetic energy**. Each magnet begins with a certain amount of magnetic energy. Over time, sometimes years, this energy decreases as the magnet pushes or pulls on other objects and magnets. The magnet loses its strength, unless its energy is recharged by an electromagnet or a magnetizer.

Uses of Magnets

Electromagnets can make bells ring. Electromagnets can make radio or stereo speakers produce sounds of music or speech. They can also cause electric door locks to open, such as door locks that you operate with a buzzer. Junkyard cranes have electromagnets for picking up and moving iron objects and cars. When the crane's electricity is switched off, the magnetism stops and the junk falls.

Permanent magnets are used in cabinet-door catches, in key holders, and on metal bulletin boards. For example, does your family use magnets to attach notes to the refrigerator? The more you investigate, the more you will discover magnets in household devices.

Stating Conclusions

Jermaine and Penh performed some experiments with electromagnets. They tested various factors that might influence the strength of electromagnets. These factors included the material of the core, the amount of electricity used, and the number of turns of wire around the core. They measured the strength of the electromagnet by counting how many brads it could pick up in each experiment. Jermaine and Penh summarized their results in the set of four diagrams you see in Figure 16-13.

1. What hypothesis is being tested by comparing experiments A and B? What conclusion can you state about these two experiments?
2. What hypothesis is being tested by comparing experiments B and C? What conclusion can you state about these two experiments?
3. What hypothesis is being tested by comparing experiments C and D? What conclusion can you state about these two experiments?
4. What can be done to make an electromagnet stronger, according to these experiments?

Figure 16-13

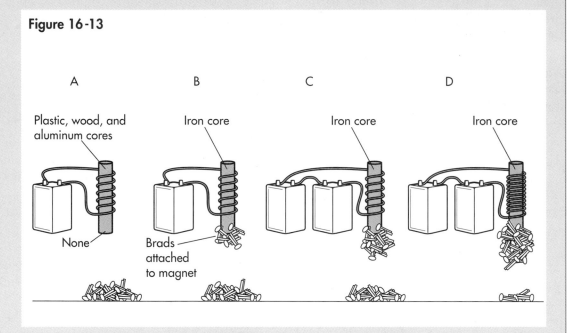

Magnets are in electric generators that produce electricity. Magnets are also needed in electric motors, which turn because of magnetic attraction and repulsion. Televisions contain electromagnets that direct electrons toward the screen, where they form a picture.

Other commonly used items that contain magnets are videotapes, audiotapes, and credit cards. The magnetic materials within these items record pictures, music, and information. Computer tapes and disks also contain magnetic materials. For this reason, it is very unwise to place any recording tape or computer disk near strong magnets. This is because a strong magnet may erase or scramble the information stored in the magnetic materials of the tape or disk.

Hospitals have strong magnets in nuclear magnetic resonance (NMR) machines. Doctors use these machines to examine the inside of organs without surgery. The machines do not operate on nuclear energy. Instead, they work by applying magnetism to the nuclei of atoms in the organs being observed. NMR machines are safer than X-rays, and they give doctors excellent images to use for diagnosis.

Scientists use magnets in many ways, such as investigating the particles found inside atoms. Engineers are planning to build trains without wheels. The trains will lift off and float above the track because of magnetic repulsion. These trains should be very fast and quiet. The ride should be smooth, too, because there will be no friction like the friction between the wheels and tracks in today's trains.

Finally, biologists have discovered that pigeons, bees, and some bacteria respond to magnetism. These organisms can detect the magnetic field of the earth, and they move along the direction of the field as they travel. Medical researchers are beginning to investigate whether people may also have some magnetic sense.

■■■ Section 16-2 Review ■■■

Write the definition for the following term in your own words.

1. **magnetic energy**

Answer these questions.

2. Three identical iron bars are lying on a table. One is not a magnet. The second is a weak magnet. The third is a strong magnet. On your own paper, draw diagrams that show how the atoms are arranged in each iron bar.
3. Why should you avoid heating a permanent magnet? Explain what would happen if you did heat one.
4. Why does a permanent magnet eventually lose its ability to attract and lift iron loads? Why does an electromagnet keep this ability?
5. Name a few common uses and a few unusual uses of magnets.

SCIENCE, TECHNOLOGY, & SOCIETY

Science and Nonsense . . . and Magnetic Jewelry

Figure 16-14

Forget about gold and gems. How would you like to wear magnets in a necklace, in earrings, or in a bracelet? Magnetic jewelry may be the next fad. Inventors of magnetic jewelry claim that magnets can relieve pain, improve your health, and give you a sense of happiness. They will sell you magnetic face masks, pins, and pillows. They claim that the north end of a magnet can stop infections and bleeding, and that the south end of a magnet can energize you.

Is this science or nonsense? That is a difficult question to answer. Sometimes popular beliefs and cures are based on some scientific truth. For example, an old belief is that copper bracelets can reduce arthritis pain. Research now shows that copper ions do partially heal arthritis. Does a copper bracelet actually promote healing? Scientists are not sure.

There are several questions you should always ask before you accept a health-related advertising claim. One way to begin this process is to investigate such advertising claims.

- Was any scientific research done to prove the claim? If so, by whom?
- Were the experiments properly controlled?
- Has the research been published in scientific or medical journals, or just in popular magazines and newspapers?
- Have other scientists repeated the experiments and gotten the same results?

Follow-up Activities

1. Gather examples from television, magazines, and newspapers for amazing advertising claims that scientists should investigate.
2. Health issues are often confusing. For instance, one day, you are advised to avoid red meat; the next day, an article implies that red meat is healthful. Who are some professionals who could discuss health issues with you to help you make decisions?

KEEPING TRACK

Section 16-1 Properties of Magnets

1. Magnets are made of iron, cobalt, and nickel metals and their alloys.
2. Most magnets are manufactured to be permanent magnets or electromagnets.
3. Each end of a magnet is a different magnetic pole.
4. Like poles repel, unlike poles attract.
5. An invisible magnetic force field exists around all magnets. The earth itself has a magnetic field. That explains why compasses point along a north-south direction.

Section 16-2 Understanding and Using Magnets

1. You can think of magnets as being made of many smaller magnets. The atoms in a magnetic material are themselves tiny magnets.
2. Magnetism is a form of energy.
3. Magnets have many ordinary applications and some special uses.

BUILDING VOCABULARY

Write the term from the list that best completes each sentence.

electromagnet, lodestone, magnetic energy, magnetic field, magnetism, permanent, poles

The ability to attract some metals with an unseen force is called ___1___. A natural magnetic rock is known as magnetite or ___2___.

An ___3___ is the most common type of temporary magnet and operates by electricity. The invisible forces of a magnet can be pictured as a ___4___ that surrounds the magnet. Because of its ___5___, a magnet can do work on an iron object. The ends of a magnet are called the north and the south ___6___. Most magnets are manufactured either to be temporary or ___7___.

SUMMARIZING

Write *true* if the statement is true. If the statement is false, change the *italicized* term to make the statement true.

1. Steel magnets used in laboratories or in home cabinets to keep doors closed are *permanent* magnets.
2. The *north pole* of a magnet points toward the north geographic pole of the earth.
3. The north pole of a magnet is *attracted to* the north pole of another magnet.
4. The north pole of a magnet is attracted to the *north pole* of another magnet.
5. *Aluminum*, cobalt, and nickel are attracted by magnets.
6. The *atoms* in a magnet are themselves magnets.
7. As more atoms line up in the same direction within a magnet, the magnetic strength *decreases*.
8. Magnetism is a form of *matter*.
9. If a magnet is cut into three parts, you will get *three* smaller magnets.
10. Magnetic trains will lift away from a magnetic track by magnetic *attraction*.

INTERPRETING INFORMATION

Examine Figure 16-15. Three magnets, A, B, and C, are placed near each other, side by side. A piece of glass is placed over them. Iron filings are sprinkled on the glass, which is gently tapped to reveal the magnetic fields. Compasses are placed near magnet A, as shown in the figure.

1. The arrow-tip end of each compass is its north pole. What does this suggest about the poles of magnet A?
2. Observe the magnetic field between magnets A and B. What can you learn about the poles of magnet B from that field?
3. Observe the magnetic field between magnets B and C. What can you learn about the poles of magnet C from that field?
4. What force will act between A and B, and what force will act between B and C? Will they come together or move apart?

5. At the bottom end of magnet C is another compass. Which way will it point, up, down, left, or right? Explain your answer.

THINK AND DISCUSS

Use the section numbers in parentheses to help you answer each question. Write your answers in complete sentences.

1. What is the difference between the north and south poles of a magnet? How do the poles interact with each other? (16-1)
2. How does a compass behave? What happens if a strong magnet is placed nearby? (16-1)
3. How can you see the field around a magnet? (16-1)
4. What facts are explained by the model of magnetism scientists use today? (16-2)
5. Why is magnetism considered a form of energy? (16-2)
6. It is not wise to lay videotapes on top of televisions. Why? (16-2)

Figure 16-15

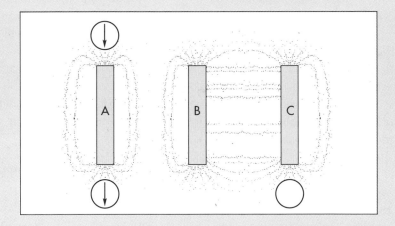

GOING FURTHER

1. Make a report on the history and development of magnets. You may want to focus on the history and importance of magnetic compasses in the growth of nations and shipping.

2. List any devices or locations at home in which there are electromagnets or permanent magnets. Make a combined list of these with your classmates.

3. Take apart an old doorbell or find a diagram of a doorbell in a library book. How does the bell work? What makes the hammer vibrate against the bell many times each second?

4. Using earth science reference books, write a report about the earth's magnetism. What creates its magnetic field? Why is earth's magnetism important, apart from making compasses work? What evidence shows that earth's magnetism varies in strength and even reverses its pole directions?

5. Using physics or technology books, investigate how the strength of a magnet is measured and how it changes with distance.

COMPETENCY REVIEW

1. Which is a temporary magnet?
   a. lodestone
   b. electromagnet
   c. compass needle
   d. magnet used in toys or games

2. What causes a compass to point toward the north?
   a. Earth's magnetism
   b. Earth's gravitation
   c. Earth's spinning
   d. the Moon's gravity

3. Which of the following objects could you pick up with a magnet?
   a. a plastic spoon
   b. a paper cup
   c. a copper penny
   d. a steel paper clip

4. Which element is used in an alloy to make magnets, but is not itself magnetic?
   a. iron   b. gold
   c. aluminum   d. nickel

5. Magnetism is a form of
   a. matter
   b. power
   c. chemical bonding
   d. energy

6. Banks, industry, and government keep computer tapes in underground rooms away from any source of magnetism. This is to
   a. prevent theft or fire damage.
   b. limit access to secret information.
   c. prevent damage to contents on the tape.
   d. recharge the magnetism on the tapes.

7. Pictures and sound are stored on videotapes as
   a. magnetic patterns.
   b. photographic film images.
   c. electric signals.
   d. microscopic bumps and ridges.

8. Magnetite is
   a. a naturally magnetic rock.
   b. a naturally magnetic metal.
   c. a manufactured magnet.
   d. an alloy used for magnets.

CHAPTER 17

ELECTRICAL ENERGY

At 5:17 PM on November 9, 1965, eight Northeastern states and some parts of Canada had a total electrical blackout. No electricity was available in these areas for a period of three to 12 hours. In New York City, 600,000 people were trapped in subway trains. Airplanes had to land at military airports that made their own electricity. You can be sure that the people who lived through the blackout realized how necessary electricity is in our society.

How many uses of electricity can you think of? For a start, consider big hospitals, factories, and office buildings, as well as the telephone, toaster, and TV at home. The more you think about it, the more you will realize just how essential electricity is in modern society. Using electricity is really a fairly recent development. Electricity became available for the first time in New York City in 1882, through the inventions of Thomas Edison. By 1900, only two percent of Americans had electricity in their homes. The United States government made an all-out effort over the following 50 years to electrify homes, industry, cities, and farms. The technology of electricity has brought about great changes not only in the United States, but all over the world.

17-1 Static Electricity

■ *Objectives*

☐ *Give examples of static electricity.*

☐ *Explain the origin of positive and negative charges.*

☐ *Distinguish between induction and conduction.*

If you have ever pulled off a cap and noticed that your hair sticks up, you have experienced **static electricity.** Static electricity results when objects gain or

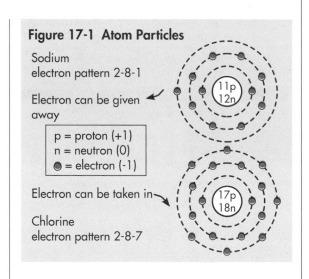

Figure 17-1 Atom Particles

Sodium
electron pattern 2-8-1

Electron can be given away

p = proton (+1)
n = neutron (0)
● = electron (-1)

11p
12n

Electron can be taken in

Chlorine
electron pattern 2-8-7

17p
18n

lose electrons, leaving the objects with a positive or negative charge. The term *static* means to stay in place or to have no motion. Static electricity does not move like regular electricity, but stays in one place.

Atomic Origin of Electric Charge

You will recall that all matter is made of atoms. Atoms themselves are made from three smaller particles. The neutron is a neutral particle in the nucleus, and the proton is a positive particle in the nucleus. The electron is a negative particle that travels around the nucleus. Look for these particles in Figure 17-1.

In a neutral atom, the number of protons equals the number of electrons. The positive and negative charges balance each other, so the total charge is zero. Look at the sodium atom and chlorine atom in Figure 17-1. If the sodium atom lost one electron, it would become an ion with a charge of +1. If chlorine gained an electron, it would become an ion with a charge of -1. Examine the table in Figure 17-2.

Placing a Charge in Objects

The concept of electrical charge applies to objects as well as atoms. A neutral object contains an equal number of protons and electrons throughout all its atoms. If electrons are transferred from one object to another, each object becomes electrically charged. The object that loses the electrons gets a positive charge, while the object that gains the electrons gets a negative charge. You can transfer electrons between objects by rubbing them together. This is the origin of static electricity. Very large numbers of electrons must be transferred to make a static charge that you can detect.

Laboratories use three devices to produce large amounts of static electricity. The devices are known as an electrophorus, a Wimshurst machine, and a Van de Graaff generator, and they, too, work by rubbing. Rubber is rubbed against some other material, sometimes using a motor, so that the rubber gains electrons. These electrons are picked up by the metal parts of the devices and used for experiments on static electricity.

Figure 17-2 Electrical Charges	
In the neutral chlorine atom	**When negatively charged**
17 protons = +17 charge	17 protons = +17 charge
17 electrons = −17 charge	18 electrons = −18 charge
Total charge = 0	Total charge = −1

Figure 17-3 shows some ways to create static charges. Study part a of the figure. When you rub a rubber rod with fur, electrons leave the fur and enter the rubber rod. The rod becomes negatively charged, and the fur becomes positively charged.

Now look at part b of the figure. When you rub a glass rod with silk, electrons leave the glass rod and enter the silk. The rod becomes positively charged, and the silk becomes negatively charged.

Protection against Sparks

Have you ever noticed metal chains or belts dragging from gasoline trucks? As trucks and cars travel, static electricity can build up on their metal bodies. It arises from the rubber wheels rubbing against the ground, and from the glass windows rubbing against the air. The static electricity can jump through the air to the ground, creating a spark, and a spark can set off a fire or explosion near gasoline. The metal chain prevents this danger by allowing the static electricity to drain off to the ground.

You may also notice that at toll booths there is a thin metal whip attached to the ground. As a car enters the toll booth, the metal whip touches it and drains off any built-up static electricity. This prevents the discomfort of sparks as the driver hands coins to the toll collector.

Types of Charge

There are positive and negative charges. Look at the experiments in Figure 17-4. In part a, the rubber rods were all rubbed with fur, and the glass rods were rubbed

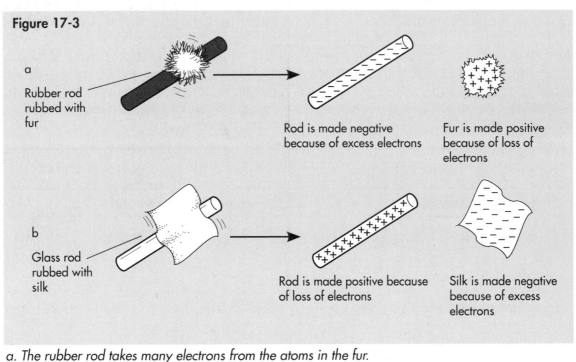

Figure 17-3

a
Rubber rod rubbed with fur

Rod is made negative because of excess electrons

Fur is made positive because of loss of electrons

b
Glass rod rubbed with silk

Rod is made positive because of loss of electrons

Silk is made negative because of excess electrons

a. The rubber rod takes many electrons from the atoms in the fur.
b. The silk takes many electrons from the atoms in the glass rod.

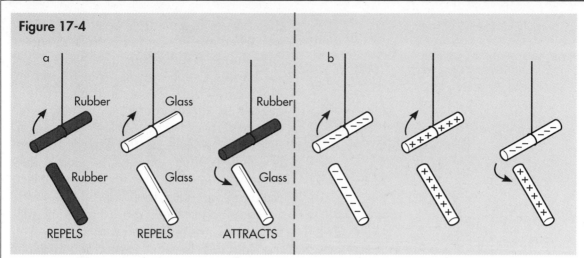

Figure 17-4

a

Rubber Glass Rubber

Rubber Glass Glass

REPELS   REPELS   ATTRACTS

b

a. Charged objects can attract or repel each other.
b. Similar charges repel, but opposites attract.

with silk. This gives each rod a static electric charge. When the charged rubber rods are brought close together, they repel each other. The same thing occurs when the charged glass rods are brought close together. A charged rubber rod and a charged glass rod, however, will attract each other.

Scientists noticed that this behavior is very similar to that of magnets, which have two different poles. They concluded that there must also be two different types of electric charge, which they named positive and negative. It is possible to explain the results above using the rule that "Objects with like charges repel; objects with unlike charges attract." Part b of the figure illustrates the rule.

Electrical Force and Energy
Charged objects attract or repel each other by exerting force. Scientists have carefully measured the strength of this force. They

found that the strength depends on two things. It depends on the amount of the charge. The strength of the force increases as the amount of charge on either object increases. The strength of the force also depends on distance. The strength decreases as the distance between the objects increases.

Objects with an electric charge can do work by pulling or pushing other objects a distance. They contain electrical energy. Electrical energy is created by the forces that act between negative and positive charges.

Induction and Conduction
When a charged object is brought near uncharged objects, there is a small force of attraction. The attraction occurs because the electrons in the uncharged object are rearranged because of the nearness of the charged object. This effect is called **electrical induction**, which means rearranging

Figure 17-5 Electrical Induction

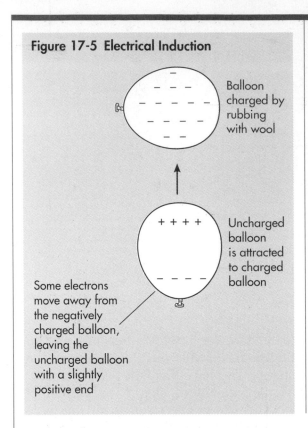

Balloon charged by rubbing with wool

Uncharged balloon is attracted to charged balloon

Some electrons move away from the negatively charged balloon, leaving the uncharged balloon with a slightly positive end

the charges in a neutral object by bringing a charged object near it. Figure 17-5 shows an example of induction.

Charges can be given to an object by **electrical conduction**, or transferring a charge by direct contact. Conduction occurs when a charged object touches an uncharged object. If the charged object is negative, then some of its excess electrons will flow into the uncharged object, making it negative, too.

Study Figure 17-6. You can demonstrate induction and conduction with a pith ball on a string. Pith is spongy wood found under the bark of some trees. It is similar to cork. You can also use small balls made of cork or aluminum foil.

In Step 1, the neutral pith ball moves toward the negative rubber rod. It is attracted to the rod by induction. The rod rearranges the charges inside the pith ball because some electrons in the ball are

Figure 17-6 Pith Ball and Charged Rod

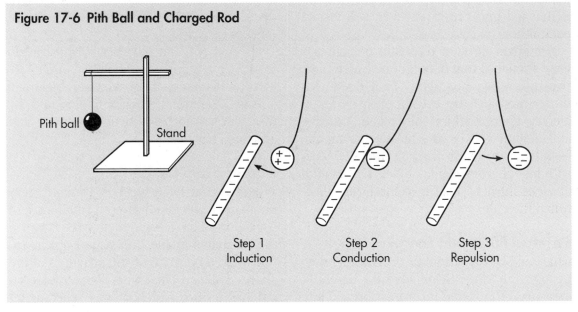

Pith ball

Stand

Step 1
Induction

Step 2
Conduction

Step 3
Repulsion

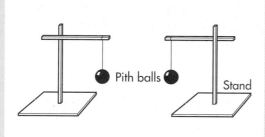

How Can You Demonstrate Electrical Charges?

Process Skills observing; stating conclusions

Materials two pith balls suspended from a string, as shown in Figure 17-7; you can also use aluminum foil squeezed into balls; rubber rod and glass rod; fur or wool; and silk or plastic wrap

Figure 17-7

Pith balls

Stand

Note Static electricity drains into the air quickly on humid or rainy days. These experiments are best done on dry days.

Procedures

1. Rub the rubber rod with wool or fur. Bring it slowly toward one pith ball. What happens?
2. Let the pith ball touch the rod and wait until it breaks away. Now try to bring the rubber rod to the pith ball. What happens?
3. Charge the other pith ball with the rubber rod in the same way.
4. Slide the two similarly charged pith balls near each other. Try to bring them together. What happens?
5. Touch each ball between two fingers to remove the static electricity.
6. Repeat steps 1-5 using the glass rod rubbed with silk or with plastic wrap.
7. Charge one pith ball negatively with the rubber rod, and charge the other positively with the glass rod. Bring them together. What happens?

Conclusions

8. Why was the neutral pith ball first attracted to the rod, then repelled by it?
9. Describe the relationship you observed between like charges, unlike charges, and a charged and uncharged object.

repelled by the negative rod and move to one side. As one side of the ball becomes slightly negative, the other side becomes slightly positive. This positive side is attracted to the negative rod.

In Step 2, the pith ball touches the negative rod. Some electrons rush into the ball, making it negative, too. This is conduction.

In Step 3, the ball flies away from the rod because it is now negative. It is repelled by the negative rod.

Electroscopes

Scientists use an **electroscope** to detect static electricity. This device consists of two thin metal leaves that fly apart when

How Can You Make an Electrophorus?

Process Skills *observing; analyzing*

Materials flat metal dish or large metal jar lid; a piece of plastic or a candle; glue; a vinyl phonograph record; a piece of glass, plastic, or rubber; wool

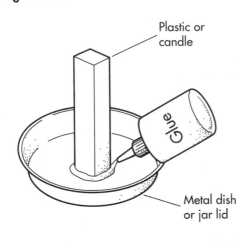

Figure 17-8

Plastic or candle

Glue

Metal dish or jar lid

Procedures

1. Glue the piece of plastic to the dish or lid, or melt the candle so that it sticks to the metal dish, as shown in Figure 17-8.
2. Place the record on a piece of glass, plastic, or rubber.
3. Rub the record with wool for 30 seconds.
4. Hold the metal dish by the plastic or candle handle you made for it, and set the dish on the record. Touch the dish with a finger of your other hand and then quickly remove your finger.
5. Quickly lift the metal plate from the record by its handle. Bring it near to a radiator or doorknob. What happens?

Conclusions

6. What occurs as you rub the record with wool?
7. What do you think happens when you touch the metal dish to the record? Your teacher may tell you some additional things that happen.
8. Why is the handle for the metal dish made of plastic or a candle?

they are given a positive or negative charge. They repel each other because they have the same charge. In Figure 17-9, the electroscope has been charged by direct contact, or conduction.

Electroscopes can also demonstrate charging by induction, as you can see by studying Figure 17-10. The rod is not touching the electroscope, yet the metal leaves fly apart. This is because the nearby rod rearranges the charges within the electroscope. This is induction.

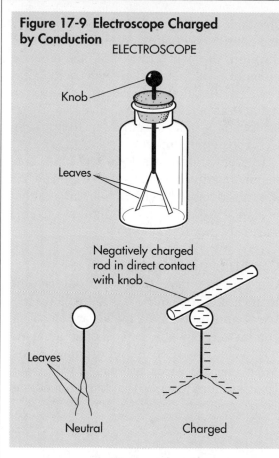

Figure 17-9 Electroscope Charged by Conduction

ELECTROSCOPE

Knob

Leaves

Negatively charged rod in direct contact with knob

Leaves

Neutral

Charged

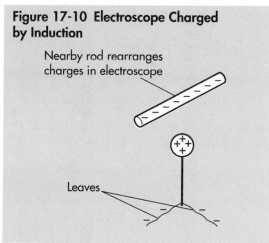

Figure 17-10 Electroscope Charged by Induction

Nearby rod rearranges charges in electroscope

Leaves

▬ Section 17-1 Review ▬

Write the definitions for the following terms in your own words.

1. **static electricity**
2. **electrical induction**
3. **electrical conduction**
4. **electroscope**

Answer these questions.

5. List examples of static electricity. Your list should include at least two examples not given in this section.
6. Why do objects become electrically charged when they are rubbed? What makes one object become positive and another become negative?
7. Do a sketch showing a positive rod brought near a neutral pith ball on a string. Describe what happens, in three separate steps. Explain what is occurring in each step.
8. How does an electroscope work?
9. Two pith balls are given charges by touching them with charged rubber and glass rods. One pith ball has a negative charge and the other ball has a positive charge. What happens when the balls are placed a few inches apart? Why?
10. Two other pith balls are given charges by touching them with charged rubber and glass rods. Both balls have negative charges. What happens when the balls are placed a few inches apart? Why?

17-2 Current Electricity

■ *Objectives*

☐ *Define electric current.*

☐ *Give examples of conductors and insulators.*

☐ *Calculate current, using Ohm's law.*

Have you ever had an electric shock? This can happen if you touch a wire without insulation, or if a wire is loose inside an electrical device. In the United States, standard electricity is 120 volts, a number that measures its strength. A shock from 120-volt electricity can give you a painful jolt. In European countries, electricity is 240 volts, which gives a very painful shock. Wherever you are, be careful with electricity!

Current

Electrons can flow through solid metal wires. Examine Figure 17-11. When the electrons are within the rubber rod, they are static electricity When they move through the wire they are an **electric current**, which is called electricity. Current is the flow of electrons through metal wires and other objects. When scientists speak of current, they mean electrons on the move!

Current only flows between spots containing unequal amounts of charge. You can see this in Figure 17-12. In this way, current is similar to heat, which only flows from a warm spot to a cooler spot.

Conductors and Insulators

Current can move easily through some materials. These materials are called **electrical conductors**. All metals are electrical conductors. Silver is the best conductor, but it is too expensive to use for ordinary wires. Most electric wires are made of copper. Heavier electric wires that hang outside on

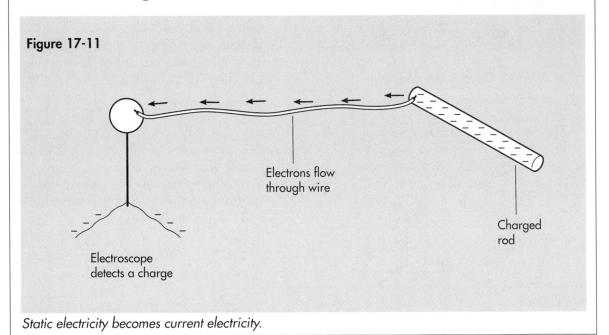

Figure 17-11

Electrons flow through wire

Charged rod

Electroscope detects a charge

Static electricity becomes current electricity.

Figure 17-12 Current Flow

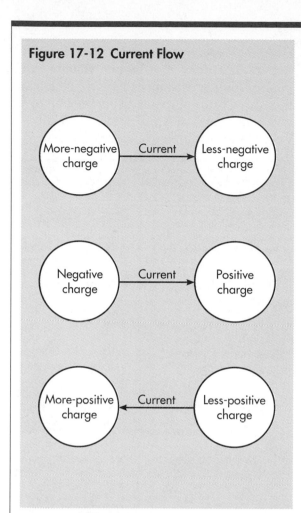

poles are made of aluminum, which is a fine conductor and is stronger than copper.

Conductors can be solids, liquids, or gases. In neon lights used for electric signs, and in fluorescent light bulbs, there are gases that conduct electricity. As the current flows through these gases, the gases glow in different colors.

Many materials do not allow electrons to flow through them easily. Such materials are poor conductors and are called **electrical insulators**. Common insulators include rubber, glass, plastic, wood, and air. Electrical wires are protected and insulated with a rubber coating, which is sometimes even wrapped with cloth fiber. These coverings protect you from getting an electric shock. **Caution** Be extra careful of electric wires that have no insulation on them. The current in them can flow into you, causing discomfort, injury, or death.

Circuits and Resistance

The path the electricity takes as it flows through wires and other devices is known as the **electric circuit** (SUR-kut). Figure 17-13 shows a large circuit. Notice that every

Figure 17-13

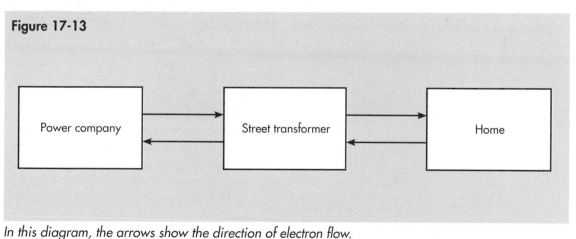

In this diagram, the arrows show the direction of electron flow.

device needs two wires. One wire delivers the current from the power company, and the other returns it after its energy is used.

Each electrical device resists the flow of the current. An object's **resistance** is a measure of how difficult it is to move electrons through the device. Electrical devices require that energy be used to "push" electrons through them. They operate by taking energy from the moving current. The simplest possible circuit consists of a source of electric current, wires, and an electrical device. A switch may be included so that the circuit can either be closed or opened. Look for these elements in Figure 17-14.

Characteristics of Electrical Current

You can compare current flowing through a wire to water flowing through pipes. Electrons are like the water molecules. The parts of an electric circuit are like the pipes, sinks, and drains through which the water flows. There are other similarities, which are discussed below.

Voltage In an electric circuit, **voltage** is similar to pressure in a water pipe. Water at a higher pressure can push through pipes better. Electricity at a higher voltage can push through resistances better. Voltage is measured by a voltmeter, using the unit of a volt. For comparison, Figure 17-15 gives voltages for examples of electricity.

Figure 17-15 Electric Voltage

Example	Typical voltage
Human nerve cells	0.07 volts
Flashlight battery	1.5 volts
Car battery	12 volts
Standard United States voltage	120 volts
Standard European voltage	240 volts
Subway and electric trains	600 volts
Lightning	100,000 to 1,000,000 volts

Amount of Current The amount of current flowing in a circuit is similar to the amount of water flowing through a pipe. Some pipes deliver only a trickle of water, but others, like those pipes supplying a fire hydrant, deliver a huge amount of water. Amount of current is measured by an instrument called an ammeter. The unit for measuring current flow is the **ampere** (AM-pir), which is usually abbreviated as amp. A circuit with

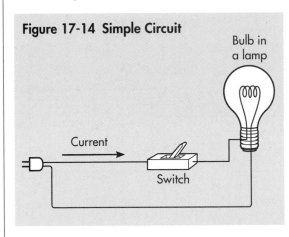

Figure 17-14 Simple Circuit

Bulb in a lamp

Current

Switch

Estimating

Like other instruments, ammeters and voltmeters come with a variety of scales. To read a meter's scale correctly, use the following four steps.

1. Study the lines on the scales in Figure 17-16. You must determine what values the lines stand for o the voltmeter.

Figure 17-16

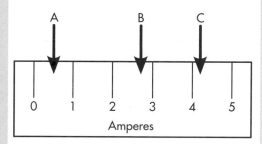

AMMETER

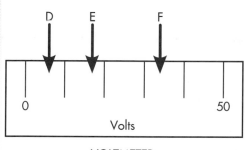

VOLTMETER

2. Look for the pointer. If it is located between two markings on the scale, note the value of the lower marking, the line the pointer moved past.
3. Look at the space around the pointer. What is the size of this space, or interval? To find out, subtract the value of the lower marking from the value of the higher marking.
4. Estimate, in tenths, how far into the interval the pointer has moved. Multiply this fraction by the size of the interval you got in step 3. This will give you the amount to add to the lower marking to obtain the correct scale reading.

The four steps listed above are applied to arrow B in the ammeter in Figure 17-16 as follows:

1. Lines stand for 0, 1, 2, 3, 4, and 5 amperes.
2. Pointer is between 2 and 3 amperes. The lower marking is 2 amperes.
3. Interval size = $3 - 2 = 1$ ampere
4. Pointer is about 7/10 of the way into the interval.
   Additional amount = 7/10 × interval size
   Additional amount = 7/10 × 1 ampere
   Additional amount = 0.7 amperes
   Scale reading = $2 + 0.7 = 2.7$ amperes for arrow B
5. Use Steps 1-4 to find readings for arrows A, C, D, E, and F.

more amperes has more electrons flowing in one second than a circuit with fewer amperes. For comparison, Figure 17-17 shows the current in some devices.

Figure 17-17 Current Flow	
Example	Typical amount of current in U.S.
A 60-watt light bulb	0.5 amperes
A 150-watt light bulb	1.3 amperes
Toaster	9 amperes
Large TV	10 amperes
Window air conditioner	15 amperes
Safe limit on most electric lines at home	15–25 amperes

Amount of Resistance Resistance is a measure of how difficult it is to move electrons through an entire circuit, or through a device within a circuit. The unit for measuring resistance is the **ohm**. A device with a high number of ohms does not let electricity pass through it easily. A device with a lower number of ohms lets electricity pass more easily.

Ohm's Law

The voltage, amount of current, and amount of resistance are related. The formula below shows the relationship. It is sometimes called Ohm's law

$$\text{Amount of current} = \frac{\text{voltage}}{\text{resistance}}$$

This formula requires that the current be in amperes, the voltage be in volts, and the resistance be in ohms. Try this example. How much current will pass through a 30-ohm electric motor in the United States?

Voltage = 120 volts (standard United States voltage)
Resistance = 30 ohms

$$\text{Amount of current} = \frac{120 \text{ volts}}{30 \text{ ohms}}$$

Amount of current = 4 amperes

Here is another example. A 12-volt battery is attached to a 1/2-ohm spotlight. What current flows through this circuit?

Voltage = 12 volts
Resistance = 1/2 ohm or 0.5 ohm

$$\text{Amount of current} = \frac{12 \text{ volts}}{0.5 \text{ ohm}}$$

Amount of current = 24 amperes

■■■ Section 17-2 Review ■■■

Write the definitions for the following terms in your own words.

1. **electric current**
2. **electric circuit**
3. **voltage**
4. **ampere**
5. **ohm**

Answer these questions.

6. What is current? What other names are used throughout this section for current?
7. Figure 17-18 shows an electrician fixing a fallen wire. Explain why each object is made of the material shown.

Figure 17-18

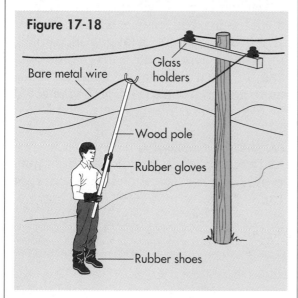

Bare metal wire

Glass holders

Wood pole

Rubber gloves

Rubber shoes

8. What characteristics does electricity have that are similar to water flowing through pipes in a house? Describe each characteristic and its units of measurement.
9. a. For what is a voltmeter used?
   b. For what is an ammeter used?
10. a. What is the current flowing through a 10-ohm resistance if the voltage is 50 volts?
    b. How much current flows through a 20-ohm motor in a subway or electric train? (See Figure 17-15.)

17-3 Characteristics of Circuits

■ *Objectives*
☐ *Compare and contrast series and parallel circuits.*
☐ *Distinguish between DC and AC electricity.*

Series Circuits

Light bulbs can be arranged in a **series circuit**. In a series circuit, the current can only flow in one path, passing through each resistance, one at a time. See the left side of Figure 17-19 on page 330. A series circuit has some unique characteristics. Any break in a series circuit stops the flow of current in the entire circuit. Therefore, if one bulb goes out, all the bulbs will go out. Furthermore, the current is the same in all parts of a series circuit. If two amperes are flowing through bulb A, two amperes will flow through bulbs B and C, through the wires, and through the electric source.

Many familiar devices, such as a string of holiday lights, are connected to each other in series circuits. If one light stops working, none of the others will work. For the string to work again, you must locate the bad bulb and replace it.

Parallel Circuits

Bulbs can also be arranged in a **parallel circuit**. In a parallel circuit, the current can flow in two or more paths. See the right side of Figure 17-19 on page 330. A parallel circuit has some unique characteristics. A break in one branch of a parallel circuit also does not stop the flow of current in other branches. If one bulb goes out, the others continue to work. Furthermore, the current can be different in different branches of a parallel circuit. The voltage is the same in each branch, however.

Most electrical appliances at home are connected in parallel circuits. In fact, you can think of each plug and wire as a branch of your home's large parallel circuit. This is why each plug can usually operate alone. If one is off, the others can be on. The voltage in each line will also be the same, 120 volts, although the amount of current will differ.

Arranging Batteries in a Circuit

Batteries store chemical energy and release electrical energy. Each battery gives out a limited current and voltage. For example, most flashlight batteries produce 1.5 volts. Batteries can be arranged either in series or in parallel, as shown in Figure 17-20.

When batteries are arranged in series, their voltages add up so that the combined voltage is larger than the voltage of each individual battery. This arrangement is used in radios and flashlights. When batteries are arranged in parallel, the total voltage is the same as the voltage of each individual battery. Many more amperes of current can be delivered by batteries in parallel circuits.

AC and DC Current

Electrons can flow in either a **direct current (DC)** or an **alternating current (AC).** Batteries produce direct current.

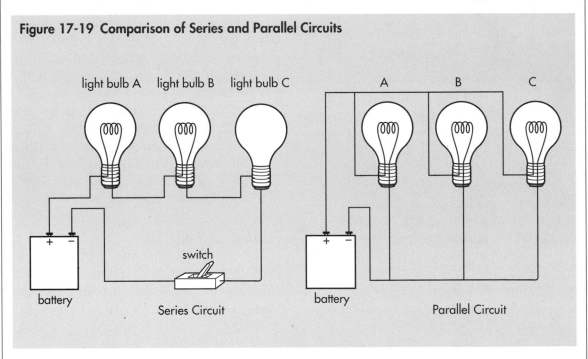

Figure 17-19 Comparison of Series and Parallel Circuits

light bulb A   light bulb B   light bulb C

battery

switch

Series Circuit

A   B   C

battery

Parallel Circuit

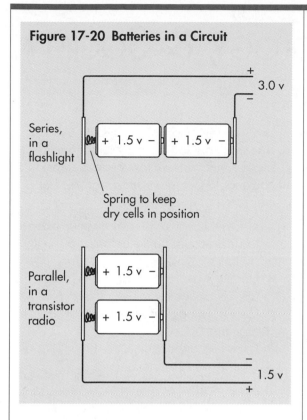

Figure 17-20 Batteries in a Circuit

In circuits carrying direct current, the electrons flow in one direction, from the negative side of the battery to the positive side of the battery. In circuits carrying alternating current, the electrons continuously change their direction of flow. In the United States, this back-and-forth flow can occur 60 times every second. This is referred to as "60 cycle" AC electricity.

Both direct current and alternating current electricity have their own advantages. Direct current is found mostly in electronic circuits like those inside radios and TVs. Alternating current is used for home and industry.

The voltage of alternating current electricity can be changed by a **transformer**.

This device increases or decreases the voltage of alternating current. The voltage of direct current cannot be changed by a transformer. Alternating current is therefore used for sending electricity over long distances. Alternating current electricity is sent as 10,000 volts over long-distance power lines. It is then transformed into 120-volt electricity for home use.

▬ Section 17-3 Review ▬

Write the definitions for the following terms in your own words.

1. **series circuit**
2. **parallel circuit**
3. **direct current (DC)**
4. **alternating current (AC)**
5. **transformer**

Answer these questions.

6. Compare series circuits and parallel circuits in regard to
   a. how many paths the current can take.
   b. the current flowing in each resistance.
   c. what happens if one resistance breaks.
7. a. What is the voltage of four 6-volt batteries that are arranged in series?
   b. What is the voltage of four 6-volt batteries that are arranged in parallel?
8. What is an advantage of using AC electricity instead of DC electricity?
9. Many products in Europe are made for 240-volt, 50-cycle alternating-current electricity. Why will these products fail to work in the United States?

Science, Technology, & Society

Systems

Below is a flow chart of a heating system that uses a thermostat to control a room's temperature. A **system** is a group of devices and actions that work together to produce results. A system can be shown by means of a flow chart like the one in Figure 17-21.

All systems have certain common features, or parts.

1. The *input* is the starting material or signals.
2. The *comparison/control* part is where decisions take place.
3. *Processing* is the part where action is taken.
4. *Output* is the result of the action.
5. *Feedback* is the information or signal from the output that helps in making further decisions and taking additional actions.

Follow-up Activities

1. What is a system? Define the five common features of all systems.

2. Draw a flow chart of the system used when calling someone by telephone. Label each part. Below are the steps you should put into your diagram, but they are not in proper order!

   - Electrical signal goes from your phone to desired phone by automatic switches.
   - Decide whether to dial, wait, hang up, speak, or redial.
   - Hear ringing, busy signal, answer machine, or silence.
   - Listen for dial tone.
   - Dial or push buttons.

3. The test marks you get in class are feedback that tells you how you are progressing. List six other examples of feedback, for both technological and non-technological systems.

4. Draw a flow chart that illustrates some of the actions involved in driving a car.

Figure 17-21 Block Diagram for a Thermostat

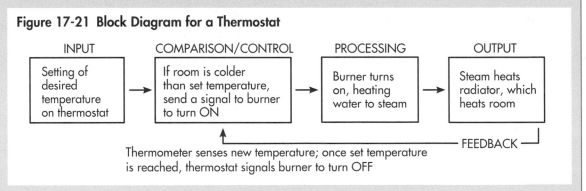

INPUT — Setting of desired temperature on thermostat

COMPARISON/CONTROL — If room is colder than set temperature, send a signal to burner to turn ON

PROCESSING — Burner turns on, heating water to steam

OUTPUT — Steam heats radiator, which heats room

FEEDBACK — Thermometer senses new temperature; once set temperature is reached, thermostat signals burner to turn OFF

CHAPTER REVIEW

Section 17-1 Static Electricity

1. Objects with a negative charge have an excess of electrons, and objects with a positive charge have electrons missing. Like charges repel, and unlike charges attract.
2. The force between charged objects increases if the amount of charge increases, or if the distance decreases.
3. Objects can be electrically charged by induction or conduction.

Section 17-2 Current Electricity

1. Current is a flow of electrons and carries electrical energy from an area of higher charge to an area of lower charge.
2. Electrical energy flows through conductors. Most metals are good conductors.
3. Some substances, such as rubber and glass, are poor electrical conductors and are called electrical insulators.
4. A simple electric circuit consists of a source of electrical energy, an electrical device such as a light bulb, and wires between them.
5. A circuit can be described by its voltage, current, and resistance. These are related through Ohm's law.

Section 17-3 Characteristics of Circuits

1. In a series circuit, there is one conducting path and one amount of current.
2. In a parallel circuit there are two or more conducting paths, each with its different current.

3. In a direct current, the electrons flow in one direction, but in an alternating current electrons can reverse direction. Alternating current electricity can be transformed into different voltages.

BUILDING VOCABULARY

Write the term from the list that best matches each statement.

circuit, electric current, electroscope, ohm, transformer

1. detects static electricity
2. a flow of electrons
3. path that electrons flow through
4. unit for measuring electrical resistance
5. changes voltage of alternating current

Explain the difference between the terms in each pair.

6. static electricity, current electricity
7. conduction, induction
8. electrical conductor, electrical insulator
9. voltage, resistance
10. series circuit, parallel circuit
11. direct current (DC), alternating current (AC)

SUMMARIZING

Write the missing word or phrase for each sentence.

1. When rubber is rubbed with fur, electrons flow into the ___ which becomes ___ charged.

2. The force between charged objects will decrease if either charge ___ or if the distance between them ___.
3. In a charged electroscope, the metal leaves or vanes will ___.
4. Rubber and glass are poor electrical ___ and good electrical ___.
5. A current of ___ amperes will flow through a TV that is rated at 12 ohms when plugged into 120 volts.

INTERPRETING INFORMATION

Figure 17-22 shows a circuit that includes an ammeter and voltmeter. Study the figure, then answer these questions.
1. Why is this called a circuit?
2. What is the voltage across the bell?
3. a. The resistance of the bell is 15 ohms. Calculate the current that will flow through the bell, using Ohm's law.
   b. Is your answer confirmed by information in the diagram? Where?

4. Are the following devices shown in series or parallel?
   a. the batteries
   b. the voltmeter
   c. the ammeter
5. What is the voltage of a single battery in Figure 17-22?

THINK AND DISCUSS

Use the section numbers in parentheses to help you find each answer. Write your answers in complete sentences.

1. What occurs when a negative object is brought near
   a. another negative object?
   b. a positive object?
   c. a neutral object? (17-1)
2. Distinguish between volts, amperes, and ohms in describing an electrical circuit. Explain each measurement by comparing

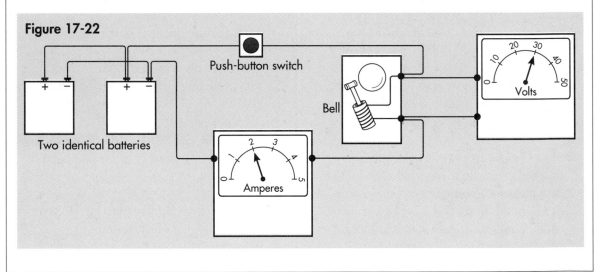

Figure 17-22

Push-button switch

Bell

Volts

Two identical batteries

Amperes

the circuit to water moving through a series of pipes and plumbing devices. (17-2)

3. Describe the characteristics of electricity made for home use in the United States.
   a. amount of voltage
   b. type of current
   c. cycles per second (17-3)
4. What is the difference between direct current and alternating current electricity? What is one big advantage of AC? (17-3)
5. What are some similarities between
   a. electric charges and magnetic poles?
   b. electric current and water flowing in pipes?
   c. electric conduction and heat conduction? (17-1, 17-2)

GOING FURTHER

Make a report on some topic related to this chapter, such as the history of electrification in the United States, production of electricity in the United States, or features of Ohm's law.

COMPETENCY REVIEW

1. What causes a neutral object to gain a positive charge of static electricity?
   a. gain of electrons
   b. loss of electrons
   c. gain of protons
   d. loss of protons
2. Some vehicles have a chain hanging underneath, touching the road. This
   a. forms an electric circuit.
   b. allows static electricity to start fires.
   c. allows static electricity to drain off.
   d. drains excess current from the battery.

3. All objects with electrical charge contain
   a. much greater mass.
   b. much less mass.
   c. magnetic poles.
   d. energy.
4. Which is not a good electrical insulator?
   a. glass   b. rubber
   c. wood   d. copper
5. What is not needed in the simplest circuit?
   a. conducting wires
   b. an electrical source
   c. a voltmeter
   d. an electrical device
6. What amount of current is in a circuit if the voltage is 10 volts and the resistance is 5 ohms?
   a. 0.5 amperes   b. 2 amperes
   c. 15 amperes   d. 50 amperes
7. Which circuit in Figure 17-23 shows *both* the batteries and the lamps connected in a series?

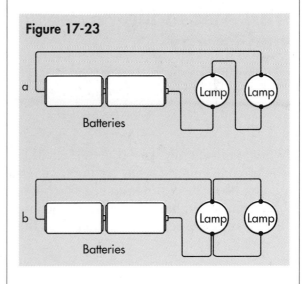

Figure 17-23

CHAPTER 18

GENERATING AND USING ELECTRICITY

Do you know where your electricity comes from? Most people know it comes from an electric company, through wires under the ground or on street poles. Still, where does the electric company get the electricity? Many electric companies just distribute electricity that they purchase from larger companies.

You may be surprised to learn that your electricity comes from several sources over various distances. People in New England get part of their electricity from Niagara Falls in New York, which may be more than 500 miles away. Part of your electricity may be produced locally, using gas, oil, or coal as fuel. Some electricity is made by nuclear plants using uranium as fuel. Wherever it is made, the electricity is sent into a network or "power pool" where it is shared, borrowed, or purchased by electric companies like yours. Through a network, a company with too little electricity can get some from a company with too much. This chapter looks at how electricity is measured, made, and used.

18-1 Measuring Power and Energy

■ *Objectives*
☐ *Calculate electrical power and energy.*
☐ *Compare devices by watts of power.*

When your family receives its monthly electric bill, you are really paying for energy you used during the month. The bill is based on a combination of three factors. These are voltage, current, and time the electricity was in use.

Figure 18-1	
Device	**Power used**
Digital clock	5 watts
Radio	25 watts
Light bulbs	15-300 watts
Toaster	600-1200 watts
TV	1500 watts

Electrical Power

Power is the rate of using energy. Power tells how quickly energy is being delivered or used in a unit of time. The units for power are the watt (W) and the kilowatt (kW). One kilowatt is equal to 1000 watts. The formula for power in electric circuits is

power = voltage × current

Imagine that a toaster requires a current of five amperes. What is its power? Remember that the standard voltage of electricity in homes is 120 volts.

power = 120 volts × 5 amperes
power = 600 watts or 0.6 kilowatts

By noting the power of electrical devices, people can compare how much energy they use in equal amounts of time. Power is often referred to as the "wattage" of a device. Look at Figure 18-1.

Measuring Electrical Energy

When you leave a light burning, you are using up energy. The amount of energy used depends on the power of the bulb and the time it is on. The formula for energy is

electrical energy = power × time

A unit for electrical energy is the **kilowatt-hour** (kWh). Assume an electric heater requires two kW of power. If it is used for three hours, how much energy does it use?

electrical energy = power × time
electrical energy = 2 kW × 3 h
electrical energy = 6 kWh

Now assume that an electric motor uses ten amperes of 240-volt electricity, and it is used for five hours. Determine the power of the motor and the energy used.

a. **power = voltage × current**
   power = 240 volts × 10 amperes
   power = 2400 watts or 2.4 kilowatts
b. **energy = power × time**
   energy = 2.4 kilowatts × 5 hours
   energy = 12 kilowatt-hours

Electric Bills

Once a month, a person from the electric company usually comes to read your electric meter. By subtracting the previous reading from the new reading, the company determines how much electrical energy you used, in kilowatt-hours. Figure 18-2 shows a typical electric bill.

Figure 18-2 A Sample Electric Bill

ELECTRIC COMPANY	
Billing date:	Feb. 15
Customer:	R. Freshman, #618-17
Feb. 1 reading:	4620
Jan. 1 reading:	−4170
Total energy used:	450 kWh
X cost/kWh	X $0.12
Total charges:	$54.00
+ 5% sales tax	2.70
Amount due:	$56.70

Write the definitions for the following terms in your own words.

1. **power**   2. **kilowatt-hour**

Answer these questions.

3. Distinguish between electrical power and electrical energy.
4. How can the watts ratings on the boxes of heaters help a customer make a choice? When would you need higher power? How might you use this information to compare the cost of two heaters?
5. An electric coffee pot operates on ten amperes of 120-volt electricity.
   a. What is its power in watts and kilowatts?
   b. How many kilowatt-hours of energy does it use in one fourth of an hour?
6. A 30-kW machine runs for five hours.
   a. What is the energy used?
   b. What is the cost of using this machine, if electricity costs eight cents per kilowatt-hour?

18-2 Generating Electricity

■ *Objectives*
□ *Give examples of the link between magnetism and electricity.*
□ *Tell how a generator works.*

One of the most amazing discoveries of the early 1800s was that electricity and magnetism are related. Later, the Scottish physicist James Maxwell showed that electric force and magnetic force are really parts of one combined force called electromagnetism. This combined force was the idea behind building machines to produce electricity.

Electricity Can Make Magnetism

You will recall from Chapter 16 that electromagnets use electricity to make magnetism. Electromagnets work because when electrons flow through a wire, they create a circular magnetic field around the wire. This circular magnetic field can be shown with iron filings or small compasses, as you can see in Figure 18-3.

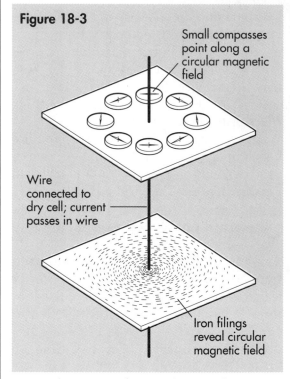

Figure 18-3

Small compasses point along a circular magnetic field

Wire connected to dry cell; current passes in wire

Iron filings reveal circular magnetic field

A straight wire conducting electricity has a magnetic field created around it.

Magnetism Can Make Electricity

The demonstration in Figure 18-4 uses a **galvanometer**, an instrument that detects small amounts of current. When the magnet is moved through the coil of wire, the galvanometer's needle moves. Electric current flows in the wire. The electrons of the atoms in the wire move because of the changing magnetic field. If you stop the magnet at any spot, the galvanometer reads zero. Current only flows as the magnet moves.

As the magnet is moved up and down within the coil, the galvanometer needle swings first to one side, then to the other. The electricity changes direction as the magnet changes direction. This device creates alternating current.

Electricity will also be formed if the coil of wire is moved up and down instead of the magnet. What counts is that there is motion between the wire and the magnet. It does not matter which one is moving.

Generators

Electricity flows in a wire when the wire cuts through a magnetic field. This is the principle behind a **generator**, a machine used to produce electricity. In a generator, a coil of wire on a shaft is made to spin within a magnet. Current flows in the wire. The voltage and amount of current increase when the magnet is made stronger, when more coils of wire are spun, or when the coils are spun faster. Look at Figure 18-5.

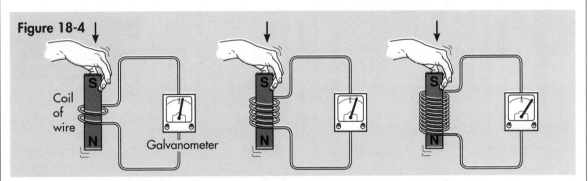

Figure 18-4

Coil of wire   Galvanometer

Moving a magnet through a coil of wire creates electricity. More coils produce more current.

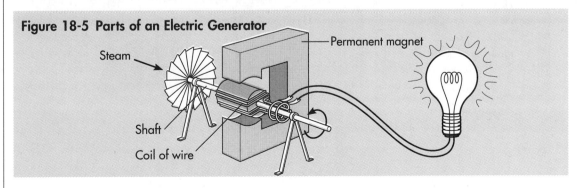

Figure 18-5 Parts of an Electric Generator

Steam

Permanent magnet

Shaft

Coil of wire

In large generators used by electric companies, the coil of wire has thousands of loops and weighs several tons! The magnet also weighs tons. It is actually an electromagnet that uses some of the electricity the generator makes. A very large force and much energy are needed to keep the heavy shaft and coil turning. These are supplied by rivers, by water from dams, by moving air, and often by high-pressure steam. To produce the steam, some fuel is needed to boil water. The diagram in Figure 18-6 shows the steps that occur in transforming energy to produce electricity.

Batteries

Batteries and dry cells are a completely different way to produce electricity. Chemical reactions between different materials and metals in the battery or dry cell release electrons. These electrons flow from the negative terminal of the battery through the circuit, and back again into the battery through the positive terminal. Look at Figure 18-7.

Figure 18-6 Energy Transformations in Making Electricity

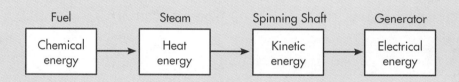

Figure 18-7

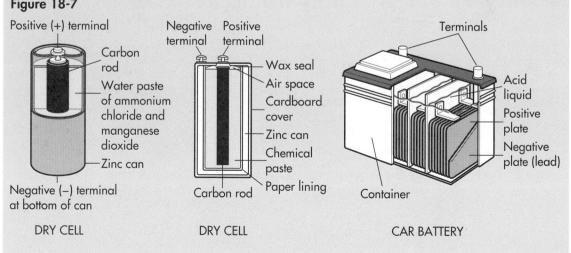

How Can Electricity Come from Lemons?

Process Skills observing; measuring; analyzing

Materials zinc and copper strips; lemons; galvanometer; electric wire

Procedures
1. Roll a lemon on a hard surface. This breaks down its tissues and releases the juice.
2. Insert the zinc and copper strips into the lemon. Put them next to each other, but do not let them touch.
3. Connect the galvanometer as shown in Figure 18-8. What happens?

Figure 18-8

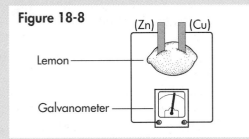

4. Connect several lemons in series, as in Figure 18-9. What happens to the reading on the galvanometer? Why?

Figure 18-9

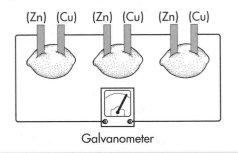

5. Insert two zinc or two copper strips into a lemon. Does current flow now?

Conclusions
6. Does the electricity really come from the lemon itself? What evidence is there that this is not true?
7. In what ways is this lemon battery similar to the batteries in Figure 18-7? What seems to be the common features of all batteries?

■■ Section 18-2 Review ■■

Write the definitions for the following terms in your own words.

1. **galvanometer**
2. **generator**

Answer these questions.

3. Describe three ways to demonstrate how magnetism and electricity can produce each other.
4. What are the key parts of a generator? What occurs to produce electricity?
5. Why is coal needed to make electricity in some generating plants?
6. What increases the voltage and amount of current made by a generator?

18-3 Safety with Electricity

■ *Objectives*

☐ *Explain the purpose of fuses, circuit breakers, and grounding.*

☐ *Describe safe and unsafe practices when using electricity.*

Have you ever been shocked by touching an electric wire? An electrical shock is uncomfortable. Although electricity provides us with comforts at home and with energy for industry, it can be dangerous. Modern society depends on electricity, but you must handle electricity with knowledge and care or else it may seriously hurt you. If you use electricity properly, it is very safe.

Fuses and Circuit Breakers

When electricity flows through the wires in your home, it makes the wires warm. As you turn on lights and appliances, you permit more current to flow. If too much current flows through the wires, they can become hot. The rubber insulation might melt, and a fire might begin. Usually, electric wires in homes are designed to carry up to 15 amperes or up to 25 amperes of current. To prevent these wires from being overloaded and becoming hot, an electrician places a **fuse** or a circuit breaker in every electric line that supplies your home. A fuse contains a piece of metal that melts if too much current flows through it. Circuit breakers contain an electromagnet that turns a switch to the off position when the amount of current

is more than it can carry. Figure 18-10 shows fuses and a circuit breaker as they are used to control the flow of electricity through wires.

If the demand for electricity is more than the limit for each line, the job of the fuse or circuit breaker is to stop the current completely. Stopping the current is for your safety because it warns you that too much electricity is being used on one line. If a fuse blows at home you should take the following steps.

■ Turn off all the lights and appliances that were on just before the current stopped.

■ Check the fuse box to see which fuse has melted and looks burned, or see which circuit breaker has flipped to the off position. Carefully unscrew and replace the blown fuse, or switch on the circuit breaker.

■ Turn some of the lights and appliances back on again, but not all of them.

Warning Handle fuses and circuit breakers with care and adult supervision. It is a good idea to wear rubber gloves. Replace the blown fuse only with a fuse of recommended size. For example, do not use a 30-ampere fuse if the previous fuse or the recommended size is 20 amperes. If you do not know the correct size, check the fuse box or ask the landlord, the superintendent, or an electrician. Do not defeat the purpose of fuse boxes by shorting them with a wire or a coin. Doing so could cause a fire.

Some appliances, such as food processors and vacuum cleaners, have their own small circuit breaker, called a reset button. Should the motor jam for some reason, the

Figure 18-10

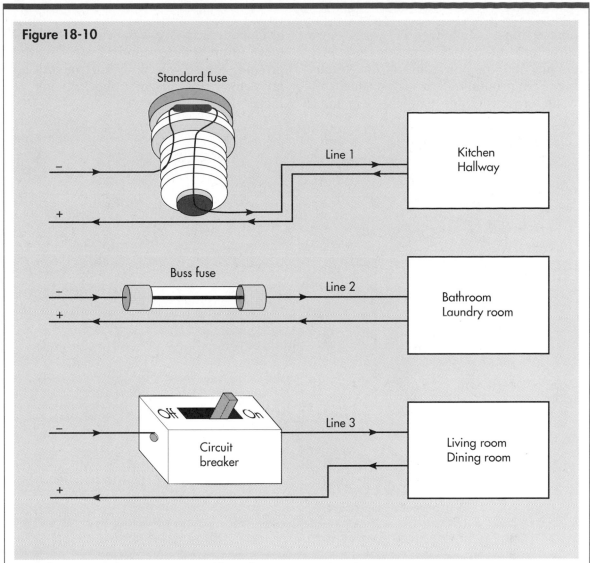

Each line in your home is protected by a fuse or circuit breaker.

button pops out. After the problem is fixed, you press the reset button to make the machine run again.

Grounding

Electrons flow very readily into any metal object that is buried in the ground. This is called **grounding**, and is used for electrical safety. Electric wires in a lamp or appliance may become loose or broken. If the wire touches the metal parts of the device, you will get a shock when you touch it.

To prevent this, most devices have a grounding wire. Newer electrical lines in homes must contain three wires. Two wires are for the electrons to flow in a circuit, and the third line is for grounding. This grounding wire is attached to a water or sewer pipe in the basement. If electrons escape from the circuit they will move through the grounding wire into the ground. A fuse will blow, but you will be protected from electric shock. Look at Figure 18-11.

Figure 18-11

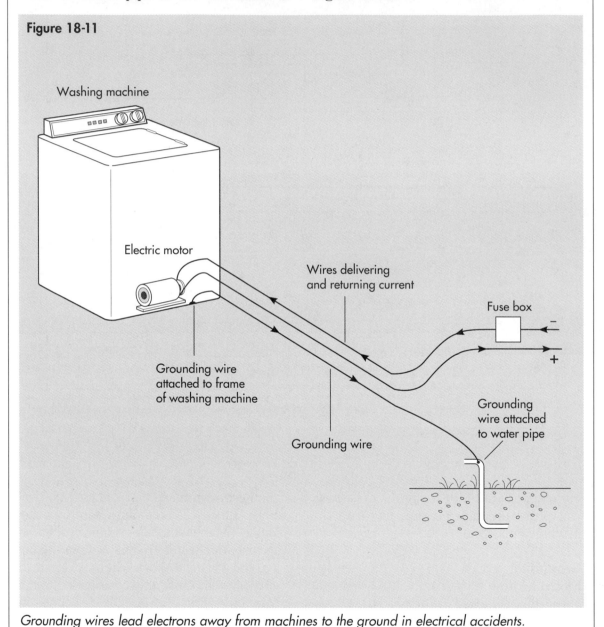

Washing machine

Electric motor

Grounding wire attached to frame of washing machine

Wires delivering and returning current

Fuse box

Grounding wire

Grounding wire attached to water pipe

−

+

Grounding wires lead electrons away from machines to the ground in electrical accidents.

Stating Hypotheses; Analyzing; Stating Conclusions

A group of students used the circuit in Figure 18-12 to measure how much current flows through different wires. For each test, they used the same voltage by using the same battery. They recorded the amperes of current going through the wire using the ammeter. Study their results in Figure 18-13.

Figure 18-13

Experiment A Three wires of identical length and thickness, but different materials

Material	Nichrome	Iron	Copper
Current	0.2	2	10

Experiment B Three iron wires of identical thickness, but different lengths

Length	Short	Medium	Long
Current	4	2	1

Experiment C Three iron wires of identical length, different thicknesses

Thickness	Thin	Medium	Thick
Current	0.5	2	8

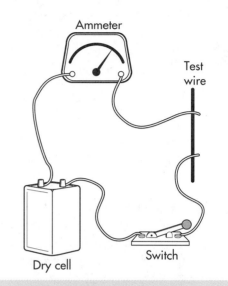

Figure 18-12

Ammeter

Test wire

Dry cell

Switch

1. What hypotheses are being tested in each experiment? Write three research hypotheses. If you need help, review the Skill Builder in Chapter 12.
2. What conclusions can you state from these experiments about conduction in a wire and the material, length, and thickness of the wire?
3. Predict which combination of material, length, and thickness will make a wire with
   a. maximum resistance, or lowest current.
   b. minimum resistance, or highest current.
4. In what ways were these experiments properly controlled?

How Does a Fuse Work?

Process Skills *observing; stating conclusions*

Materials six-volt dry cell; some steel wool; glass plate; electric wire

Procedure

1. Strip about three centimeters of insulation off both the ends of two wires. Attach the wires to the dry cell terminals as shown in Figure 18-14.
2. Place a single thin strand of steel wool on a heat-resistant surface such as a laboratory desk top or glass plate.
3. Holding only the insulated parts of the wires, touch the ends of the wires to the strand of steel wool. Place the wires about three centimeters apart, and then place them about one centimeter apart. What happens?

Conclusions

4. What path do the electrons follow when you touch the wires to the steel wool?

Figure 18-14

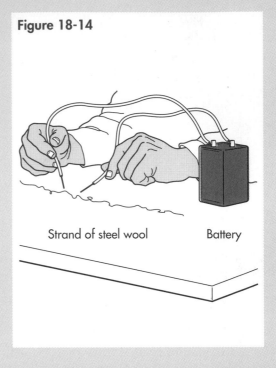

Strand of steel wool   Battery

5. What makes the steel wool get hot and melt?
6. Why do the wires from the battery remain unmelted?
7. How does this experiment show how a fuse works?

You can recognize an electric cord with grounding by the three prongs on its plug or socket. It is unwise to defeat the purpose of a three-pronged plug by placing a two-pronged adapter on it. If you must do this because the socket only has two openings, attach the green grounding wire on the adapter to the screw on the socket plate. This provides a way for the electrons to be grounded in case of problems. Study Figure 18-15.

Lightning Rods

Lightning is high-voltage electrons which make the air glow. Some homes and buildings have lightning rods. These are tall metal pipes placed on the roof, with a heavy wire connected to a metal pipe

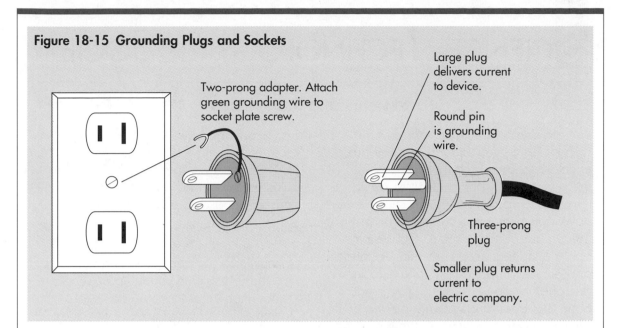

Figure 18-15 Grounding Plugs and Sockets

Two-prong adapter. Attach green grounding wire to socket plate screw.

Large plug delivers current to device.

Round pin is grounding wire.

Three-prong plug

Smaller plug returns current to electric company.

buried in the ground. Should lightning hit the building, it will most likely strike the pipe and pass down into the ground.

Water and Electricity

Water helps electricity pass through materials, especially human clothing and skin. Water also acts as a ground, since electrons can flow through the water into the pipes the water touches. For this reason, electricity and electrical devices should not be used near water. Do not plug a cord into a socket or operate electrical devices if your hands are wet or if there are spills nearby. Avoid placing electrical devices like radios and hair driers near a bath or sink.

If water does spill near or onto an electrical device, do not touch it! Carefully pull the plug from the socket with dry hands. Clean up the spill and allow the device to completely dry out.

■■■ Section 18-3 Review ■■■

Write the definitions for the following terms in your own words.

1. **fuse**   2. **grounding**

Answer these questions.

3. What is the purpose of fuses and circuit breakers? How do they work?
4. You turn on an air conditioner, and soon the lights, TV, fan, and vacuum cleaner in several rooms go out. What happened? What steps should you take to correct the situation?
5. Most new electrical devices come with a three-prong plug. Why is this a safety feature?
6. What precautions should you take regarding electrical devices near water?

SCIENCE, TECHNOLOGY, & SOCIETY

Systems and Subsystems

You are one complete person, yet your doctor might speak about your nervous, circulatory, and skeletal systems. All of these systems can be considered as subsystems inside you. Like systems, subsystems have their own input, control, processing, output, and feedback steps. Subsystems operate independently, but they do interact and provide input, output, and feedback to one another. For example, if you break a bone, your nervous system is alerted and your circulatory system sends extra blood to the injured area.

Like your body, many technological devices contain subsystems. For example, a car is a system. Its electrical subsystem involves the battery, spark plugs, and lights. The power subsystem includes the engine, crankshaft, gears, and wheels. The steering subsystem includes the steering wheel, turning gears, and axles. Subsystems operate independently, but can also operate together. The power system can provide energy for easier steering in cars with power steering. Although cruise control is part of the electrical system, it can control the gasoline valves of the power system to make the car go at a steady speed.

Advanced circuits have subsystems. These subsystems are smaller circuits that perform certain tasks, alone and together. One circuit in a television, for example, receives the picture signal and forms the image on the screen. Another circuit receives the sound signal and sends current to the speakers. Other systems keep the color balanced and prevent interference caused by passing planes and cars.

Follow-up Activities

1. Your school is a system that offers you knowledge, experiences, and skills. List some subsystems, made up of people, that enable the school to do its job.
2. Name the subsystems that are part of large systems such as electric plants, telephones, planes, and trains.

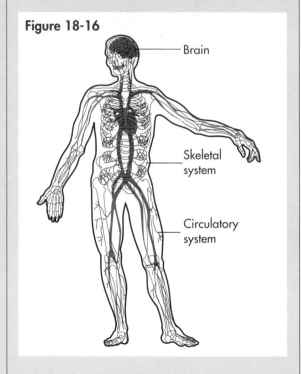

Figure 18-16

Brain

Skeletal system

Circulatory system

CHAPTER REVIEW

KEEPING TRACK

Section 18-1 Measuring Power and Energy

1. Power = voltage × current
2. Electrical energy = power × time
3. The watt is a unit that allows people to compare the rate at which various electrical devices use energy.
4. The kilowatt-hour is a unit used to measure the electrical energy used by a device during the time it operates.

Section 18-2 Generating Electricity

1. As a wire cuts through a magnetic field, electricity is produced in the wire. This is the basis for generators.
2. Batteries make electricity from chemicals.

Section 18-3 Safety with Electricity

1. Electrical energy is dangerous and can injure or kill living organisms.
2. Fuses and circuit breakers provide protection from fires that can be caused by overloaded circuits.
3. Some electrical devices should be grounded to prevent electrical shock.

BUILDING VOCABULARY

Write the term from the list that best completes each sentence.

fuse, galvanometer, grounding, kilowatt-hour, power

The rate at which a device uses up electrical energy is called its ___1___. The unit used to measure electrical energy on electric bills is the ___2___. It is better to use a ___3___ than an ammeter to measure tiny currents. A ___4___ and a circuit breaker shut off the electricity if too much current is used on one line. Many electrical devices have a ___5___ wire to prevent accidental shock.

SUMMARIZING

Write *true* if the statement is true. If the statement is false, change the *italicized* term to make the statement true.

1. Electric *energy* is measured in watts.
2. Electric energy is measured in *watts*.
3. A circuit using six volts and two amperes has *0.012 kilowatts* of power.
4. A two-kilowatt air conditioner left on for four hours uses *eight kilowatt-hours* of energy.
5. Electricity flows in a wire if the magnetic field around the wire *is not* changing.
6. A generator makes more current if the number of loops in its coil is *increased*.
7. A fuse contains a piece of *rubber* that melts if too much current flows through it.
8. Fuses work because electricity produces *magnetism* as it flows through the fuse.
9. Electrical appliances are made safer by *grounding* them.
10. Lightning is high-voltage *protons*.
11. It is *safe* to use most electrical devices near water or when they are wet.
12. Water *helps* electricity pass through materials, such as skin and clothing.

INTERPRETING INFORMATION

Some electric companies show you your energy use by putting a graph on your bill. Study Figure 18-17.

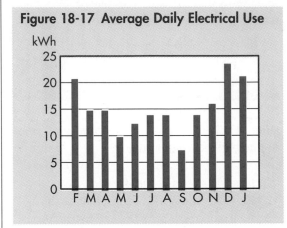

Figure 18-17 Average Daily Electrical Use

1. What was the average daily use of electricity in January?
2. If electricity is $.13 per kWh, estimate the cost for January, or 31 days.
3. During which three months does this family use greater amounts of electricity?

THINK AND DISCUSS

Use the section numbers in parentheses to help you find each answer. Write your answers in complete sentences.

1. Why can a light bulb have less power than a toaster but use more energy? (18-1)
2. A small TV uses 7.5 amperes of standard electricity at 120 volts. It is on for about eight hours a day. Calculate its

   a. power, in watts and kilowatts
   b. daily energy use, in kilowatt-hours
   c. cost of operation each day if electricity costs $.13 per kWh. (18-1)
3. How can you make electricity flow within a wire by using a magnet? How can you increase the amount of electricity made? (18-2)
4. Why is a large force needed to make electricity in generating plants? What are some of the energy transformations that occur as electricity is generated? (18-2)
5. Some people will replace a 15-ampere fuse with a 20-ampere fuse so it will not burn out so often as the 15-ampere fuse. Why might this be unsafe? How might they find out whether 20-ampere fuses are safe or not? (18-3)
6. What is the purpose of a lightning rod? (18-3)

GOING FURTHER

1. Study your family's next electric bill. List the information it contains. Try to understand each line of the bill and, if you wish, present a report about it. Especially note the meter readings, the total kWh used, and the cost per kWh. Check your meter. Is it close to the last reading on your bill?
2. An information plate is attached to every electrical device, usually on the bottom or the back. Most devices also come with consumer booklets. Read the plate or booklet of two appliances at home. What information is given? What is the power of each in watts?

CHAPTER REVIEW

COMPETENCY REVIEW

1. Here are the characteristics of a motor. Which represents its power?
   a. 120 volts
   b. 20 amperes
   c. 60 cycles AC
   d. 2.4 kilowatts

2. Electricity is produced by
   a. fuses.
   b. galvanometers.
   c. circuit breakers.
   d. generators.

3. Which is *not* true about electricity?
   a. It can injure.
   b. It can cause fires.
   c. It is safer to use in wet areas.
   d. It can kill.

4. What is the purpose of the third opening in the socket in Figure 18-18?
   a. delivers extra electricity to the device
   b. returns electricity to its source by way of pipes instead of wires
   c. provides safety by grounding
   d. goes to the fuse

5. Circuit breakers and fuses
   a. prevent too much current from flowing at one time.
   b. increase the electricity's power.
   c. boost the current flowing.
   d. prevent electric heating in wires.

6. Compared to a 300-watt bulb, a 60-watt bulb uses
   a. less energy.
   b. less energy per second.
   c. more energy.
   d. more energy per second.

Figure 18-18

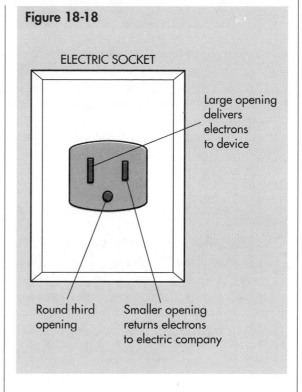

ELECTRIC SOCKET

Large opening delivers electrons to device

Round third opening

Smaller opening returns electrons to electric company

7. Your electric bill is based on the
   a. voltage of your electricity.
   b. total wattage of all your lamps and electrical appliances.
   c. kilowatt-hours of energy you used.
   d. number of electrical devices in your home.

8. If one fuse keeps burning out, it is *not* recommended that you
   a. use a fuse that passes more current.
   b. plug appliances into sockets on other electric lines.
   c. use lights and devices at different times.
   d. rewire or improve the electric lines.

CHAPTER 19

NUCLEAR ENERGY

If you examine today's newspaper, you will probably find a reference to nuclear energy or radioactivity. Your great-grandparents may never have heard of nuclear energy. Your grandparents grew up thinking of nuclear energy as atomic bombs and hydrogen bombs, weapons of great destructive power. Your parents grew up learning about nuclear energy as a fuel source and in medical advancements. Today, nuclear energy is used more than ever to make electricity. However, many people fear nuclear energy because of the safety and environmental risks it poses.

Nuclear energy is found at the core of every atom, within the nucleus. Some atoms, such as uranium, release this energy more readily than others. The amount of energy released is tremendous compared to the energy released by fuels such as coal and oil. Consequently, researchers are greatly interested in nuclear energy as a power source. In some hospitals, special machines use nuclear energy to help diagnose illnesses. In industry, machines use radioactivity to ensure consistent product quality. When nuclear energy is used, people must control it carefully because it can be dangerous. This chapter introduces you to the characteristics, uses, and dangers of nuclear energy.

19-1 Radioactivity

■ *Objectives*
☐ *Explain radioactivity.*
☐ *Compare forms of radiation.*
☐ *Use half-life in calculations.*

You have probably heard about materials that are **radioactive**, which means constantly giving off invisible particles. Radioactive materials may also constantly give off electromagnetic waves. The

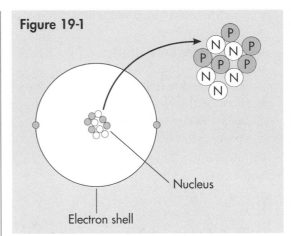

Figure 19-1

Nucleus

Electron shell

Nuclear energy holds neutrons and protons together.

waves, as well as the particles, can damage living cells and other materials. The waves and particles, however, can also serve people and science.

Unstable Nuclei

Recall that in the center of every atom is a tiny nucleus. The plural of *nucleus* is *nuclei*. The nucleus contains the protons and neutrons of the atom. Look at Figure 19-1. The protons and neutrons are held together by a powerful force of attraction that exists only within a nucleus. The energy that holds a nucleus together is called nuclear energy.

The nuclei of most elements are very stable. This means the nuclear energy in these nuclei is great enough to keep the protons and neutrons from flying apart.

The nuclei have lasted, and will last, for billions of years. A fair percentage of elements, however, contain nuclei that are unstable. After a while they break apart, releasing a proton or a neutron, or some combination of both.

This releasing of particles is what makes some materials radioactive. The nuclei are unstable. The particles and the electromagnetic waves that come out of unstable nuclei are called **radiation**. Radioactive materials give out radiation.

Types of Radiation

An unstable nucleus can eject various types of particles that contain various electric charges. Figure 19-2 describes the types of radiation given off by unstable nuclei.

Figure 19-2 Types of Radiation

Type	Comment	Electric charge
Neutron	Escapes from nucleus	0
Proton	Escapes from nucleus	+1
Alpha particle	Package of two protons and two neutrons, identical to a nucleus of helium	+2
Beta particle	A high-speed electron	−1
Gamma ray	Short burst of a high-energy electromagnetic wave	0

You can examine these different types of radiation by looking at Figure 19-3.

Figure 19-3 Types of Radiation Given off by an Unstable Nucleus

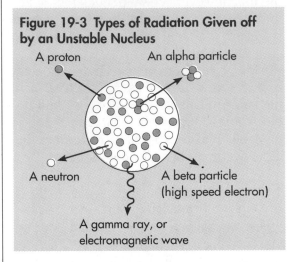

A proton

An alpha particle

A neutron

A beta particle (high speed electron)

A gamma ray, or electromagnetic wave

You may wonder how an electron can come from the nucleus of an atom. Sometimes a neutron breaks up into a proton and an electron. The proton stays in the nucleus and the electron is thrown out of the nucleus at high speed.

After the nucleus of a radioactive element breaks apart, a different atom remains behind. The remaining atom is an atom of a different element. Figure 19-4 shows some examples of this change, which is known as **radioactive decay**.

Radioactive Isotopes

Recall that isotopes are atoms of the same element that have different mass numbers. Isotopes have the same number of protons but different numbers of neutrons. Sometimes, one isotope of an element will be very stable and other isotopes will be unstable. Carbon-12 is very stable, for example, but carbon-11 and carbon-14 are

Figure 19-4 Examples of Radioactive Decay

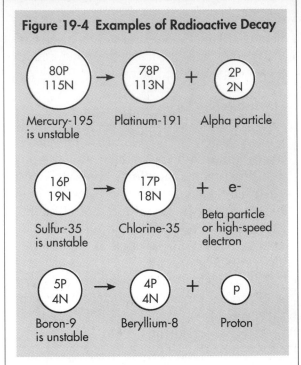

Mercury-195 is unstable → Platinum-191 + Alpha particle

Sulfur-35 is unstable → Chlorine-35 + e- Beta particle or high-speed electron

Boron-9 is unstable → Beryllium-8 + Proton

not. The unstable atoms are called **radioactive isotopes**. Study Figure 19-5.

Half-life

A sample of a radioactive element contains a huge number of atoms. Some of these atoms may decay every second. There are so many atoms in the sample, however, that

Figure 19-5 Isotopes of Carbon

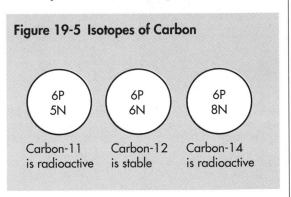

Carbon-11 is radioactive

Carbon-12 is stable

Carbon-14 is radioactive

it may take millions of years for all of them to decay.

In some ways, radioactive decay is like making popcorn. You can never be sure when any one corn kernel will pop, but you know that most of the kernels will pop within a few minutes. Similarly, no one can tell when any one nucleus will decay and release radiation. It is possible, however, to predict how much of a sample will decay in a certain amount of time.

Figure 19-6 contains some data on copper-59, a radioactive isotope. Suppose we begin with a sample of 400 grams of pure copper-59, which begins to decay and give out radiation. The amount of copper-59 left keeps decreasing as time goes by. Figure 19-6 shows the rate of its decay.

Notice that after every ten minutes exactly half of the previous amount of copper-59 remains undecayed. You can therefore say that the **half-life** of copper-59 is ten minutes. The half-life of a radioactive element is the time it takes for half of the radioactive atoms of a sample to decay.

Each radioactive isotope has a specific half-life, which can vary from less than a second to billions of years. Each radioactive isotope also gives off a specific type of radiation.

Iodine-131, for example, has a half-life of eight days. It decays by emitting a beta particle and a gamma ray. After eight days, only half of the original sample remains undecayed. In the next eight days, half of the previously unchanged amount will decay, and half will remain. The decay then continues at the same rate every eight days. Figure 19-7 illustrates this half-life process, starting with 64 grams of iodine-131.

Figure 19-6 Copper-59 Decay

Time	Grams of Cu-59 left
At start	400
10 minutes	200
20 minutes	100
30 minutes	50
40 minutes	25
50 minutes	12.5
60 minutes	6.25

Figure 19-7 Radioactive Decay of 64 grams of Iodine-131 with a Half-life of Eight Days

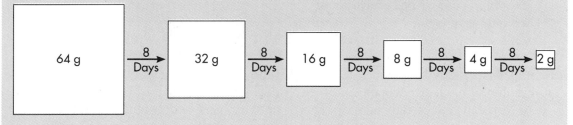

How Can You Demonstrate Half-life?

Process Skills gathering and recording data, organizing, stating conclusions

Materials 200 pennies and a covered box or can

Procedures

1. Put all the pennies in the can or box and shake them. Then pour them out on a table.
2. Separate the pennies that landed heads-up from those that landed tails-up. Remove the pennies that landed heads-up. Record the number of coins that are left.
3. Place the remaining pennies into the can and repeat procedures 1 and 2.
4. Keep repeating these procedures until all pennies have come up heads and have been removed.
5. Prepare a chart of your data that follows the format of Figure 19-8.
6. Plot your data on a graph. Place the number of the toss on the horizontal axis, and the coins left after each toss on the vertical axis.

Conclusions

7. What conclusion can you state about the number of coins that are left after each toss?

8. A heads-up penny is like an unstable atom that has decayed. How is this activity with pennies like the decay of a radioactive sample?
9. Assume that each toss represents one year in the decay of a radioactive element. Using the chart you made, state the half-life of the pennies.
10. What does the shape of your graph show about the pennies, about radioactive decay, and about half-life?

Figure 19-8

Toss number	Coins left
0 (start)	200
1	
2	
(and so on)	

Detecting Radioactivity

Photographic film, electroscopes, and Geiger counters can detect radioactivity. The radiation causes streaks on photographic film that you can see when the film is developed. In fact, this was how radioactivity was accidentally discovered by Antoine Becquerel in France in 1896.

Radiation also causes ions to form as it passes through various materials and air. Electroscopes can detect the small charge of these ions. A **Geiger counter** detects and counts the number of radiation particles that are emitted from a sample in a certain time. A Geiger counter is illustrated in Figure 19-9. Today, in addition to these detectors, very fast digital detectors can count the radiation and describe its energy, even if thousands of particles are given out each second!

Figure 19-9 Geiger Counter

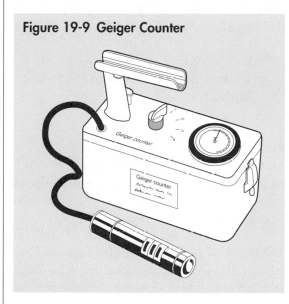

Background Radiation

Some radiation occurs on the earth, and some reaches the earth from space. Radiation from both sources is called background radiation, a level of radiation that occurs naturally. Background radiation is found in larger amounts at higher elevations. This fact concerns pilots and flight attendants, who are exposed to these strong levels of background radiation.

Science and technology have added to the radiation around us by making radioactive elements. In research and industry, people experiment with radioactive materials in order to understand nuclear energy. In doing so, they concentrate radioactive elements in purified samples and in nuclear devices.

■ Section 19-1 Review ■

Write the definitions for the following terms in your own words.

1. **radiation**
2. **radioactive decay**
3. **radioactive isotopes**
4. **half-life**
5. **Geiger counter**

Answer these questions.

6. Explain radioactivity by telling what it is, what causes it, and how it is detected.
7. Compare alpha, beta, and gamma radiation.
8. If an atom decays by emitting a proton or an alpha particle, a new element is formed. If it decays by emitting a neutron, a new isotope is formed. Explain the difference.
9. Uranium-232 has a half-life of 70 years. Beginning with 24 kilograms of pure U-232, how much will be left unchanged after 350 years? Make a table or sketch to show how much remains after each 70-year interval of time.

19-2 Nuclear Reactions

■ *Objectives*

☐ *Compare and contrast chemical and nuclear reactions.*

☐ *Distinguish between fission and fusion.*

The tremendous energy locked inside the nucleus of an atom leaks out naturally during radioactive decay. During the 1920s and 1930s, scientists studied how to release this energy. On December 2, 1942, in Chicago, a team of researchers led by Enrico Fermi achieved the first continuous release of substantial amounts of nuclear energy. This led to the development of the atomic bomb, which changed the course of World War II. Today, people use nuclear energy to make electricity and for operating ships and spacecraft.

You will recall that atoms join together or break apart in a chemical reaction, and chemical energy is released or absorbed. Similarly, when the nuclei of atoms join or break apart, nuclear energy is released or absorbed. We call this a **nuclear reaction**.

Chemical reactions occur among atoms. Nuclear reactions occur in the nuclei inside atoms. Nuclear reactions are more difficult to start and control, but they produce millions of times more energy than chemical reactions.

Natural Nuclear Reactions

Radioactive decay is a nuclear reaction that occurs naturally. Unstable nuclei react by giving off radiation, and a new product results. Just as chemical reactions are described by chemical equations, so nuclear reactions are described by nuclear equations. Look at Figure 19-10. This nuclear reaction can be written as a nuclear equation.

uranium-238 → thorium-234 + alpha particle + energy

Fission

Some radioactive elements can undergo a nuclear reaction known as **fission**. When fission occurs, a large nucleus splits into smaller nuclei. This results in the release of a large amount of nuclear energy, in the form of heat and electromagnetic energy. Fission does not occur on its own. Something must encourage fission to

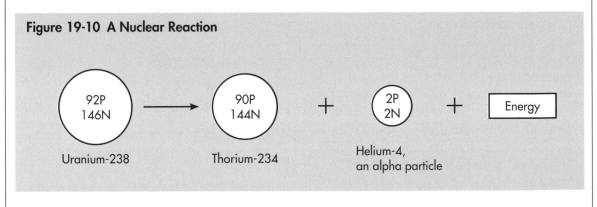

Figure 19-10 A Nuclear Reaction

92P
146N
Uranium-238

→

90P
144N
Thorium-234

+

2P
2N
Helium-4,
an alpha particle

+

Energy

start. One way to begin fission is to bombard a nucleus with a neutron.

Look at Figure 19-11. The most well-known fission reaction begins with uranium-235. The nuclear equation for this reaction is

**neutron + uranium-235 →
krypton-91 + barium-142 +
3 neutrons + energy**

This is the nuclear reaction that takes place in an atom bomb and in many electricity-generating plants.

The fission in Figure 19-11 produces three neutrons that can hit other uranium-235 nuclei, causing them to split also. This is called a **chain reaction**. In a chain reaction, the products of one nuclear reaction can trigger a reaction in other nuclei. A chain reaction may be uncontrolled, as in an atom bomb. Millions of chain reactions happen in one second, causing U-235 atoms to rapidly undergo fission. A great amount of energy is released. Figure 19-12 shows what happens in an uncontrolled chain reaction of U-235.

Most chain reactions are not as dangerous as those in atom bombs. This is because a chain reaction may be controlled so that the number of fissions per second is kept constant. This occurs in nuclear power plants.

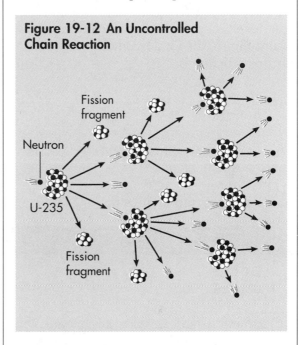

Figure 19-12 An Uncontrolled Chain Reaction

Fission fragment

Neutron

U-235

Fission fragment

Fusion

Some elements can undergo a nuclear reaction known as **fusion**. When fusion occurs, small nuclei join to form a larger nucleus. Fusion causes the release of a large amount of nuclear energy in the

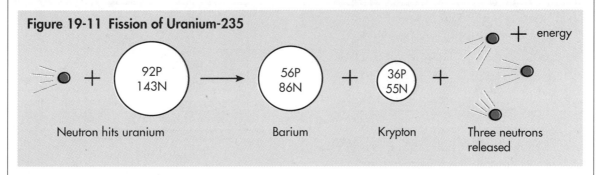

Figure 19-11 Fission of Uranium-235

92P
143N

Neutron hits uranium

56P
86N

Barium

36P
55N

Krypton

energy

Three neutrons released

form of heat. It requires extremely high temperatures to begin.

The most well-known fusion reaction occurs naturally in stars, including the sun. Hydrogen nuclei smash together in a series of steps to become helium nuclei. Study Figure 19-13. The nuclear equation for this reaction is

**hydrogen-2 + hydrogen-2 →
    helium-4**

This fusion reaction is responsible for the energy and heat of the sun. Life on the earth, therefore, depends on a nuclear reaction occurring in the sun.

Figure 19-13

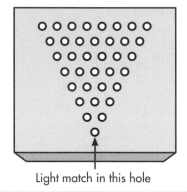

A fusion reaction is similar to reactions within the sun.

S K I L L   B U I L D E R

Making a Model

Warning Do not perform this experiment. It is a teacher demonstration.

A teacher cut 36 small holes into a wooden board, about six mm apart, in the pattern you see in Figure 19-14. Into each hole, the teacher placed a wooden safety match. Then the teacher lit the single match indicated in the figure. What do you suppose happened?

1. Why could this demonstration serve as a model of a chain reaction?
2. What does this model show about chain reactions?
3. A science model does not have to look like the real thing, it just has to imitate it. How does this demonstration show this about models? For instance, do nuclei really transfer heat and start to burn?
4. How might you set up dominoes to make a model of an uncontrolled chain reaction? It is not enough to line them up in a straight row. Why?

Figure 19-14

Light match in this hole

Increased Use of Radioactive Isotopes

People are using radioactive isotopes more and more in medicine, business, and industry. Even some fire detectors in homes use a tiny source of radioactivity. Special care must be taken to ensure that these products are handled with great care and disposed of properly. Today's applications of radioactivity involve using radioactive decay and fission to produce energy. As additional radioactive materials are used, the dangers of abuse and accidents increase. Radiation can become a serious problem if people are overexposed to it.

At present, people have applied fusion mostly in the form of thermonuclear bombs, also known as hydrogen bombs. These bombs create the same reaction that occurs on the sun, for a brief moment, as they explode. The energy of a hydrogen fusion bomb is far greater than that of an atom bomb, which works on uranium fission.

Research continues on controlling the fusion reaction so that people can use its vast energy for good. So far, however, more energy is required to make the fusion occur than the reaction itself produces. Fusion is not yet a process people can use economically.

▬ Section 19-2 Review ▬

Write the definitions for the following terms in your own words.

1. **nuclear reaction**   2. **fission**
3. **chain reaction**   4. **fusion**

Answer these questions.

5. In what ways are chemical reactions and nuclear reactions alike? How are they different?
6. What type of nuclear reactions are the first and second reactions illustrated in Figure 19-15?
7. Suppose that each of the three neutrons given out by a uranium-235 nucleus when it splits starts fission in another uranium-235 nucleus. The number of neutrons being released would rapidly multiply in the order 1, 3, 9, 27, 81, 243, 729, 2187 and so forth.
   a. Why do the numbers of neutrons increase this way?
   b. Why is this tripling important for releasing nuclear energy?
   c. What is this process called?
8. Why is uranium an important material today?

Figure 19-15

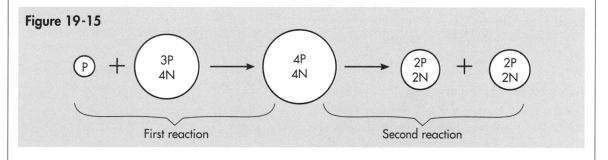

First reaction   Second reaction

19-3 Benefits of Nuclear Energy

■ *Objectives*

☐ *Give examples of uses of radioactive isotopes.*

☐ *Explain how scientists can date ancient objects through radioactivity.*

Perhaps you know people who have been tested or treated for cancer by means of radiation, which is a result of nuclear energy. About ten percent of the electricity used in the United States comes from generating plants that use nuclear energy. These are two of the benefits of nuclear energy.

Radioactive Isotopes

As discussed in Section 19-1, some elements have radioactive isotopes that researchers use to study nuclear energy. Researchers have also found ways in which radioactive isotopes can serve various beneficial purposes.

Medical Uses Doctors use some radioactive isotopes to diagnose diseases. The isotopes are given to a patient by injection or in a liquid drink. Using radiation detectors, doctors can trace the path of the radioactive material as it moves through the blood or digestive systems. Problems such as blockages can be quickly seen. When used in this manner, the radioactive isotopes are called **tracers**. For example, doctors often use radioactive barium to trace the flow and blockages in the digestive system. Doctors can use radioactive thallium to trace the flow of blood, especially through the heart.

Some radioactive isotopes are used to kill cancer cells in the body or to slow down their growth. For instance, when radioactive iodine is swallowed, it gathers in the thyroid gland, where its radiation kills cancer of the thyroid. Radiation from strongly radioactive elements can be aimed at parts of the body that have cancer cells. In this way, radiation can reduce the number of cancer cells and their rate of growth.

Quality Control in Industry The thickness of a sheet of metal can be controlled by a device like the one shown in Figure 19-16. The thicker the sheet, the more radiation it absorbs. By measuring the radiation the sheet absorbs, the device can detect very small changes in its thickness. If the sheet is a little too thick, the rollers are moved a little closer together. If the sheet is a little too thin, the rollers are moved farther apart. This is an example of feedback within a system.

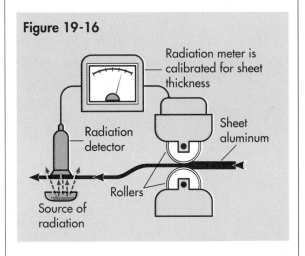

Figure 19-16

Radiation meter is calibrated for sheet thickness

Radiation detector

Sheet aluminum

Rollers

Source of radiation

There are other ways people use radioactive isotopes when they want to keep the quality of products high. Defects inside metal parts can be photographed using radiation instead of X rays. How well different oils work can be measured by making moving parts with metal that contains a tiny fraction of a radioactive element. In spots where the oil does not work well, the metal will be slightly rubbed off. This can be detected by radiation meters examining the oil. Tire manufacturers place some radioactive carbon in test tires to measure the tiny amount of rubber left on the road when the car stops.

Scientific Research Scientists use radioactive isotopes to follow the changes made to molecules during chemical reactions. They also can use detectors to trace radioactive chemicals in fertilizers as they move up a plant and help it grow. From these studies, researchers can find the proper amount of fertilizers to be used with different plants and report the information to farmers.

Carbon and Uranium Dating

The age of fossils and other ancient objects can be estimated by radioactive dating. A constant percent of all the carbon atoms in the air are radioactive carbon-14. This becomes part of trees, wood, animals, and human bodies, along with ordinary carbon-12. To date a fossil, scientists measure the amount of carbon-14 radiation that it gives off.

The half-life of carbon-14 is 5730 years. By comparing the amount of carbon-14 radiation in a fossil to the natural amount always present in the air, scientists can calculate the fossil's age. This **carbon-14 dating**, is accurate for objects less than 50,000 years old.

The age of rocks and the earth can be estimated by comparing the amount of radioactive uranium and lead isotopes in samples. This comparison is the basis of **uranium-238 dating**. Uranium-238 has a half-life of 4.5 billion years. The oldest earth rocks are 3.7 billion years old, and the moon rocks brought to the earth are 4.2 billion years old. Uranium dating helps confirm the age of 4.5 to 5 billion years for the earth and the solar system.

New Elements

When an ordinary element is bombarded with neutrons, some of its atoms capture a neutron. This creates a new isotope of the element that may not ordinarily exist. Often such isotopes are radioactive, and scientists use them for the special purposes described above.

Besides making new isotopes, scientists can use radiation to create new elements. On the periodic table, elements past uranium, number 92, are not found naturally on Earth. They have been made in laboratories by smashing nuclei together or by bombarding atoms with neutrons, protons, or alpha particles. Most of the elements between 93 and 106 exist for only a few seconds before decaying into stable nuclei with atomic numbers less than 92.

Fission Power

As you know, electricity is made in generating plants. It takes a lot of energy to turn the generator's heavy coils and shaft. In some electric plants, fission of uranium fuel is the source of this energy. The heat from the fission heats water, which turns to steam. The steam turns the blades of the turbine, which turn the generator.

Fission power runs engines on large navy ships and in spacecraft. You have probably heard of nuclear submarines, which are navy ships that run on fission. A few pounds of uranium can keep them going for years.

■ Section 19-3 Review ■

Write the definitions for the following terms in your own words.

1. **tracers**
2. **carbon-14 dating**
3. **uranium-238 dating**

Answer these questions.

4. List four beneficial uses of radioactive isotopes of elements.
5. a. Why can carbon-14 be used to date an object 20 thousand years old, but not one 20 million years old?
   b. Why can uranium-238 be used to date an object 20 million years old, but not one 20 thousand years old?
6. Describe at least one use for the radioactive isotopes of each of these elements: barium, thallium, iodine, and the carbon in rubber.
7. Element number 99, einsteinium, was named after Albert Einstein and was discovered in 1952. However, it does not exist on Earth. Explain this.

19-4 Dangers of Nuclear Energy

■ *Objectives*

☐ *List the dangerous effects and sources of radioactivity.*

☐ *State key issues about using nuclear energy to produce electricity.*

☐ *Define rads and rems.*

You have read about some of the benefits of nuclear energy. You can judge these benefits by studying some of the risks and dangers this source of energy brings with it. When you talk with other people about nuclear energy, you will probably want to weigh both sides of the issue.

Uncontrolled Radiation

Uncontrolled radiation from radioactive materials is dangerous. It can injure or kill living organisms by killing individual cells. It can also cause birth defects by making changes in the nucleus and chromosomes of each cell. Excessive radiation can promote many diseases, especially cancer.

Humans always receive a small amount of radiation from background sources such as cosmic rays from space and radioactive elements in the ground and air. The harmful effects occur when someone receives a large dose of radiation in a short period of time, or small doses over a long period of time. As you can imagine, this is the danger involved in receiving radiation treatments for diseases. It also explains the danger of using radioactive chemicals in industry or in scientific research. These chemicals often escape into the atmosphere or leak into the water system. Of course, such leaks increase the normal background level of radiation.

Recently, home owners have become concerned about naturally radioactive radon gas, Rn-222. Radon results from the radioactive decay of radium in the earth. This gas has been escaping from the ground for years, and it can accumulate in people's basements and houses. If these homes are not properly ventilated, the radon may become a health hazard.

Measuring Radiation Doses

The amount of radiation given off by radioactive sources is measured in **rads**. The ability of radiation to harm human beings is measured in **rems**. In other words, rems measure the dosage of radiation that will cause biological effects.

Alpha particles are more damaging to human tissue than other types of radiation. This means they have higher values of rems. However, alpha particles do not travel as far in the air as other types of radiation. Therefore, radioactive sources giving off alpha particles are not immediately dangerous at distances greater than 2.5 cm, or about an inch. However, if you were to swallow or breathe in a material giving off alpha particles, it would be very dangerous to your body tissue and organs.

People who work with or near radioactive materials must wear small radiometers containing photographic film. Remember that radiation causes streaking on photographic film. The radiometers are checked every week to determine the amount of radiation to which the workers have been exposed.

The average person in the United States is exposed to about 0.2 rems of radiation each year. This radiation comes from the body itself, from buildings, cosmic rays, medical X rays, and television. In radiation therapy, a person might receive a concentrated dose of 200 rems each day for weeks. Compare this to a typical X ray, which gives off about 5 to 30 millirems, or 0.005 to 0.030 rems. You can see that radiation treatment is very powerful.

People often do not feel the effects of a radiation dose under 20 rems. When the dose reaches 100 rems, however, a person is likely to experience radiation sickness. This illness causes tiredness and a loss of appetite. Larger doses of radiation can cause a person's hair to fall out. A full-body dose of 500 rems of radiation at one time will usually kill a person.

Nuclear Weapons

Nuclear weapons represent the most dangerous use of nuclear energy. Only two nuclear weapons have ever been used. Two nuclear bombs were dropped on Japanese cities near the end of World War II. The first bomb killed almost 100,000 people and destroyed about five square miles of the city of Hiroshima.

Today's atomic bombs are much more powerful. If nuclear weapons were ever used again, the climate might change because of the dust produced by the bombs. By blocking the sun, this dust cloud could prevent plants from growing well and could cool the earth drastically. Scientists have been studying this possible effect and call it nuclear winter. In addition, radiation from the explosion, and in the fallout material

Analyzing

Diagrams A and B in Figure 19-17 compare the penetrating ability of three types of radiation. Diagram A shows the distances the three types of radiation can travel in air. Diagram B shows the materials that can stop the radiation. Diagram C shows the way the radiation travels as it goes past electrically charged plates.

1. Which type of radiation can travel farthest in air? Which type can only travel short distances?
2. Which type can be stopped by very thin material? Which type may not even stop inside dense material?
3. What are some ways of stopping radiation?
4. Which particle is attracted to the positive plate and repelled by the negative plate? What does this mean about its charge?
5. What can you conclude from the paths followed by the other two types of radiation as they go past the charged plates?
6. What can you conclude from the sharp bend made by the beta particle and the wide bend made by the alpha particle? You might want to think about the fact that birds can make sharp turns, but jet planes cannot.

Figure 19-17

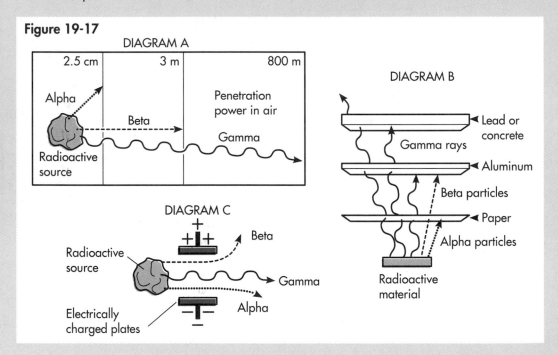

that comes after one, could cause radiation sickness and death everywhere.

Even if nuclear weapons are never again used in a war, as they were in World War II, testing them has already caused some damage. Nuclear bombs produce large amounts of radiation. This is why the United States and other nations now test nuclear weapons underground, not in the atmosphere as they used to. It was discovered that radioactive isotopes such as strontium-90 were part of the fallout from atmospheric testing. Strontium-90 got into plants, into the food chain, and into human bodies, posing a serious health hazard.

Nuclear Energy Safety Issues

Society must rigidly control nuclear power plants to prevent the accidental or deliberate release of dangerous radioactive materials. People in each nuclear plant must account for every gram of radioactive material that goes in or out. A serious nuclear plant accident like the one that took place in Chernobyl, Soviet Union, in 1986 could release large amounts of radiation. Therefore, many safety systems and emergency procedures must be prepared before such plants are permitted to operate. In the United States, the Nuclear Energy Commission has the authority to review and license plants that use nuclear energy. Safety is the agency's top concern.

People have recently found out about many nuclear accidents and deliberate dumping of nuclear materials in this country. Such activity has been occurring since 1950. The government has pledged never again to permit the release of radioactive materials. Accidents do happen, however.

The accidents at Three Mile Island in Pennsylvania, in 1979, and at Chernobyl have been the worst accidents so far involving nuclear power plants.

Disposing of radioactive materials is a current problem for society to handle. By law, such materials must be collected, transported, and buried in underground sites established by the government. Transporting them by truck or by train often causes great alarm among people who live along the route. What if there were an accident?

▬ Section 19-4 Review ▬

Write the definitions for the following terms in your own words.

1. **rads**   2. **rems**

Answer these questions.

3. What are some natural and artificial sources of radiation?
4. What are the dangers in overexposure to radioactive materials?
5. Explain these statements in rems.
   a. One chest X ray may be ten percent of the average radiation dose expected for a whole year.
   b. A 1000-rem dose in a local spot may kill cancer cells, but a similar full-body dose will kill the patient.
6. What are some safety issues connected with using nuclear energy to produce electricity?

SCIENCE, TECHNOLOGY, & SOCIETY

Making Up Our Minds on Nuclear Energy

Figure 19-18

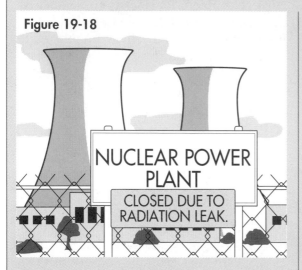

NUCLEAR POWER PLANT
CLOSED DUE TO RADIATION LEAK.

People in society weigh the pros and cons of new technologies. Some technologies, such as VCRs, become almost instant successes. Others, such as video discs, fail to catch on. Usually people will wait for a product to be improved before they start buying it. Home computers were difficult to use at first. When they came with simple instructions and increased memory, they gained popularity.

The United States has a similar wait-and-see attitude toward nuclear energy. In Europe, about 50 percent of all the electricity comes from nuclear power plants, but in the United States it is only about ten percent. Since 1970, no new nuclear power plants have been built because groups have questioned or protested the use of nuclear fuel instead of coal, oil, and gas.

People who favor using nuclear power plants say that we should not depend on oil from other nations, and that burning our country's coal creates more pollution than uranium does. Besides, we will eventually run out of other fuels. They claim that safety devices and procedures are so good that a major accident is less likely in nuclear plants than in ordinary power plants.

People opposed to nuclear power plants say they are just too dangerous. Accidents have happened. The radioactive waste produced by the plants must be shipped and then buried for thousands of years. Also, the price of electricity made by using uranium has been higher than electricity made by using other fuels because of safety and insurance costs.

Both sides argue strongly and are politically active. At this time, the public's fears about nuclear energy are stronger than any desire for the possible benefits. This attitude could change through improvements in the technology and in the safety records of nuclear plants.

Follow-up Activities
1. Gather material for or against using nuclear energy to make electricity. List the major issues raised. What is your opinion? Why?
2. What improvements might make the public accept nuclear energy?
3. What themes about technology and society are illustrated in this topic?

KEEPING TRACK

Section 19-1 Radioactivity

1. Some isotopes of atoms are unstable and give off radiation. They are radioactive.
2. Radiation may consist of particles or electromagnetic waves in rays.
3. The half-life of a radioactive element is the time needed for half of the atoms in a sample to decay.

Section 19-2 Nuclear Reactions

1. In nuclear reactions, atomic nuclei change, releasing a great amount of energy.
2. In fission, a large nucleus breaks apart into smaller nuclei. In fusion, small nuclei join and turn into one larger nucleus.
3. Some nuclear reactions keep going because they become chain reactions.

Section 19-3 Benefits of Nuclear Energy

1. People use radioactive isotopes in medicine, industry, and research.
2. Researchers can find the age of ancient objects and rocks by using carbon-14 or uranium-238 dating.
3. People can use energy from the fission of uranium atoms to produce electricity.

Section 19-4 Dangers of Nuclear Energy

1. Excessive radiation can injure or kill people and can cause birth defects and cancer.
2. Nuclear energy poses problems in the areas of safety, health, and waste disposal.

BUILDING VOCABULARY

Write the term from the list that best matches each statement.

Geiger counter, half-life, radioactive, radioactive decay, radioactive isotope, tracers

1. gives off invisible particles and rays
2. process in which unstable nuclei give off particles and become stable
3. the form of an element that is unstable
4. time during which 50 percent of a radioactive sample changes by releasing particles
5. radioactive chemicals used to study people, plants, and chemical reactions
6. an instrument that detects radiation

Explain the difference between the terms in each pair.

7. fission, fusion
8. nuclear reaction, chain reaction
9. carbon-14 dating, uranium-238 dating
10. rads, rems

SUMMARIZING

Write the missing term for each sentence.

1. A ___ particle is a high-speed electron.
2. Alpha particles have a ___ charge.
3. Gold-192 has a half-life of four hours. If you start with a sample containing 80 grams, you will have ___ grams left after four hours.
4. During radioactive ___, unstable nuclei change into stable nuclei by giving off ___.

5. One neutron can make a uranium-235 atom break apart, releasing ___ neutrons.
6. In a ___ reaction, the products of one nuclear reaction can trigger a reaction in other nuclei.
7. In the fusion reaction of the sun, the element ___ is changed into the element ___.
8. The ___ of ___ fuel produces energy for nuclear power plants and for the atom bomb.

INTERPRETING INFORMATION

The graph in Figure 19-19 shows the origins of radiation exposure for the average person in the United States.

1. What is the total percentage of radiation caused by people? What percentage occurs naturally?
2. Which source of radiation contributes the least to the total amount of radiation in our environment?

3. Which source of radiation might be easily reduced, thus making a big cut in radiation exposure?
4. Does the low percentage for nuclear power plants mean they pose only minor problems? Why or why not?

THINK AND DISCUSS

Use the section numbers in parentheses to help you find each answer. Write your answers in complete sentences.

1. What happens to the atomic number of an atom, meaning the number of protons, if it gives off
   a. a proton?   b. a neutron?
   c. an alpha particle? (19-1)
2. Suppose you have a sample containing four grams of Na-24, a radioactive isotope of sodium with a half-life of 15 hours. Make a chart showing the amount of Na-24 left in the sample after 15, 30, 45, 60, and 75 hours. (19-1)

Figure 19-19

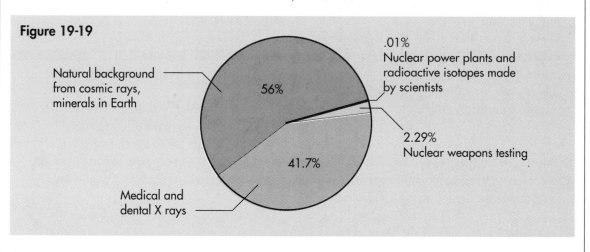

Natural background from cosmic rays, minerals in Earth — 56%

.01% Nuclear power plants and radioactive isotopes made by scientists

2.29% Nuclear weapons testing

Medical and dental X rays — 41.7%

3. Compare fission and fusion by telling
   a. what occurs.
   b. what elements are used.
   c. how each is started.
   d. where they occur.
   e. their energies. (19-2)
4. Give examples of radioactive isotopes used as tracers. (19-3)
5. What factors might determine how harmful a dose of radiation is to humans? (19-4)
6. What issues about the waste products of nuclear reactions concern society? (19-4)

GOING FURTHER

1. Find newspaper and magazine articles that deal with the topics of radioactivity and nuclear energy introduced in this chapter. What issues of safety, health, and the environment are most often discussed in the articles? If you wish, write a report or make a poster highlighting the main issues in each category.
2. Make a detailed list of the radioactive isotopes used in industry, scientific research, and medicine. Tell why and how the isotopes are used in a brief report.

COMPETENCY REVIEW

1. Radioactivity comes from
   a. an atom's nucleus.
   b. an atom's electron shells.
   c. a molecule's chemical bonds.
   d. a material's heat energy.
2. Which type of radiation is an electromagnetic wave?
   a. proton
   b. alpha particle
   c. beta particle
   d. gamma ray
3. A radioactive element has a half-life of one day. Beginning with ten grams, how much will be left after two full days?
   a. 10 grams
   b. 5 grams
   c. 2.5 grams
   d. 1.25 grams
4. Radiation can be detected by
   a. photographic film.
   b. electroscopes.
   c. Geiger counters.
   d. all the above
5. Obtaining energy from uranium involves all of the following *except*
   a. a chemical reaction.
   b. a nuclear reaction.
   c. a fission reaction.
   d. a chain reaction.
6. Hydrogen is used as
   a. a fuel for fission reactions.
   b. a fuel for fusion reactions.
   c. a tracer in medical examinations.
   d. a radiation cancer cure.
7. Which is *not* true about radioactive materials?
   a. They can cause cancer.
   b. They can cure cancer.
   c. They all become harmless in a very short time.
   d. They release energy.
8. The damage done by radiation depends on the
   a. strength and rate of emission.
   b. type of radiation.
   c. exposure time.
   d. all the above
9. What radioactive gas leaks from the ground?
   a. nitrogen
   b. radon
   c. uranium
   d. hydrogen

CHAPTER 20

ENERGY SOURCES: PAST AND PRESENT

In this book, you have learned about types of energy like electromagnetic energy, mechanical energy, and heat energy. You have seen how people can apply energy to do work. Figure 20-1 shows an ancient way and a modern way in which people have applied energy to do work. Much energy comes from fuel, and you know about several fuels. How do people use fuels to do work? Where do they find these fuels? Most importantly, are there enough fuels for people to use in doing more and more work? Scientists, historians, and politicians are studying questions such as these today.

When you watch television, cook food, or light a bulb, you are using energy. You may take energy for granted and barely give it a thought. The amount of energy you use every day, though, is far greater than the amount used by your ancestors or, for example, people living in rain forests today. This chapter examines the energy sources and energy demands of the past and present.

20-1 Energy Trends

■ *Objectives*

☐ *Trace the developments in energy sources and demand.*

☐ *Illustrate how changes in energy sources and technology have changed society.*

There have been three trends in the use of energy. First, each new generation has used more energy than the previous one did. Second, new sources of energy have been developed. Third, there has been a shift from **renewable energy sources** to

Figure 20-1 Two Ways Energy is Applied to do Work

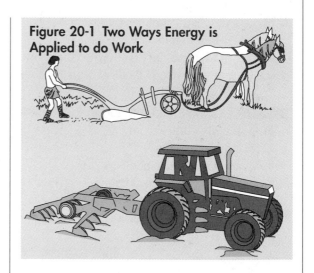

nonrenewable energy sources. A renewable source of energy is one that nature can replace within a short period of time. Fresh water, wind, sunshine, and plants are renewable energy sources. A nonrenewable source of energy is one that is not replaced easily after it has been used up. Coal, oil, gas, and uranium are nonrenewable energy sources. These four are the major sources of energy today.

Before the Seventeenth Century

Ancient food-gathering societies relied mainly on the sun for energy. The sun enabled the growth of plants and trees, which provided food. People spent much time searching for fruits and grains. The discovery of fire provided a new source of heat and light. With fire came a way to make tools and hunting weapons, so hunting societies developed next. The meat of animals provided more concentrated energy than fruits and grains so people spent less time searching for food.

The development of agriculture allowed for more-efficient use of solar energy in raising crops and animals for food. The water wheel was invented in the first century B.C. Early civilizations used water wheels for pumping water to irrigate crops, and for grinding grain. Flowing water provided the energy.

A few improvements in the use of energy occurred before the fifteenth century. There were advances in water wheel construction, and windmills appeared throughout Europe in the twelfth century. Wood remained the main source of heat as well as fuel for trades such as glass-blowing and metalworking.

The Seventeenth and Eighteenth Centuries

As wood was used up and became harder to find, people began to use coal. This substitution marked a change from a renewable energy source, trees, to a nonrenewable energy source. The first steam-operated pump was invented in 1698 to drain water from deep coal mines. This machine led to the invention of the steam engine, shown in Figure 20-2. During the 1700s, the steam engine, fueled by wood and coal, gradually improved.

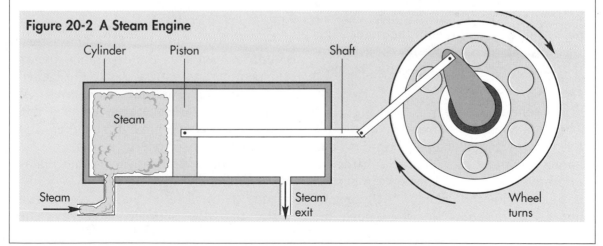

Figure 20-2 A Steam Engine

Cylinder   Piston   Shaft

Steam

Steam

Steam exit

Wheel turns

The steam engine permitted the Industrial Revolution of the mid-1700s to occur. Manufacturing and commerce grew because engines now did the work of muscles, and the engines were faster and stronger than people. At home and on the farm, though, animals and wood were still the main sources of energy. This was soon to change.

The Nineteenth Century

In the 1800s, the use of steam engines increased, and the engines were improved in design and efficiency. People now preferred using coal to using wood. Entire industries, like the textiles industry and farming, became mechanized with these engines. Mechanized industries used machines to do work instead of people or animals. Mechanization was part of the Industrial Revolution that continued to spread in Europe and in America.

The invention of the battery and generator prepared the way for the electrical age. Thomas Edison, who invented the light bulb in 1879, supervised setting up the first electricity system in New York City in 1882. Electricity put a clean and reliable source of energy into the home. This was a time when people used oil just for lubricating machines and lighting lamps.

The Twentieth Century

The automobile was invented in the latter part of the nineteenth century and soon began to use the new **internal combustion engine**. This system burned gasoline inside the cylinder of the engine instead of injecting steam to push the piston. At first, cars were so expensive that only wealthy people could afford them. By 1929, car sales in the United States totaled over 4,455,000. Gasoline engines were used everywhere, so oil exploration boomed. Consumption of electricity also increased by seven times. Factories started to use electric motors and electric lamps more and more.

Between World Wars I and II, 1918–1939, electricity became available throughout the United States. New generating plants and transmission lines brought electricity to cities, towns, and farms. Every ten years, the consumption of electricity doubled. Electricity that came from burning coal and oil became the chief source of energy during this period.

People also built long-distance pipelines. These pipelines brought natural gas to most homes, where people used it for heating and cooking. Natural gas was also used in industry and to generate some electricity.

After World War II, oil replaced coal as the major source of energy. People used oil to make gasoline, heating oil, diesel fuel, and kerosene. Americans used so much oil that the United States grew dependent on other nations for fuel. Middle Eastern countries, such as Saudi Arabia, began shipping oil and natural gas to the United States to run cars and industries. At various times, international political and economic events have threatened the availability of these energy sources.

As the population of the United States has continually grown, the energy demands in the United States have grown faster than the sources of energy. The result has been increased importing of oil and rapidly rising energy costs. In 1950, a gallon of gasoline cost about 30 cents; today it costs about $1.30.

As you read in Chapter 19, uranium as a source of tremendous energy was developed in the twentieth century. Since 1970, however, problems with safety, waste disposal, and costs have halted the construction of nuclear power plants in the United States. Although Europe produces 50 percent of its electricity from nuclear sources, the United States produces only about ten percent this way. Nuclear energy remains a controversial issue at the present time. However, changes in nuclear technology and energy costs may settle the nuclear issue in the future.

S K I L L   B U I L D E R

Interpreting a Table and Making a Graph

Examine the table in Figure 20-3. Keep in mind that *per capita* means "for each person."

1. Explain what the numbers given in this table represent.

Figure 20-3

Society	Energy needed per capita per day, in kilocalories
Primitive	2000
Early civilizations	12,000
Nineteenth century	70,000
Twentieth century	250,000

2. How many times greater is the energy use of people today as compared to the use in primitive society?
3. What possible factors have caused the changes in energy shown in the table?
4. Explain whether or not you think this trend will continue. What energy per capita might you predict for the twenty-first century?
5. Make a bar graph of this data, naming the society types across the bottom axis. You will need a full sheet of graph paper to properly show the energy on the vertical axis. If each line represents 10,000 kilocalories, you will need 25 lines. If each line is 5000 kilocalories, you will need 50 lines.
6. Assume the data for the twentieth century in the table is for the United States. Assume also that in 1990, the population of the United States is 250,000,000 people. How would you calculate the yearly use of energy in the United States in 1990 from these data? Show your work.

Section 20-1 Review

Write the definitions for the following terms in your own words.

1. **renewable energy source**
2. **nonrenewable energy source**
3. **internal combustion engine**

Answer the following questions.

4. Give examples of the three trends in energy described at the start of this section.
5. List the energy sources people have used in historical order.
6. Give examples of an energy source or technology that changed society.
7. What are the key issues about energy facing the United States today?
8. Why has oil become increasingly important since 1900?

20-2 Today's Energy Sources

■ *Objectives*

☐ *Outline how electricity is generated.*

☐ *Describe fossil fuels, hydroelectric energy sources, and nuclear sources of energy.*

☐ *Explain what occurs in a nuclear-reactor core.*

Electricity

Electricity is the major form of energy used in homes and industry today, but it is not a primary source of energy.

Electricity must be made from other, primary sources of energy, such as coal. The following primary energy sources are used to make electricity:

■ Coal, oil, and natural gas were formed by the decay of ancient animals and plants. These are called **fossil fuels**.
■ **Hydroelectric energy** refers to moving water, which is used to turn the generators that make electricity. Rivers and dams provide the water for hydroelectric energy to make electricity; *hydro-* means "water."
■ Nuclear energy is fueled by uranium.

As you know, electricity is produced in heavy coils of wire that spin rapidly within a large electromagnet. The large coil of wire, called the armature, is located at the end of a heavy steel shaft that rotates rapidly. Energy is needed to turn the shaft and armature, both of which weigh many tons. Moving water can do this by spinning a set of fan blades called a **turbine**, which turns the shaft.

Fossil fuels and uranium release heat that is used to make high-pressure steam. The steam makes the blades of a turbine spin. Without these primary energy sources, generators would not turn and electricity would not be made.

The electricity produced by generators is at a high voltage and a very high current. It is sent through thick aluminum wires, known as transmission lines, to distribution centers and cities that may be several hundred miles away. There, transformers reduce the voltage into the standard 120 volts or 240 volts used in homes and businesses.

How Do Turbines Work?

Process Skills *making a model*

Materials cardboard circle, toothpick, thread, and paper clips; flask with one-hole stopper, short glass tubing, and rubber hose; medicine dropper; ring stand and Bunsen burner; boiling chips, which are small stones; paper towels; safety goggles; a pinwheel; clay.

Procedures

1. Cut four slits into a cardboard circle and fold back their edges as shown in Figure 20-4. Push the toothpick through the center of the circle and attach the string with paper clips to it.
2. Place the edge of the wheel under a stream of water. What happens?

Figure 20-4

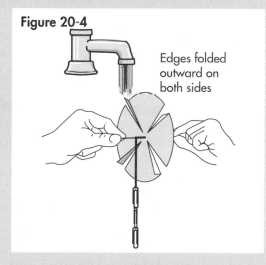

Edges folded outward on both sides

3. **Caution** Use safety goggles. Place a small amount of water in the flask. Add boiling chips to prevent the water from spurting out. Insert the glass part of the medicine dropper at the free end of the tubing, as shown in Figure 20-5.

Figure 20-5

Rubber stopper with glass tube

Pinwheel and clay

Medicine dropper end

4. Using the Bunsen burner, gently boil the water in the flask. Do not produce steam any faster than it can escape from the medicine dropper at the end of the tubing.
5. Hold the end of the tubing with paper towels. Direct the steam at the pinwheel. What happens?
6. Turn off the burner.

Conclusions

7. What energy changes occurred in both experiments?
8. In what ways do these experiments show how electricity is made?
9. Which method wastes more energy?

Crude Oil

Crude oil is a liquid fossil fuel. It is a source of energy for generating electricity, heating, and transportation. Crude oil is a mixture of organic chemicals, mainly hydrocarbons. These were formed by chemical changes in dead plants and animals caused by heat and pressure underground.

Crude oil, which is also called petroleum or oil, is trapped in soft rocks and large pools underground. More than half of the world's oil is located in Middle Eastern countries such as Saudi Arabia and Iran. The United States has only about five percent of the total oil in the world. Pennsylvania, West Virginia, Oklahoma, and Texas have the main supplies of petroleum.

People remove oil from underground or from under the ocean floor by pumping it from wells. An oil well is a hole drilled into the ground that contains a pipe to reach the oil. In some states, there are oil wells everywhere, even in people's backyards. An oil well may go down 5000 feet and can cost from $100,000 to several million dollars to drill. Most wells turn out to be dry. Only a small percentage produce enough oil to make drilling profitable.

After the oil is pumped, it is transported to **refineries** by pipeline or by ship. In refineries, workers separate the oil mixture into its many compounds by boiling. Each compound is purified and further processed. These compounds are known as petroleum products. They are sent by pipelines, trucks, and barges for storage and distribution to all parts of the nation.

Crude oil contains materials that can burn. It also contains the basic materials people use to make synthetic compounds, such as plastics and medicines. These are called the **by-products** of oil. Oil is known as black gold because its energy and compounds are so valuable. Figure 20-6 shows some products that are made from crude oil.

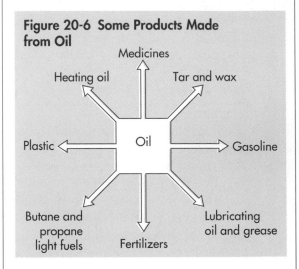

Figure 20-6 Some Products Made from Oil

Medicines
Heating oil
Tar and wax
Plastic
Oil
Gasoline
Butane and propane light fuels
Fertilizers
Lubricating oil and grease

Natural Gas

Natural gas is a fossil fuel used for home cooking and heating and also for making electricity. Natural gas consists mainly of methane, which has the formula CH_4. Gas companies add an odor to the gas so that people can smell the gas if it is leaking.

Natural gas is found underground in the same areas that contain crude oil. People pump the gas from wells to gas refineries, to storage tanks, and eventually to the customer. Natural gas changes into a liquid at $-160°$ C. In this form, as Liquified Natural Gas (LNG), the gas has much less volume, and it is easy to ship around the world inside high-pressure metal tanks.

One big advantage of natural gas is that it burns cleanly, producing no smoke. It also contains no sulfur compounds, so no sulfur dioxide (SO_2) is formed. Burning natural gas (CH_4) only forms carbon dioxide (CO_2) and water, as you can see in the following equation.

$$CH_4 + 2\ O_2 \rightarrow CO_2 + 2\ H_2O$$

Some by-products of natural gas are rubber, drugs, and dyes.

Coal

When ancient plants and animals were buried under swamps, lakes, or mud, they decayed slowly. They eventually formed a soft material known as peat. Peat can burn, and people dig it up and use it for fuel in some places. Over time, as peat beds sink deeper, the pressure and temperature underground change them into coal. Coal contains plenty of pure carbon. Coal, oil, and natural gas are often found close together.

The United States has a large supply of coal, mainly in the Rocky Mountains, the Midwest, and the Appalachian Mountains near Pennsylvania. Figure 20-7 shows some types of coal and their characteristics.

People mine coal using two methods. In strip mining, they remove coal after bulldozing away the soil and rock that covers it. In underground mining, they remove coal by digging tunnels. Conveyor belts bring the coal to the surface, as shown in Figure 20-8.

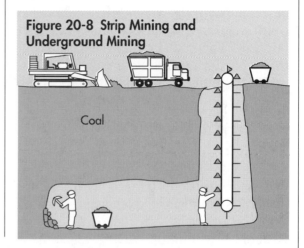

Figure 20-8 Strip Mining and Underground Mining

Coal

Figure 20-7 Types of Coal

Type	Hardness	Heat content	Sulfur content
Lignite	Soft	Low	High
Bituminous coal	Hard	High	Low
Anthracite	Hardest	Highest	None

The coal is then washed and crushed into pieces of certain sizes. Trucks and trains carry the coal to many storage places and then to the customers. Coal can be stored in piles in the open air.

Coal is used mainly to produce electricity. Many homes and industries still burn coal for heat, although this practice has been decreasing for many years. Coal may be baked in air-tight ovens to remove coal tar and oil materials. The material left behind is called coke, which is very pure in carbon. People use coke in manufacturing steel. They also use coke in furnaces that require very clean burning. The coal tars and oils removed from coal during baking are used to make many by-products, such as plastics, dyes, and detergents.

Hydroelectric Power

The kinetic energy of moving water can be used to spin the giant turbines that generate electricity. This is hydroelectric power. Hydroelectric plants are located along major rivers, especially in California, Oregon, Washington, Arizona, and New York. Dams are built to hold large amounts of water in deep artificial lakes high above the power plants. As you can see in Figure 20-9, water released from the dam makes energy available to the hydroelectric plant.

As water in larger pipes drops hundreds of feet from the dam, it gains kinetic energy. At the lowest level, the water strikes the blades of a turbine. River water forced to collect in such lakes ensures a steady flow of water through the plant even if the river flow decreases.

Uranium

People mine uranium ore mainly in the Rocky Mountain states and in Texas. Ordinary uranium contains one percent of the isotope U-235 and nearly 99 percent of the isotope U-238. However, only nuclei of U-235 are **fissionable**, meaning they can split apart to release nuclear energy as heat. To maintain a controlled chain reaction, the uranium ore must be processed so that it contains at least three percent U-235. The final form, called enriched uranium, is shaped into pellets, like

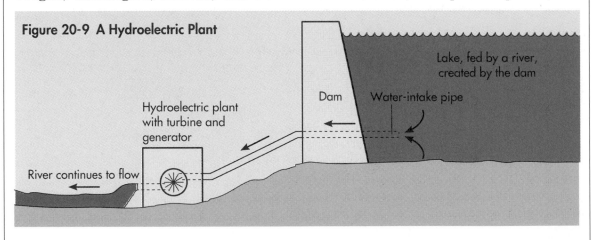

Figure 20-9 A Hydroelectric Plant

Lake, fed by a river, created by the dam

Dam   Water-intake pipe

Hydroelectric plant with turbine and generator

River continues to flow

checkers. These pellets are used in nuclear power plants.

Figure 20-10 is a diagram of a nuclear power plant.

Figure 20-10 A Nuclear Power Plant

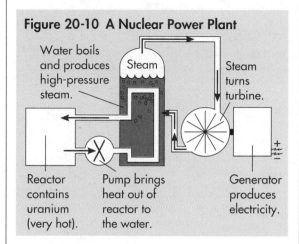

Water boils and produces high-pressure steam.

Steam

Steam turns turbine.

Reactor contains uranium (very hot).

Pump brings heat out of reactor to the water.

Generator produces electricity.

The heart or center of a nuclear plant is the **reactor core**, where the chain reaction occurs. The core is very hot. Figure 20-11 shows a diagram of a reactor core.

There are three parts in the core. Fuel rods are long metal tubes containing uranium pellets. As fission proceeds, the trillions of neutrons produced each second cause the U-235 nuclei to split apart inside fuel rods. Neutrons pass between neighboring fuel rods. The moderator material surrounds the fuel rods. It slows the neutrons so that they are more effective in splitting more uranium atoms.

The control rods are made of materials, such as boron, that stop neutrons. They are positioned between the fuel rods. When someone lowers the control rods into the core, the fission decreases. Control rods control the rate of the reaction and can instantly stop the reaction in an emergency.

Figure 20-11 Reactor Core

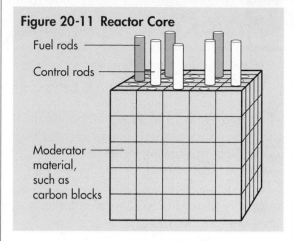

Fuel rods

Control rods

Moderator material, such as carbon blocks

■ Section 20-2 Review ■

Write the definitions for the following terms in your own words.

1. **fossil fuels**
2. **hydroelectric energy**
3. **turbine**
4. **refineries**
5. **reactor core**

Answer these questions.

6. What are the important primary sources of energy today? Why is electricity not included among them?
7. Why is crude oil known as black gold?
8. Why is it not surprising that coal, oil, and natural gas often are found in the same areas?
9. Why are dams part of hydroelectric plants?
10. What are the purposes of the fuel rods, moderator material, and control rods inside a nuclear reactor's core?

SCIENCE, TECHNOLOGY, & SOCIETY

Building Dams and Making Trade-offs

A trade-off tries to balance the good and bad sides of choices when making a decision. Building dams involves trade-offs. On the good side, dams increase the production of electricity while reducing the cost of electricity for millions of people. Building a dam and a generating plant provides thousands of jobs. The lake formed behind the dam offers recreation such as boating, fishing, and swimming. Dams reduce flood damage because the extra water fills the lake instead of overflowing the river.

On the bad side, dams sometimes require that farms, and even whole towns, be relocated, or else they would be flooded. For example, monuments in Egypt were submerged when a dam was built on the Nile River. Dams also reduce the amount of water in a river, so farms and cities may not get the water they need. Situations like this lead to lawsuits. The Colorado River is so dammed up that Los Angeles, California, has lost much of the water it used to get from that river. Boat traffic on a river is blocked by a dam, unless expensive canals are built to bypass the dam. Sometimes, dams disrupt fish and animal life. For instance, salmon must swim up a river to lay eggs. A dam can prevent salmon from swimming up river and cause an entire population to die out.

A dam raises tough political questions. Who will own and operate the dam, the government or a company? Who will pay for its construction? Who must be ordered to move, and by whom? During the 1930s these issues were addressed by the Army Corps of Engineers, which was responsible for building dams. The largest dam and electrification project at that time was the Tennessee Valley Authority, often called TVA, which made hard choices and difficult trade-offs.

Figure 20-12

Follow-up Activities

1. Laws are stricter today than they were when many of the dams in this country were built. Dams that permanently harm fish life may not get approved. What is your opinion on this?
2. Gather information on building dams. Focus on the issues and trade-offs that are being faced today.

■■■■■ KEEPING TRACK ■■■■■

Section 20-1 Energy Trends

1. The consumption of energy to meet human needs has increased with passing time.
2. Societies have shifted from renewable to nonrenewable sources of energy.
3. Nonrenewable energy sources, such as fossil fuels, cannot be replaced.
4. The steam engine, internal combustion engine, and electricity have changed society and its energy needs.

Section 20-2 Today's Energy Sources

1. Electrical energy is a key part of our society. It is used to heat and cool buildings, run machines and appliances, and provide lighting. Electrical energy must be produced from other energy sources, such as fossil fuels, uranium, and falling water.
2. Fossil fuels are energy sources formed over millions of years from dead plants and animals. Some fossil fuels are oil, coal, and natural gas.
3. Gasoline and heating fuel come from oil and they are the main oil products used for energy.
4. Natural gas is used chiefly to produce heat for industrial needs and for heating and cooking in the home.
5. Plastics, fertilizers, and drugs are products made from fossil fuels.
6. Coal is used mostly to produce electricity and to provide heat for industrial needs.
7. The energy of moving water can be used to produce electrical energy and is known as hydroelectric power.

8. The nuclear energy contained in uranium is used to produce heat, which in turn produces electricity.

■■■■■ BUILDING VOCABULARY ■■■■■

Write the term from the list that best completes each sentence.

by-products, fissionable, fossil, internal combustion, nonrenewable, reactor core, refinery, renewable

Trees are a ___1___ source of energy because, after they are cut down, others grow in their place. Engines that run on gasoline or diesel fuel are ___2___ engines. Coal, oil, and natural gas are all ___3___ fuels and are examples of ___4___ sources of energy. Plastics and medicines are ___5___ of fossil fuels. Uranium-235 is the ___6___ material in the ___7___ of a nuclear plant. Oil is separated into its components in a ___8___.

■■■■■ SUMMARIZING ■■■■■

Write *true* if the statement is true. If the statement is false, change the *italicized* term to make the statement true.

1. *Hydroelectric* plants use water.
2. The use of energy has *slowly* increased in the last one hundred years.
3. The *electric motor* permitted the eighteenth-century Industrial Revolution.
4. Electricity is a *primary* energy source.
5. The United States has a *small* percentage of the world's total oil.

6. Coal, oil, and natural gas *are often* found near one another.
7. Of the fossil fuels, *natural gas* burns cleanest.
8. The *moderator material* in a reactor's core can stop neutrons and therefore can stop the reaction.

INTERPRETING INFORMATION

The graph in Figure 20-13 shows the energy sources used for making electricity in the United States.

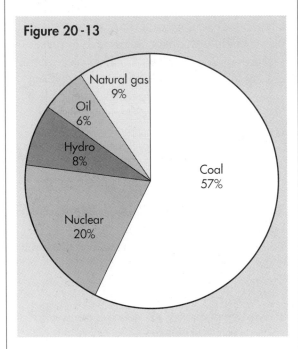

Figure 20-13

Natural gas 9%
Oil 6%
Hydro 8%
Coal 57%
Nuclear 20%

1. What is the major fuel for making electricity?
2. What percentage of the sources shown are fossil fuels?

3. What percentage of the sources shown are nonrenewable?
4. The lowest percentage on this graph is for oil. However, more oil is used in the United States than any other fuel. Is this a contradiction? Why or why not?
5. Is it correct to say that coal supplies about one half, and nuclear fuel about one fifth, of the energy used to make electricity? Why or why not?

THINK AND DISCUSS

Use the section numbers in parentheses to help you find each answer. Write your answers in complete sentences.

1. Compare the energy needs and sources of people 2000 years ago and those of society today. (20-1)
2. Why is the United States called an energy-dependent country? What disadvantages come from this situation? (20-1)
3. Except for uranium, all energy sources can trace their energy to the sun. Explain this. (20-1, 20-2)
4. Outline the key steps in making electricity from coal. (20-2)
5. Compare the amount of processing that oil, coal, and natural gas require after people take them from the earth. (20-2)
6. Describe the fuel used and the safety features needed in the core of a nuclear reactor. Why do slowing neutrons and stopping neutrons have completely different effects in nuclear reactors? (20-2)

CHAPTER REVIEW

GOING FURTHER

1. Investigate and list at least 20 products derived from fossil fuels.
2. Find out in detail how oil or natural gas gets from the well to your house.
3. Make a detailed sketch of a nuclear plant. Describe where problems can arise in its operation.

COMPETENCY REVIEW

1. Which statement is *not* correct regarding energy in human history?
   a. Energy demand has increased greatly.
   b. New energy sources were needed.
   c. Renewable sources replaced nonrenewable sources.
   d. New inventions used more energy.
2. Which is a renewable energy source?
   a. uranium   b. river water
   c. coal   d. oil
3. Which is a fossil fuel?
   a. natural gas   b. wood
   c. uranium   d. animal oil or fat
4. Which one of the following is not a primary energy source?
   a. electricity   b. moving water
   c. coal   d. uranium
5. Which graph in Figure 20-14 shows how the need for energy changed in the last 200 years?
6. In making electricity, the energy source is needed to
   a. supply electrons.
   b. turn a turbine.
   c. conduct electrical energy.
   d. create sparks.
7. The main purpose of a hydroelectric dam is to
   a. prevent river floods.
   b. provide lake recreation.
   c. store water for drinking.
   d. provide moving water whenever it is needed.
8. What is the fuel in nuclear reactors?
   a. neutrons   b. moderators
   c. uranium   d. boron control rods

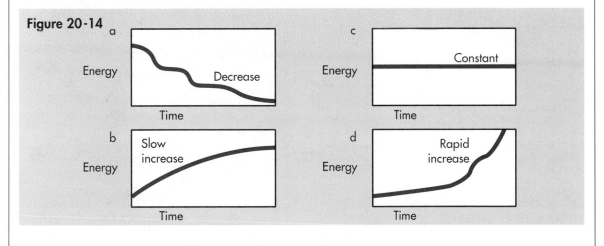

Figure 20-14

a. Energy / Time — Decrease
b. Energy / Time — Slow increase
c. Energy / Time — Constant
d. Energy / Time — Rapid increase

CHAPTER 21

ENERGY ISSUES FOR THE UNITED STATES

During the early 1970s, Americans faced an energy crisis for the first time. Ask your parents, older relatives, or older friends about the oil crisis in those years. Heating oil for homes became very expensive. The price of gasoline tripled, and there were gasoline shortages. People would wait in line for an hour to buy gasoline. To reduce the wait, some states tried an odd/even system for buying gasoline. Cars with odd-numbered license plates could get gasoline only on odd-numbered days, and those with even-numbered plates could get gasoline only on even-numbered days.

Since then, oil is easier to get, but at much higher prices. The basic problems, however, have not gone away. The first problem is demand. The demand for energy keeps growing faster than energy sources. Second, the United States depends on foreign countries for its major sources of energy, which are oil and natural gas. You will learn more about these two problems in this chapter.

21-1 Patterns of Energy Use

■ *Objectives*

☐ *Describe the United States energy crisis.*

☐ *Survey where energy is used and obtained in the United States.*

☐ *Describe the advantages and disadvantages of electricity.*

☐ *List the causes of our increasing demand for energy and the decreasing supply of energy.*

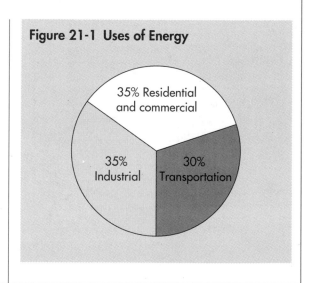

Figure 21-1 Uses of Energy

35% Residential and commercial

35% Industrial

30% Transportation

Where does the United States get its energy? How does the United States use energy? Why is the demand for energy increasing faster than the supply? These questions have very complex answers involving the patterns of energy use in this country.

Energy Use in the United States

The United States has only about six percent of the world's population but uses about 30 percent of the world's total energy. That is a remarkable fact. What might you think about that fact if you lived in another country?

Energy use in the United States can be broken into three **sectors**. These three sectors are the key parts of our economy. The transportation sector involves moving people and goods by cars, trucks, planes, trains, and boats. The residential and commercial sector includes energy for homes, public buildings, and workplaces. This energy can be used in cooling, heating, cooking, refrigeration, lights, and electrical appliances. The industrial sector involves building things and making products. Figure 21-1 compares energy use in the three sectors. The percents shown are approximate.

Within the three sectors, Americans use energy in various forms. Electricity has become the leading form of energy. In fact, 40 percent of the total energy used in the United States goes into making electricity. The rest, sixty percent, of the energy used in this country involves the direct use of various fuels such as gasoline, coal, natural gas, heating oil, jet and diesel fuel, wood, kerosene, and alcohol.

Electricity Is Costly

People prefer to use electricity because it is convenient and clean. Electricity, however, is very costly to make. As stated previously, 40 percent of all the energy used in the United States goes into making electricity. What may surprise you is that only about 25 percent of that amount actually becomes electrical energy. The other 75 percent is lost as heat in making or transmitting the electricity. This means that about 30 percent of all the energy used by the United States is lost in making electricity! Study the following equations.

percent of all energy used to make electricity $\times$ **percent lost as heat**

= percent of energy wasted

$$40\% \times 75\% = 30\%$$

As the equation indicates, electricity is less energy efficient than primary sources of energy, such as oil and natural gas. For example, cooking and heating with electricity uses four times more energy than cooking and heating with natural gas and heating oil. This is because about 75 percent of the energy used to make the electricity is lost as heat in the generator and transmission lines.

Where the United States Gets Energy

Oil is the largest source of energy in the United States today. The United States obtains energy both from its own natural resources and by buying it from other countries.

Sources of the United States' energy are shown in Figure 21-2. As you can see, oil, natural gas, coal, and nuclear energy are the major sources of energy in the United States. Figure 21-3 shows how the United States uses these major sources of energy.

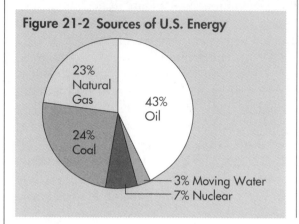

Figure 21-2 Sources of U.S. Energy

Personal Usage Of Energy

About 40 percent of all the energy used in the United States is used for personal transportation and in homes. If Americans reduce their personal use of energy, the country would need much less energy. Individuals can begin saving energy, and if this catches on, it will greatly reduce the energy used in the United States.

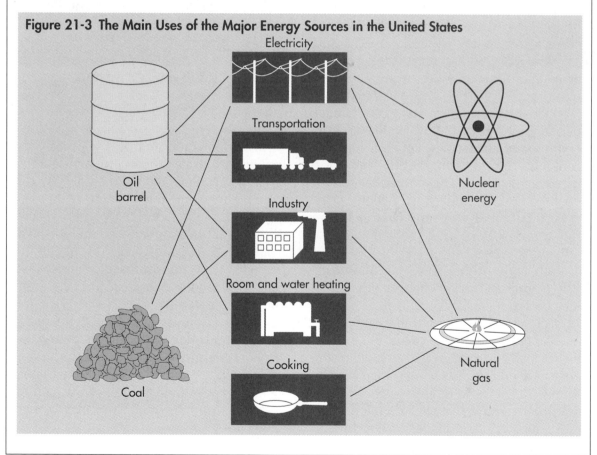

Figure 21-3 The Main Uses of the Major Energy Sources in the United States

Increasing Energy Demands

The general trend is that energy use is increasing. On a few occasions, the energy use per person has decreased because of shortages, conservation campaigns, a poor national economy, or an increase in price. Energy reduction in these cases shows that energy usage can be modified. Energy usage rises and falls but has generally been rising. Reasons for the rise are increased population, an improved economy, availability of energy, and lowered costs.

People are using more energy for a variety of reasons. Many goods and appliances require a lot of energy either to make or to operate. Advertising encourages people to buy these products. For example, people today desire air conditioning, several radios and TVs, and plastic products. In offices there are computers, FAX machines, and photocopy machines. These modern products require energy to make and operate.

Americans use cars a lot. A large system of interstate highways has encouraged automobile travel during the last 30 years. This has led to an increased use of energy.

When companies and people increase their use of natural gas or electricity, they are charged less for each unit of energy. The reduced cost, however, encourages additional energy use. Tax credits also encourage industries to expand and modernize, usually by adding or enlarging equipment. These changes are not necessarily made with a goal of efficient energy use. In many cases, energy use increases.

The number of people living in the suburbs has been increasing. A suburb is a smaller community near a city. The housing in suburbs is mostly for single families. The growth of the suburbs increases energy demands because heating single-family homes requires more energy per person than heating apartment buildings. In addition, people who live in suburbs often have jobs in the city. Many of them use cars to drive to work, resulting in increased energy use.

Finally, technology has provided society with machines and labor-saving devices that need energy to work. Elevators, power tools, kitchen appliances, and snow blowers are examples of **energy-intensive** equipment. Energy-intensive devices consume much more energy than the equipment they replace. Other energy-intensive technological devices include recreational equipment such as power boats, snowmobiles, and motorcycles. Figure 21-4 illustrates energy trends, which are being observed around the world as more nations develop.

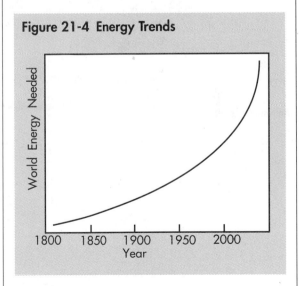

Figure 21-4 Energy Trends

How Can You Use the Library to Research a Topic?

Process Skills *gathering and recording data; organizing*

Materials card catalog, paper, pencil, *Reader's Guide to Periodical Literature,* other library references

Visit your school or local library with your classmates. Ask your teacher or the librarian to show you how to research a technical topic.

Procedures

1. Use the card catalog to locate books on technology themes. Both the topics and book titles are listed alphabetically. Use various wordings to find as many references as possible. For example, look up "energy," "energy crisis," "nuclear energy," "energy, nuclear," and so on.

2. List five books you find about a topic in technology that interests you. Include the title, author, library call number, copyright date, and number of pages. Include any brief summary that may be given about the book in the card catalog.

3. Use the *Reader's Guide to Periodical Literature* to find five articles from magazines or journals on your topic. List each article's title, author, the magazine name, its volume number and issue, and its date of publication.

4. Use any other library references to locate more information on your topic. For example; ask the librarian how to use the library's computer directory, the *New York Times Index* of news articles, the encyclopedias, and the almanacs. You can find many articles on microfilm or in computer information networks that are linked to larger libraries by telephone. Try to use at least three special sources of information like these.

5. What scientific and technical magazines does your library get? Ask for their yearly indexes and find three articles related to your topic.

Conclusion

6. Read three of the references you found. Write a one-paragraph summary of each article or part of a book. Tell some of the interesting facts you learned in each reading. Did you find this reference easy or hard to read? How did the author get your interest?

Figure 21-5

Decreasing Energy Supplies

The need for energy keeps growing faster than old and new sources of energy. In some recent years, the rate of fossil-fuel production has actually gone down. Various factors work to decrease the supplies of energy available to the United States and to the rest of the world.

One factor has to do with economics. It is very expensive to look for oil and natural gas and to drill wells for these energy sources. It is cheaper to search for and drill oil in Middle Eastern countries than in America because of our labor, materials, insurance, and safety costs. Some people feel that tax laws in the United States encourage American companies to locate oil and natural gas outside rather than inside the United States. This could eventually decrease the amount of exploration for oil and natural gas in the United States.

A second factor in decreasing sources of energy has to do with the environment. People wish to reduce or avoid the pollution problems that come with using certain fuels. For example, coal was a leading source of energy for many years. However, burning coal causes air pollution, so the use of coal has been greatly reduced.

Finally, the **reserves** of fossil fuels are being used up. A reserve is the amount of fossil fuels or uranium in an individual area, in an individual country, or in the world. It is now harder to find new places containing oil, and it is harder to remove oil from places that previously were full. In the United States, about 80 percent of the known oil reserve will have been used up by the year 2000. By 2000, natural gas will also be in short supply in United States. These are serious facts to face. Soon the United States may have to buy even more oil and natural gas from other nations.

Much of the hydroelectric energy possible within the United States is already being used. Most large rivers are dammed and have hydroelectric plants. Even if the number of hydroelectric plants could be doubled by the year 2000, only ten percent of our electrical needs could be met by the additional plants. Uranium ore reserves are expected to provide enough U-235 to last into the middle of the next century, about 50 or 60 years from now. As you can see, long-range planning regarding energy is needed now.

■■■ Section 21-1 Review ■■■

Write the definitions for the following terms in your own words.

1. **sectors**
2. **energy-intensive**
3. **reserves**

Answer these questions.

4. The United States faces an energy crisis. Give some reasons to justify this statement.
5. What are the advantages and disadvantages of electrical energy?
6. Give some reasons for the rising need for more and more energy.
7. Give some reasons for the decreasing energy supplies.

21-2 Energy Issues and Implications

■ *Objectives*

☐ *Describe environmental issues connected with energy.*

☐ *Suggest how energy problems may affect our standard of living.*

☐ *Give examples of how oil gets involved with international relations.*

Most people only think about how the United States energy crisis affects them personally. Electricity and gasoline prices are up. This car saves more energy than that car. Sometimes, gasoline or heating oil is in short supply, and people have to wait for it. These are some individual issues connected with energy. There are also bigger issues for our environment, for our nation, and for international relations.

Environmental Issues

Environmental issues have led to the decreased use of some fuels because of the problems caused by their use. Also, environmental issues are always present when companies and governments try to open up more sources of energy.

Air Pollution Burning fuel, especially coal, causes air pollution. Sulfur and sulfur compounds that are in with fossil fuels are especially damaging. During burning, poisonous sulfur dioxide (SO_2) is produced along with carbon dioxide (CO_2).

Attempts to clean up the air have made other environmental problems worse. For example, industry is using less coal because of standards set by the new, tough **Clean Air Act** passed by Congress. This law has led to a switch to oil and natural gas, which are less available and more expensive. No one wants to use up all the oil and natural gas. Returning to a wider use of coal, however, will increase the pollution in the air.

Antipollution devices in cars have helped to reduce air pollution caused by car exhaust. The devices, however, cause cars to get fewer miles for each gallon of gasoline. Cars then need more gasoline, which leads to increased use of oil.

Land and Water Pollution News reports seem to regularly tell of oil spills. Spills can occur when the crude oil or fuel is transported by pipelines, ships, or trucks. Activities such as coal mining and drilling for oil have increased pollution on land and in water.

Oil spills and illegal dumping of chemicals and garbage in water have caused many rivers to become unattractive, even unhealthy. The nation depends on those energy supplies, however. Perhaps improved procedures and heavy fines for polluting will reduce the number of accidents and illegal dumpings.

The construction of dams and pipelines can damage the ecology of an area. Steps must be taken to reduce any harmful effect. Today, any project proposal for dams, roads, nuclear plants, and buildings must contain an **environmental impact statement**. This statement lists the possible negative effects the project can have on the environment. The environment includes people, animals, plants, air, land,

and water. The impact statement lists methods of lowering or avoiding negative effects. An environmental impact statement is available to the public, and people have the right to debate it before work on the project begins.

The United States is now committed to searching for oil in the ocean. It takes time for people to debate environmental impact statements, however, and the objections of environmentalists and public officials must be addressed. Even with great care, ocean pollution will probably increase, causing further damage to valuable fishing areas and beaches.

Most possible new hydroelectric sites in the United States exist only in national parks and forests. Building dams and generating plants along these rivers would result in a great loss of beauty and recreational areas. People need parks, away from modern technology, as part of their enjoyment of life.

Thermal Pollution Electric power plants produce large amounts of heat. This heat is released into the air or into river water, where it may harm animals and plants. Cars, air conditioners, and other machinery contribute to thermal, or heat, pollution. Thermal pollution is worst on hot summer days in cities. One solution to this problem is to make generating plants more efficient so that more fuel energy winds up as electricity, not as heat.

Nuclear Hazards Thirty years ago, people believed that by now 40% of the electricity used in the United States would come from nuclear power. The actual number is about 10%. Why did the promise of nuclear energy fail? During the past 30 years, scientists and the public became very concerned about the tremendous dangers of nuclear energy. Processing uranium ore and operating a nuclear reactor always present the dangers of explosion and radiation leakage. The cost for safety devices and procedures has made electricity from nuclear plants as expensive as from non-nuclear plants. Thus, nuclear power has not resulted in any great savings.

Nuclear power plants also produce dangerous radioactive wastes that must be transported over long distances for burial. Many communities have resisted the construction of nuclear plants in their areas, as well as the storage and transport of nuclear wastes. The economic, political, and legal battles over nuclear energy have made it unattractive to electric companies at this time. Although Europe produces about 50% of its electricity at nuclear power plants, the United States may have to wait for an even bigger energy crisis before building more nuclear power plants.

Health Hazards to Energy Workers Energy workers such as coal miners, oil-well operators, and nuclear technicians are protected by strict health and safety laws. These laws are very welcome. Procedures that help to protect the health and safety of workers, however, can reduce productivity, especially in underground coal mines. For example, miners must take long breaks from work after a certain number of hours in the mine. This trade-off benefits the health of miners, but yields less coal.

Analyzing

A diagram can be useful for helping you understand some kinds of data. Look at Figure 21-6.

1. What percentage of oil is used to make gasoline?
2. What activity uses the greater amount of oil products in total, heating or transportation?

3. What is the major use for coal?
4. What percentage of coal is not used in the United States?
5. The United States uses about 700 million gallons of oil every day!
   a. How many of these gallons are "lost," or used up, during the refining process?
   b. How many of these gallons are used for transportation?

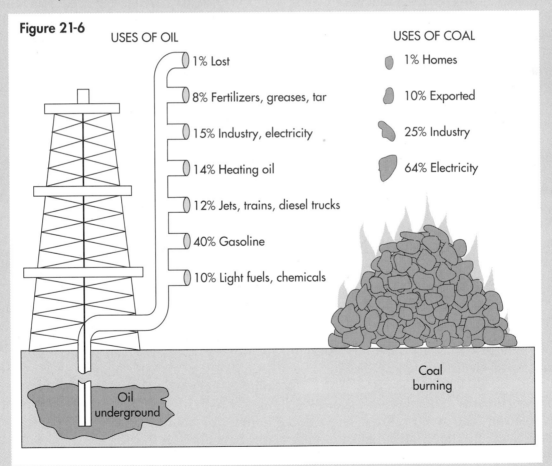

Figure 21-6

USES OF OIL

- 1% Lost
- 8% Fertilizers, greases, tar
- 15% Industry, electricity
- 14% Heating oil
- 12% Jets, trains, diesel trucks
- 40% Gasoline
- 10% Light fuels, chemicals

USES OF COAL

- 1% Homes
- 10% Exported
- 25% Industry
- 64% Electricity

Oil underground

Coal burning

Standard of Living

The United States enjoys a high **standard of living** compared with other nations. The standard of living is the quality and quantity of food, shelter, education, health, jobs, and leisure time available to most people. A country's standard of living depends on the energy sources that are available to it. Some people fear that the energy crisis will result in a reduced standard of living. They may be right. The cost of the electricity and fuel to operate luxury items may make them too expensive to own.

Several nations enjoy a standard of living like our own, yet consume far less energy per person. Switzerland, Sweden, Denmark, Norway, and Australia are examples. Therefore, if Americans can stop wasting energy, it appears likely they can maintain a good standard of living. Society, however, must take energy issues seriously. One approach to conserving energy is to drastically raise energy costs. When the price of energy goes up, people usually cut their use of energy.

International Relations

The need for imported oil has made the United States dependent on the Middle East, a politically unstable area. Vital oil supplies can be cut off by events that the United States cannot control. In recent years, the U.S. Navy has patrolled Middle Eastern waters to make sure oil-shipping routes are kept open for all nations.

Many of the countries that produce large amounts of oil belong to OPEC, the Organization of Petroleum Exporting Countries. OPEC is an international business organization that represents some of the countries that produce oil. OPEC sets prices and production quotas for its members. Since the United States depends on OPEC nations for much of its oil, it tries to maintain friendly relations with the member countries.

The United States must also consider the energy needs of small, developing nations. These nations believe their technological advancement will be hurt if they get less oil because the United States takes so much. This is a reason for being a good neighbor on the planet and sharing energy resources fairly.

■■■ Section 21-2 Review ■■■

Write the definitions for the following terms in your own words.

1. **Clean Air Act**
2. **environmental impact statement**
3. **standard of living**

Answer these questions.

4. Briefly describe the main environmental issues related to United States energy use.
5. Give two examples that show how trying to solve one energy-related problem can cause another problem.
6. What evidence exists that Americans can cut back on energy use and still keep a good standard of living?
7. What happened to the promise of cheap and abundant electricity from nuclear energy?

SCIENCE, TECHNOLOGY, & SOCIETY

America's Worst Oil Spill

Figure 21-7

Prince William Sound is in the circled area.

On March 24, 1989, an American oil tanker hit a reef in Prince William Sound, Alaska. About 11 million gallons of oil leaked from a gash in the ship's hull. The oil spread over rich fishing areas, killing many fish. It washed ashore for hundreds of miles, making the shore dangerous for people, birds, seals, and other living things. Important growths of sea plants were destroyed, as were many marshes that were wildlife refuges.

The oil company, the state, and the federal government spent millions of dollars to clean up, but recovered only half the oil. Hundreds of lawsuits were filed against the captain and the company for damages to property and business.

Several proposals have been made to prevent future oil spills. For example, tankers and barges should have double hulls. A hole in the outer hull would not cause a leak of oil contained within the inner hull. Also, radar control of ships near ports needs to be as good as the control of planes near airports. A radar system, which might be combined with satellite photos, could cost over $450 million. Special pilots steer a ship in channels near the shore. The pilots leave the ship and turn it over to the captain once it is out at sea. New rules may make a pilot stay with tankers longer. Pilots and captains might be given joint training on more safety measures. Finally, the fines and money for damages that a captain and a company might have to pay for any accidents should be sharply increased.

These proposals will take time to debate and enforce. One question is, Who pays for the special equipment and policing needed? New technology can provide some answers to the problem, but it needs the assistance of a well-informed society, new laws, and safety procedures as well.

Follow-up Activities
1. Investigate an accident involving fuels that happened in the last five years. Prepare a report.
2. Look for other safety suggestions regarding energy. Tell which are related to technology and which are not.

KEEPING TRACK

Section 21-1 Patterns of Energy Use

1. The demand for energy in the United States has been growing faster than present sources can supply.
2. Most of the United States energy demand is met by fossil fuels. The United States imports much of the oil it uses. United States reserves of coal are much greater than those of oil and natural gas.
3. Worldwide reserves of fossil fuels and uranium are running out.

Section 21-2 Energy Issues and Implications

1. The environmental problems connected with energy production include air, water, land, and thermal pollution and the disposal of nuclear energy wastes.
2. Burning fossil fuels, especially coal, contributes to air pollution.
3. Activities like mining for coal, drilling for oil and gas, and transporting fuel by ship may increase pollution on water and land.
4. Construction of dams and pipelines may damage the ecology of an area.
5. Nuclear power plants produce dangerous waste that must be stored somewhere.
6. Energy issues play a large part in the standard of living and international relations of the United States.

BUILDING VOCABULARY

Write the term from the list that best matches each statement.

Clean Air Act, energy-intensive, environmental impact statement, reserves

1. requires a lot of energy to make or operate
2. natural supplies of fuel in a country
3. sets standards to reduce pollution
4. lists possible harmful effects of a project

Explain how the terms in each pair are related.

5. Clean Air Act, reserves
6. energy-intensive, standard of living
7. sectors, Clean Air Act

SUMMARIZING

Write the missing term for each sentence.

1. Transportation needs consume about ___ percent of United States energy.
2. Making electricity takes about ___ percent of United States energy, and only ___ percent of that becomes electricity.
3. ___ is the fuel used to supply the highest percentage of United States energy.
4. The growth of suburbs ___ the energy requirements of the United States.
5. While coal is abundant in the United States, it contributes a lot to ___.
6. Most hydroelectric sites not yet dammed in the United States are located in ___.
7. ___ has not yet lived up to its promise of generating safe, inexpensive, abundant electricity.
8. Smaller nations now demand their share of ___ in order to develop.

INTERPRETING INFORMATION

Figure 21-8 compares the energy costs for different means of transportation.

1. What is the least energy-intensive means of transportation
   a. between cities?
   b. within cities, excluding bikes and walking?
   c. for shipping freight?
2. About how many times more energy is needed within a city to drive a car than to walk?
3. Which is the least energy-intensive means of travel?
4. A bus needs more fuel per mile than a car. Why is it listed as less energy-intensive than a car in this table?
5. Why do people tend to use more energy-intensive means of travel?
6. How many BTU's are needed to transport 40 tons of freight for 300 miles by truck?

THINK AND DISCUSS

Use the section numbers in parentheses to help you find each answer. Write your answers in complete sentences.

1. What are some problems that arise in finding energy sources? (21-1)

Figure 21-8

Between cities		Within a city		For shipping freight	
Method	Energy in BTUs per mile per person	Method	Energy in BTUs per mile per person	Method	Energy in BTUs per mile per ton shipped
Train	2900	Walking	300	Boat	680
Bus	1600	Bicycle	200	Pipeline	1900
Car	3500	Train	2300	Train	700
Plane	8400	Bus	3800	Truck	2800
		Car	8500	Plane	62,000

2. Since industry and business use so much energy, individuals cannot have much of an effect on reducing United States fuel needs. Explain why you agree or disagree. (21-1)

3. List some energy-intensive things or activities in your own life that you could change without changing your standard of living much. (21-1, 21-3)

4. What role does energy play in international relations? (21-2)

GOING FURTHER

Plastic garbage bags are bigger and stronger than paper bags. However, these bags come from crude oil and require far more energy to manufacture than paper bags. List other common and convenient items that are energy-intensive.

COMPETENCY REVIEW

1. Most of United States energy comes from
   a. oil.   b. moving water.
   c. coal.   d. uranium.

2. The United States has a large reserve of
   a. oil.   b. moving water.
   c. coal.   d. uranium.

3. Burning fossils fuels contributes primarily to
   a. radiation hazards.
   b. air pollution.
   c. water pollution.
   d. land pollution.

4. Water pollution other than that caused by dumping of wastes is usually connected with obtaining and shipping

   a. oil.   b. natural gas.
   c. coal.   d. uranium.

5. Waste material from nuclear plants is
   a. no longer a controversial issue.
   b. safe after a short storage.
   c. easy to get rid of.
   d. dangerous for a long time.

6. Dams and pipelines can
   a. make electricity more expensive.
   b. increase air pollution.
   c. harm animal and plant life in an area.
   d. increase thermal pollution.

7. Examine Figure 21-9.
   What percentage of United States energy comes from fossil fuels?
   a. 10 percent   b. 40 percent
   c. 25 percent   d. 90 percent

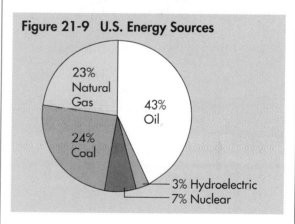

Figure 21-9 U.S. Energy Sources

23% Natural Gas

43% Oil

24% Coal

3% Hydroelectric
7% Nuclear

8. Which is *not* a major factor in the United States energy crisis?
   a. increasing need for energy
   b. decreasing supplies of energy
   c. importing oil
   d. exporting coal

CHAPTER 22

CONSERVING ENERGY

Do you like to take long, hot showers? Do you leave the TV on and lights burning when you leave a room? Do you automatically turn on the air conditioner in warm weather? Do you toss bottles and cans into the garbage instead of returning them for recycling? If so, you are like millions of Americans. Unfortunately, these practices contribute to the energy crisis. When you think over what you have learned in this course, you can see that such practices are unwise. They waste resources that keep rising in price and cannot be replaced. If people in the United States want to save energy, what they need to toss out are energy-wasting habits and practices and form new habits, like the one shown in Figure 22-1, which will save energy.

As you become aware of the energy people use in everyday life, you begin to see how much energy is wasted without even a second thought. Extend your awareness further by thinking about the way in which many ordinary products are made. Use your awareness to help you change your own behaviors and the behaviors of your family and friends. The wise use of energy is not an issue for America to face in the future. It is a challenge and a necessity to deal with now.

22-1 Ways to Conserve

■ *Objectives*
☐ *State ways to conserve energy.*
☐ *Discuss goals for developing new sources of energy.*

Many people in the United States want to learn to **conserve** energy. To conserve means to save something that is in limited supply. Conserving, therefore, means saving what you have. Conserving also means finding substitutes to use in place

Figure 22-1

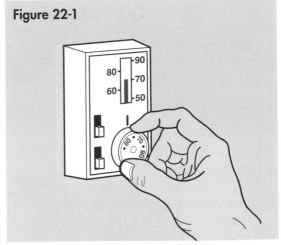

Turn your thermostat lower in the winter.

of what you are saving. There are actually only a few ways to conserve energy, but there are thousands of examples of each way.

Reasons to Conserve

One reason for saving energy is so that your children, and theirs, can enjoy at least the same standard of living you have. On the present course of our nation, the same standard of living will not be possible. People are using more and more energy while energy sources are rapidly disappearing. If people do not plan now, there may be nothing for future generations.

Your family can save money by saving energy. Supply and demand set the price of energy as they do with other products. Prices increase when the supply of a product decreases or when the demand and production costs increase. One result of the energy crisis, therefore, is the rising price of energy. The prices of gasoline, electricity, and natural gas have gone up over the years. All of these resources are in great demand because they are increasingly needed to make products. Learn to save energy, and you and your family can cut your energy bills.

How to Conserve Energy

The larger white circles in Figure 22-2 suggest 4 main ways to conserve energy. How many specific ways can you think of?

Use Less The first and most important way to conserve energy is to use it less. For instance, you can use energy more wisely. Why heat rooms at home that people barely use? Why drive a short distance when you can walk or ride a bicycle? Walking and riding also provide exercise and promote fitness. You can save energy by reducing waste. Why heat a room with a window open? Why let water run before you step into the shower? Practices like these waste energy.

You can save energy by **curtailment**, which means making do with less. Some families keep their room thermostats set at 62°F or 65°F when they are using heat, instead of at 70°F, so they save both

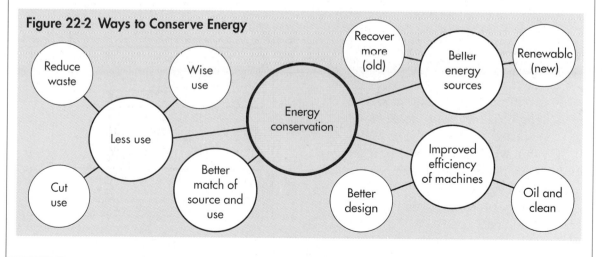

Figure 22-2 Ways to Conserve Energy

energy and money. On cold days, these families wear warm clothing. On hot days, deal with the heat by using fans instead of air conditioners.

Improve Efficiency Another way to conserve energy is to improve the efficiency of machines. As you know from the second law of thermodynamics, all machines waste some energy as heat. Machines can, however, be made to produce more useful work and waste less energy. These are the benefits when machines are efficient. For example, two cars of about the same size may have very different efficiencies because of engine design, body shape, and total weight. One car may get 35 miles per gallon, the other only 20 miles per gallon. The first is more energy efficient than the second. One machine is more **energy efficient** than another if it uses less energy to do the same amount of work. Buyers can compare similar products to find out which is most energy efficient. Many products, such as cars and refrigerators, must have labels stating how efficient they are as compared with other products of the same kind. Efficiency increases when you keep machines in good running order. You can help a machine be efficient by cleaning and oiling it.

S K I L L   B U I L D E R

Analyzing

Federal law requires manufacturers to place an "Energy Guide" label on many electrical devices and cars. Look at Figure 22-3 to see an example.

1. For what appliance has this Energy Guide been prepared?
2. What is the expected yearly cost of electricity to run this model?
3. How does its cost compare with the cost of operating other models?
4. What national average electric rate was used to predict the cost?
5. What would be the cost of operating this appliance in an area where electricity is
   a. eight cents per kilowatt hour?
   b. nine cents per kilowatt hour?

Figure 22-3 An Energy Guide

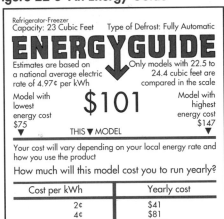

Refrigerator-Freezer
Capacity: 23 Cubic Feet   Type of Defrost: Fully Automatic

ENERGYGUIDE

Estimates are based on a national average electric rate of 4.97¢ per kWh

Only models with 22.5 to 24.4 cubic feet are compared in the scale

$101

Model with lowest energy cost $75 ▼

THIS ▼ MODEL

Model with highest energy cost $147 ▼

Your cost will vary depending on your local energy rate and how you use the product

How much will this model cost you to run yearly?

Cost per kWh	Yearly cost
2¢	$41
4¢	$81
6¢	$122
8¢	$163
10¢	$203
12¢	$244

Ask your salesperson or local utility for the energy rate (cost per kWh) in your area

IMPORTANT Removal of this label before consumer purchase is a violation of federal law (42 U.S.C. 6302)

Matching Source and Use Another way to conserve energy is to improve the match between energy uses and energy sources. For example, traveling between cities by bus or train instead of by car or plane would save energy. Buses and trains use less energy per person than cars. Therefore, buses and trains are a good match between transportation and efficient use of energy.

Steel mills were located close to coal mines, since coal was needed to heat and purify iron. Why ship coal long distances to steel mills? It helped to match the use of coal with the sources of coal by building certain industries, like steel mills, near coal deposits.

Developing Better Energy Sources Some industries are trying to **recover** fuel from old sources by improved methods. To recover means to obtain even more fuel from a used-up well or mine. For example, many oil wells no longer produce gushing streams of oil. People think of them as dry even though there may still be much oil trapped in soft rocks and small pools. By pumping hot water and steam down into the holes in these large underground areas, people can recover the remaining oil. Study Figure 22-4.

Another goal in the United States is to find replacements for nonrenewable energy sources such as oil and natural gas. The United States depends more on nonrenewable energy sources than on renewable energy sources. Developing renewable energy sources, such as the wind and plants, would be helpful. These renewable sources of energy are discussed in detail in the final chapter of this book.

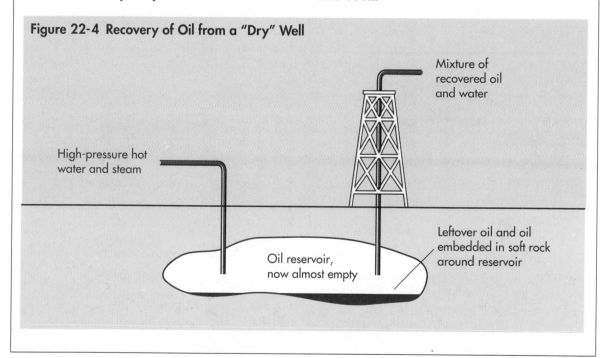

Figure 22-4 Recovery of Oil from a "Dry" Well

Mixture of recovered oil and water

High-pressure hot water and steam

Leftover oil and oil embedded in soft rock around reservoir

Oil reservoir, now almost empty

Section 22-1 Review

Write the definitions for the following terms in your own words.

1. **conserve**
2. **curtailment**
3. **energy efficient**
4. **recover**

Answer these questions.

5. What reasons can you think of for people to learn to conserve energy?
6. Explain why the price of energy reflects and affects the energy crisis.
7. Present your own new ideas for conserving energy by
   a. curtailment,
   b. reducing waste,
   c. wise use of energy, and
   d. improving efficiency.
8. What two main goals should the United States aim toward in developing its energy sources?

22-2 Examples of Conserving Energy

■ *Objectives*

☐ *Name ways to save energy in transportation, industry, and homes.*

☐ *Discuss the advantages and disadvantages of conserving energy.*

There are ways to reduce the amount of energy you use without making much change in your way of life. Everybody can reduce their energy use while changing their standard of living very little.

Transportation

Transportation uses about 30 percent of this country's energy. Petroleum is the energy source used for almost all transportation. If people cut back on traveling and use more efficient cars, a reduction in petroleum usage is possible. Planning ahead, making fewer trips, and bicycling or walking whenever possible will also help.

Recently, gasoline prices have been as high as $1.40 per gallon. Look at Figure 22-5. What serious changes would people make if gasoline cost $3.00 or $5.00 a gallon?

What were your answers? If people would make changes now, they might avoid such steep prices in the near future. Some experts, however, believe that sharp rises in fuel prices are the only way to change American habits regarding energy and transportation. These experts point out that, in many countries, gasoline costs more than $5.00 a gallon. High gasoline prices would bring about less use of gasoline.

Figure 22-5

What would happen if gasoline were to cost more than $3.00 per gallon?

Americans often travel by the least efficient means. For greatest efficiency, people could ride trains and buses within and between cities. Other ways to decrease fuel usage include riding in groups in vans and car pools. To encourage energy conservation in transportation, many communities give special privileges and reduced tolls to cars with three or more people riding to work. After being forgotten because of the automobile, vehicles such as trolleys, subway trains, ferries, and commuter trains are making a comeback in many cities. These are signs of using energy wisely.

Cars can be made more efficient by reducing friction, improving their aerodynamic designs, using lighter alloys, and modifying the engine design. Small cars today can get 45 miles per gallon on the highway. Further improvements may bring this to 60 miles per gallon. That is an amazing improvement since the 1950s, when gasoline was cheap, cars were big, and efficiencies of 12 miles per gallon were normal.

The way a driver handles a car also affects its energy efficiency. Observing the 55- or 65-mile-per-hour highway speed limit is more energy efficient than driving faster. In fact, driving at 40 mph gives the greatest fuel efficiency. Driving faster or slower can mean using more gasoline for the same trip. Drivers should also accelerate smoothly and try to drive at a steady speed. In city driving, they can reduce the amount of braking and idling time. Furthermore, a well-lubricated car that has regular oil changes and tune-ups is more efficient than one that does not.

Tires should be inflated with the right amount of air pressure to save fuel and money. Car manuals give the recommended tire pressure, usually in **psi**, or pounds per square inch. There is often writing on the side of a tire that indicates the highest pressure that is safe. Most tires call for about 28–30 psi. Radial tires are more expensive than standard tires, but they increase fuel efficiency, which means getting more miles per gallon.

Engineers constantly redesign airplanes to make them more efficient. For example, planes today weigh less and are shaped to move through the air with less friction than older planes. Examine Figure 22-6. Recently, there has been interest in switching from jet engines to propeller engines. Propeller engines push the plane more efficiently than jet engines do.

Diesel fuel comes from petroleum, just like gasoline, but it is less expensive to obtain and purify. Greater use of diesel

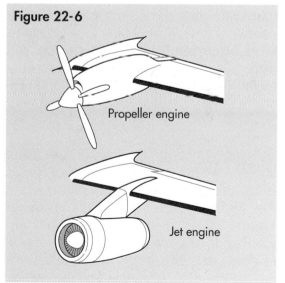

Figure 22-6

Propeller engine

Jet engine

Two types of airplane engines

engines in cars and trucks could result in energy savings. If you have ever smelled the strong odor of the exhaust from a diesel-powered engine, however, you know that using diesel fuel raises questions about air pollution.

Industry

Industry uses about 35 percent of the energy consumed in the United States. Efforts to conserve energy in industry have already yielded savings. Thermostats can sometimes be set at lower temperatures, and lighting can be reduced, without hurting the productivity, comfort, or health of the workers. Good maintenance of industrial equipment keeps it energy efficient.

Recently, some industries have considered recycling and using waste materials to decrease their energy needs. For example, Italy produces steel efficiently by melting scrap metal with new iron. Japan has modern steel plants that use recycled materials and have furnaces that lose relatively little heat. Many United States steel plants are old and inefficient and were not designed to use recycled scrap metal. American steel is therefore expensive and faces stiff competition. Only the Soviet Union produces steel that costs more to make.

A lot of energy gets wasted in many industries. For example, the hot gases from most furnaces are just poured into the air by way of chimneys. Look at Figure 22-7. Today, some companies are starting to use **cogeneration**. In cogeneration, the hot gases produced by a factory's furnace are used to make steam.

Figure 22-7 Cogeneration

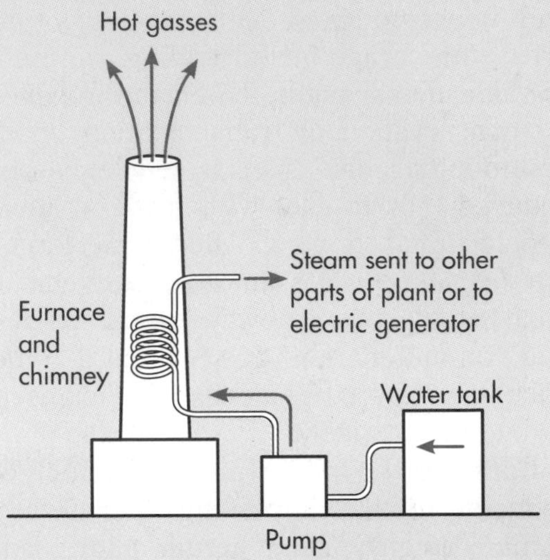

Steam produced in this way can be used for warming the buildings or for other uses, such as turning a small turbine to produce electricity from an electric generator. In any case, energy that was wasted as hot gases can be partly recovered and used for other purposes.

Using computers in factories can also help improve energy efficiency. People can use computers to watch and control industrial processes. Factory operators can more quickly spot and correct problems that cause energy loss.

Homes and Businesses

Homes, businesses, and schools use about 35 percent of the energy in the United States. It is hard to make big energy improvements in these places, because they depend heavily on electrical energy. As you know, producing electricity is a very inefficient process and requires large amounts of oil and natural gas.

Keeping homes and buildings cooler in the winter and warmer in the summer is one way to conserve energy. People can also conserve energy by improving building insulation to keep heat in or out. Many electric and gas companies will examine homes to find where heat is leaking out. They suggest ways for the owner to correct any problems. For example, a company might suggest installing storm windows, weather stripping, or adding fiberglass insulation. Weather stripping means sealing openings around doors and windows with strips of cloth or another material. In addition, planting shade trees may reduce the cost of cooling a home in summer by as much as 20 percent.

The process of **retrofitting** adds modern devices or materials to old buildings in order to reduce heat loss and improve energy efficiency. For example, replacing an old furnace with a more efficient one is retrofitting. Putting drapes over previously uncovered windows decreases energy use, because less heat now escapes through the glass. A good example of retrofitting is installing a solar heater on the roof and making it part of the heating or hot water systems of the house.

Families and businesses can also save energy by reducing the temperature of the hot water they use. To do this, they adjust the thermostat of the hot water heater. Reducing the temperature from 140°F to 110°F can cut bills for heating water by 20%.

There are other, small ways to save energy in the home. Hot water pipes can be insulated. When washing clothes or dishes, fill machines to capacity. When people purchase major appliances, such as dishwashers, it is helpful to compare models by using their Energy Guides, also called an Energy Efficiency Rating label, or EER. Look for the most efficient model, with a yearly energy cost near the low end of the range for similar models.

Examine Figure 22-8. It is more economical to use area lamps, which illuminate one area in a room, than to flood an entire room with light. Fluorescent lights are more energy efficient than

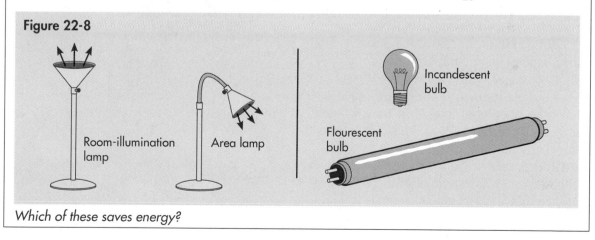

Figure 22-8

Room-illumination lamp

Area lamp

Incandescent bulb

Flourescent bulb

Which of these saves energy?

incandescent lights. Incandescent lights contain a glowing-hot wire. Fluorescent lights contain a gas. Fluorescent lights make more light and less heat.

People also make personal decisions that affect energy use. A person whose hobby is building things with power tools will have larger energy bills than a person whose hobbies include reading and stamp collecting. People who grow their own vegetables and livestock save money and save energy used for shipping.

Advantages and Disadvantages

Energy improvements often pay for themselves, which means that the energy savings may be greater than the cost of the improvements. Most improvements not only pay for themselves after several years, they result in energy savings right away.

The energy saved by conservation is available for use elsewhere, without the great expense of having to find new energy sources. In the United States, lower oil consumption strengthens the economy and national security because we become less dependent on other nations for energy. Using less energy also reduces environmental problems.

Conserving energy does have a few disadvantages. It may reduce some personal comfort, convenience, and freedom. Short, warm showers are not as pleasant to some people as long, hot ones. People may have to decide that some "freedoms" are really unnecessary luxuries.

Conservation has caused some job changes and unemployment. For example, robots have replaced people in some

steps in making cars. See Figure 22-9. *In certain tasks* robots are more efficient, faster, and do better work than people.

For a building to be more energy efficient, it should be constructed so that there is little chance for air to escape. People inside, however, may miss having fresh air. Air quality is reduced. Sometimes, expensive, energy-consuming fans and air conditioners must be installed to take care of the problem. The extra costs can undo the energy savings gained by making the building airtight.

When energy costs go up, people conserve energy. Price increases, however, affect the poor to the greatest extent. A monthly electric bill that is $10.00 higher is not a great burden for many electricity users, but it can be difficult for poor families. Increased prices can be hard on retired people, who may live on a fixed amount of money.

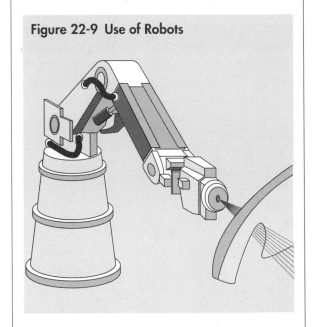

Figure 22-9 Use of Robots

How Do You Insulate an Ice Cube?

Process Skills *making a model; evaluating; stating conclusions*

Materials ice cubes of identical size; assorted materials to use in constructing ice-cube boxes with various types of insulation; thermometer

Procedures

1. This activity is a contest, so form teams with your classmates. Each team should discuss how to construct a container that will keep an ice cube from fully melting over a three-hour period. Since this is a contest, keep your team's discussions secret.
2. Consider the following points in designing your container.
   - Use lightweight materials.
   - Use insulating materials.
   - Air spaces insulate well.
   - Prevent heat from entering through openings.
   - Light-colored surfaces reflect heat, and dark-colored surfaces absorb heat.
3. When you are ready to build your container, ask the teacher for the materials you need.
4. Build your container.
5. Each team will receive an identical cube of ice. Put the ice in the container you have built. The team that has the most ice left after three hours is the winner. If no team has ice left, the team whose water from the melted ice cube is coldest wins.
6. After the three-hour period, check the containers each half hour to see which can keep some ice unmelted for the longest time.

Conclusions

7. What was special about the containers in which ice cubes lasted longest?
8. How can you apply what you learned from this experiment to home heating and cooling?

■■■ Section 22-2 Review ■■■

Write the definitions for the following terms in your own words.

1. **psi**
2. **cogeneration**
3. **retrofitting**

Answer these questions.

4. List four ways to reduce energy use in transportation.
5. Name four ways a driver can spend less for gasoline.
6. Why are some industries considering cogeneration and computerization?
7. How can home owners use less energy?
8. List three advantages and three disadvantages of conserving energy.

SCIENCE, TECHNOLOGY, & SOCIETY

The Fuelwood Crisis

Of the 5 billion people in the world today, nearly 2 billion use wood as a principal fuel. Often, they find the wood on the ground or cut down local trees. Almost 70 percent of these people lack secure supplies of wood, and five percent cannot find enough wood to meet their health needs. This five percent go cold and eat uncooked food. Both conditions threaten the health of nearly 100 million people!

By the year 2000, 2.7 billion people may be in a similar situation, because trees that have been cut down have not been replaced. This is the fuelwood crisis. This crisis is unknown in the United States, but it is a tragic fact of life in many African, Asian, and South American countries.

The fuelwood crisis worsened in the 1970s as a result of a worldwide increase in petroleum prices. Poor families could no longer afford kerosene and had to start using wood. The wood ran out, so they started using animal dung and vegetable waste as fuel. In India, roughly 35 percent of all energy comes from these materials, which are dried and burned instead of being used to fertilize crops. In some areas, trees are being replanted with the United Nations' help, but the fuelwood crisis will not go away soon.

One energy-related need in poor nations is for stoves. Many people cook over stones, and this wastes energy. Simple, inexpensive stoves could help the people save about half of the fuel they now use.

Figure 22-10

Stoves would change their standard of living. Now, in some places, people spend half of each day finding food and fuel.

Developed countries could also provide kerosene or coal to such developing nations. Small electric generators that operate by wind or streams could be built to provide some energy. Huge education programs about trees and energy are also essential. Developed countries can provide these things and help poorer nations overcome their energy crises.

Follow-up Activities
1. How does this reading illustrate the global effects of technology?
2. How can developed nations help developing nations?
3. Should people in the United States worry about the fuelwood crisis?
4. What might the worldwide energy situation be in 50 years? What might it be in 500 years?

Section 22-1 Ways to Conserve

1. Conservation means using energy wisely by reducing waste and improving efficiency.
2. Saving energy saves money.
3. The United States must recover more energy from its old sources and develop new, renewable sources.

Section 22-2 Examples of Conserving Energy

Reductions of energy use are possible in transportation, industry, businesses, and homes.

Write the term from the list that best completes each sentence.

cogeneration, conserve, curtailment, energy efficient, psi, recover, retrofitting

One way to ___1___ energy is to use less of it. This is ___2___. Another way is to use machines that are more ___3___. There is a lot of fuel left in old mines and wells, which the United States can ___4___ using new equipment. A car gets better mileage if its tires are set with the correct amount of air pressure, usually measured in ___5___. The technology of ___6___ involves using the heat that goes up a chimney for more useful purposes in a factory. Adding a solar heater to an old house is an example of ___7___.

Write *true* if the statement is true. If the statement is false, change the *italicized* term to make the statement true.

1. As supplies of energy decrease, the price of energy will *decrease.*
2. A car that goes 30 miles per gallon of gasoline is *more* efficient than one that goes 20 miles per gallon.
3. Using mass-transit vehicles such as buses and trains instead of cars *increases* energy consumption.
4. Driving at steady speeds, with less braking and less accelerating, is one way to *reduce* fuel use and costs.
5. Industry can become *more efficient* by recycling old materials.
6. An economical and adequate setting for hot water heaters is *160° F.*
7. Incandescent bulbs give *more light with less heat* than do fluorescent lights.

Figure 22-11 shows the percent of energy saved by making some things with recycled

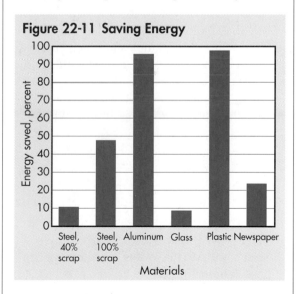

Figure 22-11 Saving Energy

material rather than by producing them brand new, from scratch.

1. What percent of energy is saved by using old newspaper to make new newspaper?
2. What percent of energy is saved by using 100 percent scrap instead of 40 percent scrap in making steel?
3. For which recycled material is the energy saving largest?
4. For which material is the energy saving lowest?
5. Why is there a difference in energy savings?

THINK AND DISCUSS

Use the section numbers in parentheses to help you find each answer. Write your answers in complete sentences.

1. What reasons would you give for conserving energy if you were asked to speak on that topic? (22-1)
2. Give some real examples in which energy is wasted, used unwisely, or used inefficiently. (22-1)
3. Most aluminum companies are located near electric generating plants. Why is this a good match between an energy source and its use? You may want to review the STS feature in Chapter 6 to see how aluminum is made. (22-1)
4. List some possible short-term and long-term consequences of a gasoline price of $10.00 per gallon. (22-2)
5. List some practical ways to reduce energy use at home. (22-2)

6. Why is it harder to save energy in an all-electric house than in a house that also uses natural gas, coal, or heating oil? (22-2)
7. Give two new examples of retrofitting. (22-2)
8. What can you learn from an EER label? (22-1, 22-2)
9. Name some luxuries or activities that you could do without in order to reduce your energy use and energy bills. (22-2)

GOING FURTHER

1. Figure 22-12 shows 13 target areas for energy conservation in a house. Investigate ways to conserve energy in at least five of the areas shown. Write a report on your findings.
2. Obtain or copy an Energy Guide or EER label from an appliance in a neighborhood store. Bring it in for sharing and discussing in class.

COMPETENCY REVIEW

1. Conserving energy
   a. makes the United States more secure.
   b. helps preserve a way of life for future generations.
   c. saves money.
   d. all the above
2. Which can cause a reduction in fuel prices?
   a. fewer sources of energy
   b. less demand for energy
   c. higher cost of producing fuel
   d. all the above

Figure 22-12 Targets for Energy Conservation

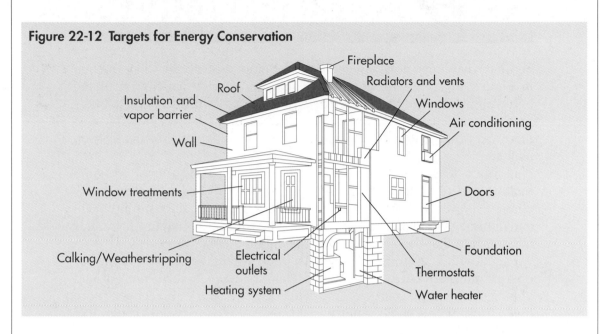

3. Which temperature settings will mean air conditioners and a furnace will use less total energy over year?
   a. 70°F in summer, 70°F in winter
   b. 65°F in summer, 75°F in winter
   c. 75°F in summer, 65°F in winter
   d. 85°F in summer, 62°F in winter
4. The least costly and most energy-efficient way to go to nearby stores is by
   a. car.   b. walking.
   c. bus.   d. cab.
5. Which one of the following does *not* save energy?
   a. curtailment   b. reducing efficiency
   c. cogeneration   d. retrofitting
6. What should a car driver do to get more miles for every gallon of gasoline?
   a. Drive slowly on highways.
   b. Drive fast on highways.
   c. Drive at steady speeds and use the brake less.
   d. Keep tires underinflated, with less air pressure than suggested.
7. Which one of the following indicates that a car is energy efficient?
   a. large number of miles per gallon
   b. large gasoline tank
   c. low price
   d. large engine
8. What practice regarding lights will reduce your use of electrical energy?
   a. use of fluorescent bulbs
   b. use of area lights
   c. turning off lights when they are not needed
   d. all the above

CHAPTER 23

ENERGY ALTERNATIVES

Some day you may drive an electric car. Before electric cars, like the one in Figure 23-1, become common, engineers must learn how to make lightweight batteries to store electricity. Engineers will also need to know how to make electric engines more powerful. In the future, you may also drive cars that use fuels besides gasoline, such as gasohol made from plants.

In years to come, electricity will probably be used more for home heating and cooking. The energy sources used to make electricity will change, however. Homes may be partially heated using sunlight. Oil and natural gas will probably be replaced by sulfur-free coal products. It is most urgent that people develop nonpolluting sources of energy.

Faced with a growing energy crisis, the United States is looking for energy alternatives. Energy alternatives are ways to save energy and obtain new energy sources. One alternative is to try to recover more fuels from the ground. Another alternative is to develop renewable sources of energy. A third alternative is to improve the efficiency of energy use, especially in producing electricity. This final chapter examines some of these alternatives.

23-1 Recovering Additional Energy

■ *Objectives*

☐ *Describe ways to increase recovery of oil and natural gas.*

☐ *Tell how to make coal a clean and convenient energy source.*

☐ *State the benefits and burdens of breeder reactors.*

As you learned in Chapter 22, worldwide reserves of oil, natural gas, and uranium

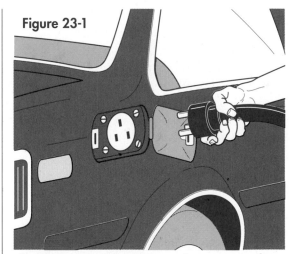

Figure 23-1

The electric car is an energy alternative.

are decreasing. This does not mean the old wells and mines are completely empty. New technologies will allow companies to recover some fuels from the old locations.

Recovering More Oil

In many old oil wells, oil is trapped in small pools or within the pores of the soft rock around the oil reservoir. One way to recover more of this oil is to pump steam and hot water into the wells. Hot water and steam force the oil out. The mixture of oil and water that results is pumped to the surface, where the oil and water separate. This method is known as water flooding.

Another way to recover trapped oil is by dissolving it. In this method, liquified propane (C_3H_8) or liquified carbon dioxide (CO_2) is pumped into old wells. These two substances dissolve the remaining oil, which is then pumped to the surface. This procedure must be carried out under high pressure and is known as **miscible drive**. The word *miscible* means one liquid is soluble in another liquid. In this case, the trapped oil is soluble, or miscible, in the liquified propane or carbon dioxide.

Oil Shale

In the western United States, there are large amounts of a soft rock known as **oil shale**. Oil shale contains solid organic material mixed in with the stone. When shale is heated, the organic material decomposes into an oily material that is very similar to petroleum.

Technicians can crush and heat shale close to where it is mined. Most of the gases given off during heating can be condensed into crude oil. Gases that do not condense can be used as fuel for the furnaces that heat the shale.

Scientists have also explored the possibility of heating the shale in place, underground. Study Figure 23-2. A hole is dug into shale, through which air and a flammable material are pumped. A fire starts underground, within the shale. Heat from the fire causes the organic

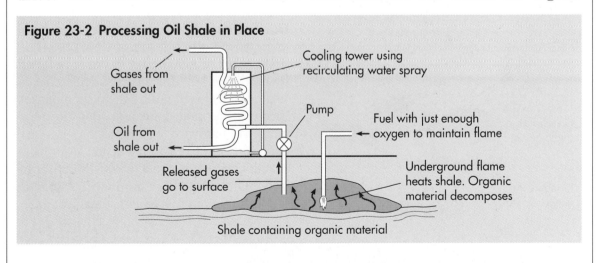

Figure 23-2 Processing Oil Shale in Place

Gases from shale out

Cooling tower using recirculating water spray

Pump

Oil from shale out

Fuel with just enough oxygen to maintain flame

Released gases go to surface

Underground flame heats shale. Organic material decomposes

Shale containing organic material

material in the shale to decompose into oil. The newly made oil is pumped out through other holes nearby. This saves on the cost of mining and treating shale.

Recovering More Natural Gas

In many old underground reservoirs of natural gas, a lot of gas is still trapped in small pockets or within the surrounding soft rock. Using explosives or water under high-pressure, people can crack the rock. Trapped gases then can escape and be collected by the pipes of the old well. Another approach is to drill wells that follow natural fractures, or cracks, in rock. Examine Figure 23-3. Drilling wells helps release trapped pockets of natural gas.

Making Coal More Acceptable

The United States has a vast amount of coal that is not being mined today. There are two reasons. First, sulfur burns along with the coal and produces sulfur dioxide (SO_2), a major air pollutant. Second, coal is difficult to transport and use. Years ago, people had to keep coal bins in their homes and had to shovel the coal into furnaces every hour or two. Oil and natural gas are much easier to use. Objections to coal may be overcome by three new technologies.

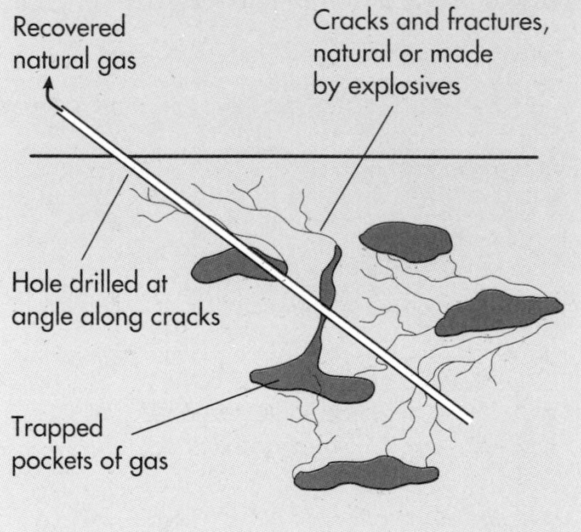

Solvent Refining   Some special liquids, known as solvents, can dissolve the ash and sulfur found in coal. Treating coal with solvents makes the coal pure, or refined. It is now a clean-burning solid with high heat content. This process of treating coal is called **solvent refining**.

One problem with solvent refining is that the solvents are expensive and toxic. The sulfur and ash dissolved in the solvent must be removed, so the solvent can be reused. Dumping the solvent would be illegal and unsafe and would cause more pollution problems than burning the unpurified coal.

Gasification   It is possible to obtain methane gas and other gas fuels from coal. The process for producing these gases is called **gasification**. In this process, the coal is heated in a closed tank. Steam then passes through the tank. Depending on the temperature, pressure, and catalysts that are used, different reactions occur. All these reactions work because coal is mostly carbon. One reaction produces methane (CH_4), the main ingredient of natural gas.

$$2C + 2H_2O \rightarrow CH_4 + CO_2$$

Figure 23-3 Getting More Natural Gas out of an Old Field of Gas

Recovered natural gas

Cracks and fractures, natural or made by explosives

Hole drilled at angle along cracks

Trapped pockets of gas

Like natural gas, methane produced from coal and water can be used for cooking and heating. Manufactured methane can also be sent through ordinary gas lines.

Another gasification reaction produces two gases, both of which burn:

$$C + H_2O \rightarrow H_2 + CO$$

The products made in this reaction are very dangerous. Hydrogen (H_2) is explosive, and carbon monoxide (CO) is poisonous. This mixture provides great energy when burned in furnaces, however. The chemical equation for burning the mixture is

$$CO + H_2 + O_2 \rightarrow CO_2 + H_2O + \text{heat}$$

Although the reactants are deadly, the products are safe. The mixture of H_2 and CO gases is used in industries where safety can be closely watched to prevent explosions or release of poison gas.

The advantage of gasification is that coal and steam make new fuels that are convenient to use. Coal may be more widely used because of this process. Of course, the coal must be sulfur-free, or burning it will produce air-polluting gases.

Liquefaction It is possible to start with coal and, through a series of complicated reactions, produce convenient liquid fuels such as heating oil, diesel fuel, and gasoline. This process is called **liquefaction**. So far, the process is too expensive to replace petroleum products. In the future, though, liquefaction may be used more. Once again, sulfur-free coal must be used. This requirement adds considerable cost to the gasification and liquefaction processes.

Recovering More Nuclear Energy

Recall that uranium-235 atoms are the fuel for nuclear reactors, but they make up only one percent of all the uranium atoms in uranium ore. Uranium-238 atoms, which do not undergo fission and do not release energy, make up the other 99 percent. When neutrons from the chain reaction of U-235 hit U-238 atoms in the fuel rods, however, they produce a new isotope called plutonium-239. Study Figure 23-4.

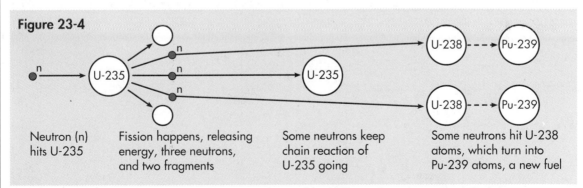

Figure 23-4

Neutron (n) hits U-235

Fission happens, releasing energy, three neutrons, and two fragments

Some neutrons keep chain reaction of U-235 going

Some neutrons hit U-238 atoms, which turn into Pu-239 atoms, a new fuel

In the breeder reaction, uranium isotopes work together to form plutonium, a new fuel that can undergo fission.

How Can Coal Be Converted to Fuel Gas?

Process Skills observing; stating conclusions

Materials two test tubes; delivery tubes as shown in Figure 23-5; Bunsen burner; beaker with cold water; one-hole and two-hole rubber stoppers; ring stand and clamp; ceramic tile; some crushed coal

Procedures

1. Set up the equipment as shown in Figure 23-5. Ask your teacher for permission to continue.
2. Heat the coal in the test tube for three minutes. After one minute, the air in the equipment will have been pushed out, so put a match to the outlet tube. The emerging gas will burn. It is called coal gas.
3. Turn off the burner. Allow the equipment to cool down.
4. Observe, do not touch, the liquid that condensed inside the test tube.
5. Dump the coal from the test tube onto a ceramic tile. It is now a very pure form of carbon known as coke.
6. Direct the tip of the outer cone of a Bunsen burner's flame on the coke. Notice how the coke glows and burns without smoke.

Conclusions

7. Why do some gases coming from the heated coal become a liquid in the cold test tube while others escape from the outlet tube?
8. The gases that come out of the coal are from organic material that is in the coal. How does this activity differ from gasification and liquefaction?
9. Why touch the tip of the outer cone of the flame, instead of the tip of the inner cone, to the coke?
10. Investigate the uses of coal gas, coal tar, and coke.

Figure 23-5

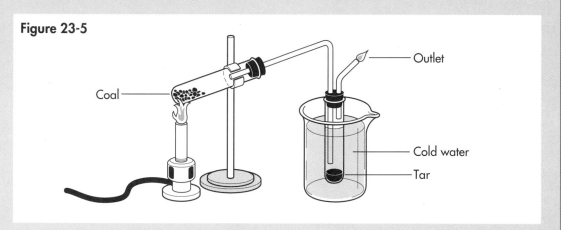

Coal — Outlet — Cold water — Tar

Plutonium-239 is radioactive and has a half-life of 24,000 years. It can be used as a reactor fuel, for it can undergo fission, like U-235, and release energy. After being separated and purified, the plutonium is placed in fuel rods near the uranium fuel rods in a reactor's core.

In nuclear reactors, about four Pu-239 atoms are made for every three U-235 atoms destroyed. The reactor is multiplying, or "breeding," new fuel from old fuel. This is called a breeding reaction and is used in a **breeder reactor**. Look at the flow diagram in Figure 23-6.

The advantage of using breeder reactors is that they create plutonium-239 fuel from uranium-238, which is not a fuel. Breeder reactors enable a country to extend the use of its uranium reserves by thousands of years. In ten to 15 years, a breeder reactor can create enough fuel to run itself and another reactor for ten to 15 more years!

Like ordinary nuclear reactors, breeder reactors produce heat, which boils water into steam. The steam turns a generator that makes electricity. Breeder reactors may be able to supply a large amount of electrical energy in the future.

Plutonium-239 is very dangerous, however, even more dangerous than uranium-235. It gives off strong radiation for hundreds of thousands of years. Another concern is that thousands of tons of plutonium will be bred each year if breeder reactors are widely used. It is possible to take some of the plutonium and make nuclear bombs. Security around breeder reactors is very strong, to prevent any theft or misuse of this dangerous element.

Breeder reactors are harder to build than ordinary nuclear reactors. Instead of water, liquid sodium metal must be used to transfer heat from the core, increasing costs and safety problems. Therefore, breeder technology may not be widely used in the United States.

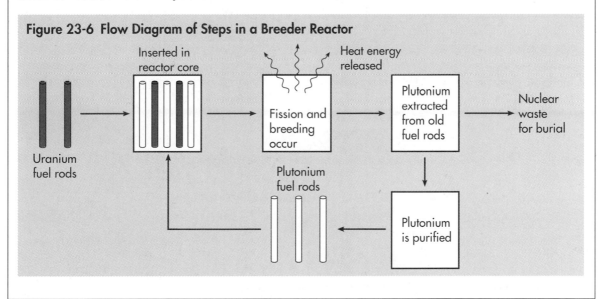

Figure 23-6 Flow Diagram of Steps in a Breeder Reactor

Section 23-1 Review

Write the definitions for the following terms in your own words.

1. **miscible drive**
2. **oil shale**
3. **solvent refining**
4. **gasification**
5. **breeder reactor**

Answer these questions.

6. Distinguish between the energy alternatives of improved recovery, improved efficiency, and new sources.
7. How can oil be obtained from shale rock? American oil companies see great possibilities for oil shale. Why?
8. How can new technologies make coal a cleaner-burning, more convenient source of energy than at present?
9. What are the roles of the following nuclei within breeder reactors?
   a. U-235
   b. U-238
   c. Pu-239

23-2 New Sources of Energy

■ *Objectives*

☐ *Describe systems that collect solar energy for heating and electricity.*

☐ *Identify types of biomass and biofuels.*

There are sources of energy that have not yet been widely used but which may change the energy picture in the United States in the future. These sources are rich in energy, and they are renewable. Their energy is resupplied, after a fairly short period of time, by natural processes.

Solar Energy

The heat and light of the sun is **solar energy**. Deep within the sun, nuclear reactions send out electromagnetic energy in the form of waves. A tiny part of this electromagnetic energy hits the earth, mainly as infrared waves, light, and ultraviolet waves. Look at Figure 23-7.

Figure 23-7

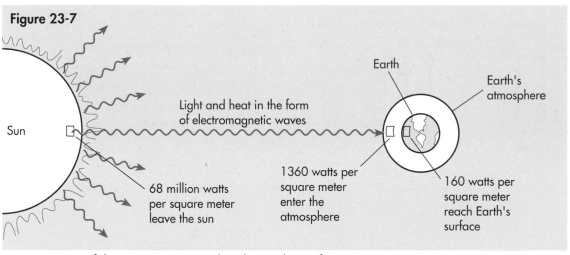

Light and heat in the form of electromagnetic waves

Sun

68 million watts per square meter leave the sun

1360 watts per square meter enter the atmosphere

Earth

Earth's atmosphere

160 watts per square meter reach Earth's surface

A tiny portion of the sun's energy reaches the earth's surface.

The sun is the primary source of energy for the earth. The sun warms the atmosphere and ground during the day. At night, this energy is returned to space by heat radiation, cooling the air and land. If people could save all the light and heat of the sun that reaches the earth's surface, it could provide thousands of times the energy the world needs.

It is hard to capture the sun's energy because it is spread out over the earth. Besides, the amount of solar energy changes according to the time of day, season of the year, geographic location, and weather. Any technology for capturing solar energy must adjust to these factors. Still, it is worth the effort to capture solar energy. In the United States, the daily solar energy shining on an area just 10 feet long by 10 feet wide is equal to the energy in one gallon of gasoline!

People can use the sun's energy to heat buildings, homes, or water. One way is to design buildings and homes so that they collect solar heat without special equipment. Look at Figure 23-8. This is called a **passive solar heating system**. Most windows can be positioned on the sunny side of a house. The section of the roof that faces the sun can have glass skylights. Glass lets light waves into a room and then traps the heat they make.

Building materials are an important consideration in passive solar heating. Bricks, adobe clay, ceramic tiles, and even tanks of water placed in a room will warm up during the day and release heat slowly at night. The windows and skylights must have shades and drapes to prevent heat from escaping through them during the night. Awnings and roof overhangs can allow the sun to warm a building during

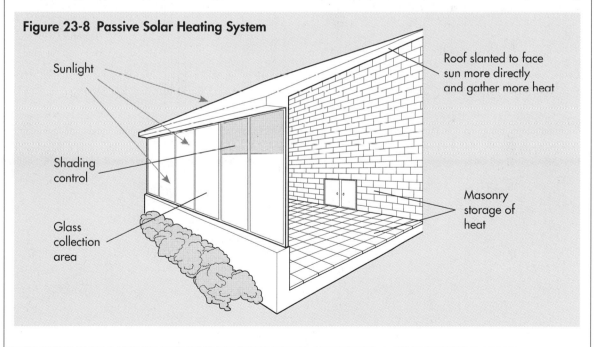

Figure 23-8 Passive Solar Heating System

Sunlight

Shading control

Glass collection area

Roof slanted to face sun more directly and gather more heat

Masonry storage of heat

the winter and provide shade during the summer. Trees can do this, too. Leafy trees can provide cooling shade in the summer. During winter, without leaves, the trees let the sun provide warmth. Trees also serve as windbreaks, reducing drafts that enter a building.

In an **active solar heating system,** special collectors placed on the roof or ground gather heat from the sun. The collectors have air or water in them that is heated by the sun. Study Figure 23-9. Warmed air circulates through vents and fans. Some heat can be stored for the night by passing the warmed air over rocks and bricks. Warmed water can be pumped to radiators or sent to insulated tanks for use at night or on cloudy days. Water warmed by the sun can also be used as part of the hot-water system of a house.

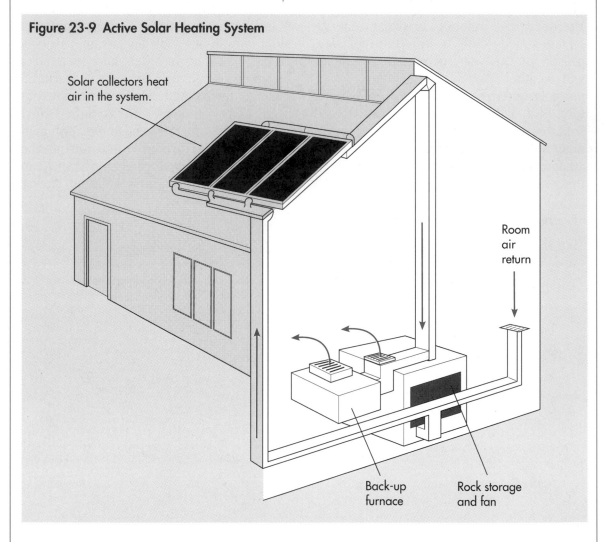

Figure 23-9 Active Solar Heating System

Solar collectors heat air in the system.

Room air return

Back-up furnace

Rock storage and fan

Electricity from Solar Energy

There are three promising techniques for obtaining electricity from solar energy. They involve changing one form of energy into another.

Thermal Energy Conversion Large, concave mirrors can focus sunlight on a tank of water, as shown in Figure 23-10. The intense heat boils the water into steam. The steam is used in the usual way to make electricity in generators.

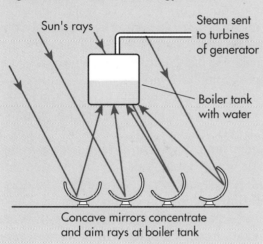

Figure 23-10 Thermal Energy Conversion

Sun's rays

Steam sent to turbines of generator

Boiler tank with water

Concave mirrors concentrate and aim rays at boiler tank

Ocean Thermal Energy Conversion Lakes and oceans make up about 70 percent of the earth's surface. They act as solar collectors. The water on top of a lake or the ocean can be near 80° F (27° C) because of solar heating, while the water below is at 40° F (4° C). Scientists are trying to use this difference in temperature to make electricity. This is called **ocean thermal energy conversion** (OTEC).

One method involves pumping warm surface water into an evaporator that is kept at a low pressure. Under low pressure, water will boil at 70° F to 80° F. The steam this makes can turn turbines to generate electricity. Note that the boiling occurs without heating. The warmth of the water itself provides the energy to make the steam.

About 1% of the water changes to steam. The rest cools because of the energy removed by boiling. It is pumped back to the cooler level of the ocean.

Another method involves using liquids with low boiling points, such as ammonia, Freon, or propane, to turn the turbines. Examine Figure 23-11. Warm ocean water provides the heat to turn these liquids into a gas. After the gas

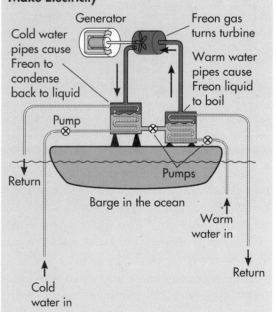

Figure 23-11 Ocean Heat Used to Make Electricity

Generator

Cold water pipes cause Freon to condense back to liquid

Freon gas turns turbine

Warm water pipes cause Freon liquid to boil

Pump

Return

Pumps

Barge in the ocean

Warm water in

Return

Cold water in

spins a turbine, it is condensed back into a liquid using cold ocean water.

The basic idea behind both methods is that the sun warms the water on top of the ocean, which is then used to make electricity. Some day, large barges may float in the ocean, making electricity. The electricity would be sent to the land by cables. Research is needed to learn whether the cost of this technology is worth the amount of electrical energy it produces.

Solar Cells   Solar cells are made of silicon, cadmium sulfide, and other materials that produce electricity from light. The electrons in their atoms are knocked loose by sunlight and can be made to flow as electricity. Study Figure 23-12. This is known as **photovoltaic conversion**.

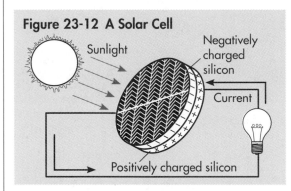

Figure 23-12 A Solar Cell

Sunlight

Negatively charged silicon

Current

Positively charged silicon

Many devices, such as solar-powered calculators and toys, use solar energy today. Spacecraft have large wings that contain solar cells for operating electric circuits and recharging batteries. The sun shines strongly on these solar cells in space. Solar cells that can produce electricity on the earth are still being developed. They convert only a small fraction of light energy into electricity. Some day, however, buildings may have roof-top panels of solar cells to supplement electricity from the electric company.

Biomass

As you know, plants trap the sun's energy in a chemical reaction called photosynthesis. Plant material is known as **biomass** because it is mass formed by biological reactions. In the future, biomass may be grown specifically to provide energy.

Since plants can be grown again after being cut, biomass represents a renewable source of energy. Growing plants for energy is not efficient, however. Only one to three percent of the solar energy that strikes a plant is converted into chemical energy in its biomass. Still, biomass is a supply of energy people can use today.

Some plants grow fast. Huge amounts can be grown and harvested within a short time. They can be burned as a fuel, like wood, or they can be converted into other useful forms of energy, such as methane, alcohol, oil, and charcoal. These products are called **biofuels**, for they are made from biomass. In recent years, plants have been used to make millions of gallons of alcohol, both methanol and ethanol, which is mixed with gasoline. The mixture is known as **gasohol**. It is an inexpensive substitute for gasoline for some cars.

As fossil fuels become more expensive, people may return to burning wood or charcoal. Charcoal is made by heating wood without air for several days. This removes gases and tars from the wood, and only pure carbon remains as charcoal.

Fast-growing trees will have to be planted to supply this energy source.

Fuel can also be made from the "left-overs" in forests, farms, and garbage. A lot of garbage can burn when it is dried. The same is true of the sludge produced by waste treatment plants. Leaves, stalks from crops, and animal waste can also burn. These sources of energy can produce steam, either for heating or for making electricity.

Solid waste can also be converted into fuel. Heating certain waste materials in the absence of air causes them to decompose into methane gas and oil products. Some wastes, like garbage, sewage, and manure can be digested by bacteria. The products of the digestion are methane gas or alcohol, two fuels.

Using biomass for energy helps eliminate waste and produces energy. It would mean that garbage dumps could be reduced in size or closed. The biomass could replace fossil fuels and reduce air pollution. Obviously, the production of biofuels must be watched carefully to make sure that it does not reduce food supplies, soil fertilization using manure, and wood for building purposes.

■■■ Section 23-2 Review ■■■

Write the definitions for the following terms in your own words.

1. **solar energy**
2. **ocean thermal energy conversion**
3. **photovoltaic conversion**
4. **biomass**
5. **gasohol**

Answer these questions.

6. What are the potential benefits and problems of solar energy?
7. Distinguish between passive and active solar heating systems.
8. Briefly describe three ways to make electricity from solar energy.
9. What are the potential benefits and problems of using biomass to make biofuels?

23-3 Additional Sources of Energy

■ *Objectives*

☐ *Explain how wind, geothermal hot spots, tides, and fusion can be used as energy sources.*

☐ *Discuss the potentials and problems of fuel cells.*

The United States is starting to investigate further sources of energy. Some of these are easy to find but do not generate much energy. Other energy sources are hard to use but can provide a lot of energy. Whether these additional sources can provide enough energy at reasonable cost remains to be seen.

Wind Energy

You have probably seen pictures of windmills. They are used throughout the world to pump water up from underground wells and to grind grain into flour. Today, there is a new interest in using wind energy to produce electricity. The mechanical energy of the wind makes the blades of a windmill spin, and these turn a generator that makes electrical energy. New windmills are called wind generators. They are designed to be much more efficient than old windmills. Study the wind generator in Figure 23-13.

Figure 23-13 A Wind Generator

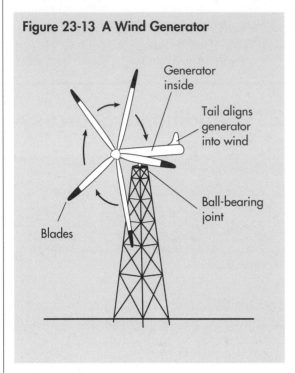

Generator inside

Tail aligns generator into wind

Ball-bearing joint

Blades

The power output of a wind generator, or the energy it makes per second, increases with wind speed, radius of the blades, and efficiency of the generator. It is important to place a wind generator where the wind is usually strong, as on the top of a hill or tall building. The best wind speeds for making electricity are between ten and 25 miles per hour. Because wind keeps changing direction, the wind generator must be able to turn quickly to catch the wind at its greatest speed. Wind

generators have large propeller-type blades. Their efficiency can be increased by improved blade design, by placing the generator directly behind the blade, and by using ball bearings on all moving parts.

Wind generators are about 20 percent efficient today. This means about 20 percent of the wind energy that moves past the blades is transformed into electrical energy. Theoretically, efficiency can be improved to about 50 percent. A recent study estimated that by the year 2000, wind energy could meet up to five percent of this country's energy needs. California already has about 10,000 wind generators.

Wind energy is actually a form of solar energy. The sun warms different places on the earth's surface differently. This uneven heating causes changes in air pressure that produce wind.

One obvious advantage of wind energy is that it is constantly being renewed. Wind energy is also free. Operating wind generators has no bad environmental effects. It is economical in areas where fuel costs are high and the wind blows steadily.

There are some problems associated with wind generators. Because the wind often blows gently or changes direction, wind generators produce a full amount of electricity only about 25 percent of the time. People have to design energy-storage systems to provide energy when the wind dies down. To build such systems could mean using large sets of batteries, which are recharged when the wind blows again. People could use wind energy to pump water up into tanks or ponds. When the wind stops, the water can flow downhill to turn electric generators. Also, people can compress air and store it in tanks or sealed caves. The compressed air can turn electric generators when needed.

Because wind generators turn at various speeds, the AC electricity they produce does not have a steady number of cycles per second. Recall that, in the United States, electrical equipment is made to operate with 60 cycles per second AC electricity. This technological problem can be overcome using electrical converters or constant-speed gears in the generator. Unfortunately, these reduce the efficiency of the wind generator.

Geothermal Energy

Deep within the earth, temperatures are high enough that rock and iron melt. Some of this heat is left over from a time when the earth was melted material. Heat is also produced by radioactive decay of elements within the earth's core.

In some places, underground heat produces hot water and steam, which come out through cracks in rocks. These are known as hot spots. One famous hot spot is Old Faithful geyser in Yellowstone Park, Wyoming. Look at Figure 23-14.

Figure 23-14 A Geyser

The heat inside the earth is known as **geothermal energy**. In some places, geothermal energy is used to heat buildings, make hot water, and generate electricity. There are many hot spots in the United States. Most are located in the western states and Alaska, where volcanic activity is high and there are many cracks in the earth.

There are some problems with geothermal energy. When hot water and steam is brought out of holes drilled near a hot spot, other polluting gases and minerals may also come out. These include sulfur, boron, and hydrogen sulfide, a poisonous gas that smells like rotten eggs.

Geothermal electric plants must be located very close to hot spots. Many of these are in scenic areas, which could lose their beauty to unsightly plants and transmission wires. Finally, no one knows what happens as water and steam are removed from deep underground. The earth may shift or cave in at such spots.

Tidal Energy

Tides result from the gravitational pull of the moon on the oceans of the earth. About every 6.5 hours, the level of ocean water at any one place goes up or down. This is the **tide**. The flow of water in the ocean and along the ocean shore is called the tidal flow. In some places, the tide is high, and the tidal flow is fast. This occurs where bays and rivers connect to the ocean through narrow channels. For example, in the Bay of Fundy, Canada, the tides rise and fall by 30 feet and the tidal flow can be 15 miles per hour, which is very fast for water. Examine

Figure 23-15. In such places, the tidal flow of water can be used to make electricity.

The entrance channels to these bays or rivers can be dammed so that the tidal flow is forced to go through turbines. Construction costs would be high and, therefore, would reduce the economic benefits of using tidal energy to make electricity. Because there are a limited number of places where tidal flow is fast, tidal energy will probably not be a major source of electricity in the future.

Fuel Cells

You have learned that making electricity is not an efficient process. In burning fuel, making steam, turning generators, and transmitting current, nearly 75 percent of the fuel's energy is lost as heat. A **fuel cell** is a more direct way to produce electricity from fuels. In a fuel cell, hydrogen and oxygen gases cause a solution containing hydroxide ions (OH) to produce electricity. The other product of the reaction is water.

$$2H_2 + O_2 \rightarrow 2H_2O + \textbf{electrical energy}$$

Notice that this equation and the action of a fuel cell are the reverse of the electrolysis of water discussed on page 139. In electrolysis, electricity is used to decompose water into its elements. In a fuel cell, electricity is made as water is formed from its elements.

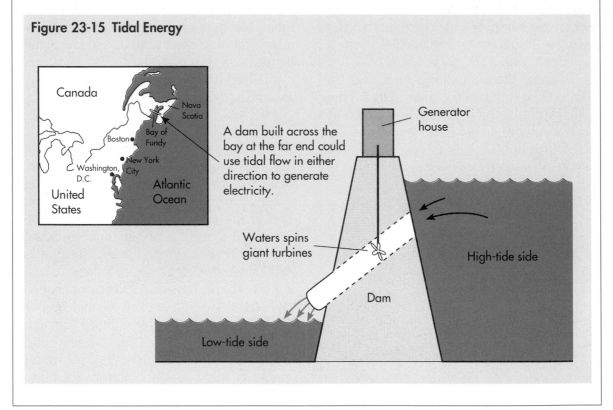

Figure 23-15 Tidal Energy

Canada

Nova Scotia

Boston

Bay of Fundy

New York City

Washington, D.C.

United States

Atlantic Ocean

A dam built across the bay at the far end could use tidal flow in either direction to generate electricity.

Generator house

Waters spins giant turbines

High-tide side

Dam

Low-tide side

Fuel cells were used in space to make electricity for the Apollo and Gemini missions to the moon. Some day, every building and house might have a fuel cell. Because fuel cells are small and can be located close to the electricity user, there is little loss of energy due to transmission. Fuel cells are very efficient, and they operate quietly.

Unfortunately, at this time, fuel cells are expensive to build. Can large fuel cells be made to supply enough electricity for a home at a reasonable cost? That is still a big question. Fuel cells also tend to break down and need repair because of the chemicals and hot temperatures inside them. Someday, technology may improve fuel cells so that they contribute a substantial percentage to America's energy needs. At this time they have limited use.

Micro-hydroelectric Energy

Hydroelectric plants are placed near huge dams that form lakes containing billions of gallons of water. A lot of water must drop down hundreds of feet to turn the generators. Engineers have designed generators, however, that make reduced amounts of electricity from smaller amounts of flowing water. By means of pipes and tanks, even the water in a small stream can be used. Electricity produced in this way is micro-hydroelectric energy, the prefix *micro* meaning small in relative size.

Use of micro-hydroelectric generators can be important in isolated areas where there are streams or small ponds. This method is vital in nations that must supply cheap energy to villages and towns. China, for example, has built more than 100,000 micro-hydroelectric units to make electricity.

Fusion Energy

In fission, a large nucleus, like uranium-235, is made to split. During splitting, much nuclear energy is released. In fusion, small nuclei such as those of hydrogen are made to combine. Fusion also releases much nuclear energy. In fact, the sun shines because of fusion reactions. Some of the fusion reactions that occur are illustrated in Figure 23-16.

Figure 23-16 Examples of Fusion Reactions

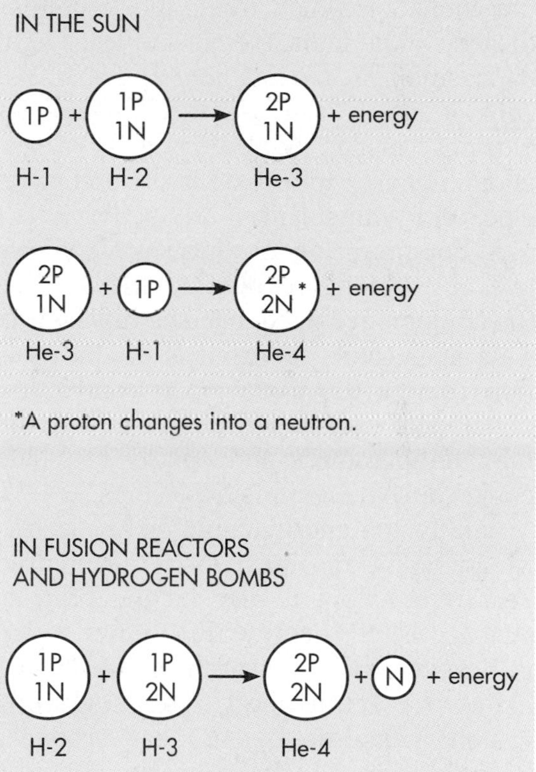

IN THE SUN

$$\left(\begin{array}{c}1P\end{array}\right) + \left(\begin{array}{c}1P \\ 1N\end{array}\right) \longrightarrow \left(\begin{array}{c}2P \\ 1N\end{array}\right) + \text{energy}$$

H-1　　　H-2　　　　He-3

$$\left(\begin{array}{c}2P \\ 1N\end{array}\right) + \left(\begin{array}{c}1P\end{array}\right) \longrightarrow \left(\begin{array}{c}2P \\ 2N\end{array}\right)^* + \text{energy}$$

He-3　　H-1　　　He-4

*A proton changes into a neutron.

IN FUSION REACTORS
AND HYDROGEN BOMBS

$$\left(\begin{array}{c}1P \\ 1N\end{array}\right) + \left(\begin{array}{c}1P \\ 2N\end{array}\right) \longrightarrow \left(\begin{array}{c}2P \\ 2N\end{array}\right) + \left(\begin{array}{c}N\end{array}\right) + \text{energy}$$

H-2　　　H-3　　　　He-4

The big advantage of fusion is that it uses hydrogen-2, called deuterium, and hydrogen-3, called tritium, as fuels. Oceans contain large amounts of hydrogen-2 from which hydrogen-3 can be made. Fusion also does not produce radioactive products, as fission does. The big disadvantage of fusion is that it only occurs at extremely high temperatures, like 50 million° C! No one yet knows how to handle and contain such temperatures. It is like trying to handle a tiny piece of the sun inside a container.

One experimental device keeps the extremely hot hydrogen gas circulating inside a magnetic field. The magnetism keeps the gas from touching the sides of the container, which would melt. Another device shoots intense beams of laser light on a small spot containing H-2 and H-3 compounds. It is hoped that the tiny explosions made by fusion will release enough energy to fire the lasers and make electricity with steam.

Although fusion produces no radioactive or polluting products, it does cause the equipment used to become radioactive. This is because neutrons from the reaction change the iron, copper, and other metals into radioactive isotopes. Also, the high temperatures and energy of fusion could cause dangerous explosions.

Still, fusion energy could make electricity abundant and inexpensive. A hundred years from now, it may be the primary source of all energy. For now, fusion is a key area of research in the United States. A **controlled self-sustaining fusion reaction** is one that produces enough energy to keep itself going at a steady rate. So far, no controlled self-sustaining fusion reaction has occurred. It still takes more energy to make a controlled fusion reaction happen than the reaction itself produces. Perhaps, in the near future, there will be a self-sustaining reaction which might lead to further improvements that would produce extra energy for the public.

Section 23-3 Review

Write the definitions for the following terms in your own words.

1. **geothermal energy**
2. **tide**
3. **fuel cell**
4. **controlled self-sustaining fusion reaction**

Answer these questions.

5. What are the advantages and disadvantages of using wind energy to make electricity?
6. How might people overcome the disadvantages of wind energy?
7. Why are geothermal, tidal, and microhydroelectric energies available only in a limited number of locations?
8. Explain why fuel cells are more efficient for making electricity than are standard generators.
9. Many believe that fusion energy will solve all the energy problems of the world. Why might they be right? Why might they be wrong?

Analyzing

Often data can be made more interesting and understandable by using pictures. The pictures in Figure 23-17 show various energy sources. The order of the pictures indicates how much energy is supplied by that source. In each row, the source that supplies the most energy is on the left and the source that supplies the least energy is on the right. These three sets of pictures compare the actual sources of energy in 1988 with the expected sources in 2000, as predicted by the experts in the Department of Energy and by the public.

1. Name the top three sources of energy in 1988, in expert predictions for 2000, and in the public's prediction for 2000.
2. What are a few big differences between the experts' view and the public's view about energy in the year 2000?
3. How do these pictures show that the experts are cautious and conservative about future changes, while the public is enthusiastic, if unrealistic, about future changes?
4. What surprises you most in the prediction made by the experts for 2000?

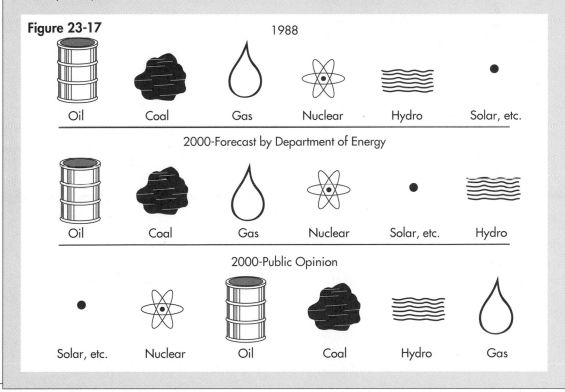

Figure 23-17

1988

| Oil | Coal | Gas | Nuclear | Hydro | Solar, etc. |

2000-Forecast by Department of Energy

| Oil | Coal | Gas | Nuclear | Solar, etc. | Hydro |

2000-Public Opinion

| Solar, etc. | Nuclear | Oil | Coal | Hydro | Gas |

SCIENCE, TECHNOLOGY, & SOCIETY

Learning to Conserve, Reuse, and Recycle

Science and technology can develop new energy sources and recover more fuel from old sources. Science and technology, however, cannot solve the energy crisis alone. Laws and government can help. People can also help by changing some poor habits. Americans need to learn how to conserve energy and reuse or recycle materials.

Many schools are helping with this effort. They have started environmental and energy studies and activities. In one school, students learned songs about these topics and the earth. Some schools run auditorium programs on the environment and energy, with shows or speakers from the government, citizens' groups, or companies. Posters, letter writing, and compositions have been used to give students the facts. Students have investigated problems in their neighborhoods and interviewed people about the problems. Students often discuss these topics at home and families have learned to save energy, money, and materials. Teachers now attend workshops on energy and the environment.

In some schools, students learn by doing. In the lunchroom, they separate trash into garbage; cans, bottles, paper items, and plastics to be recycled; and trays, plates, and other objects to be reused. Some students collect cans and bottles to raise money for school projects. They can sell cans and bottles for recycling or return them for a deposit.

Follow-up Activities

1. What suggestions do you have for educating people in your school or neighborhood about the environment and energy?
2. Identify groups that educate people about saving energy and recycling. Look for posters, advertisements, articles, and programs about these groups. If you wish, write for information.
3. List other situations in which science and technology alone cannot solve society's problems.

Figure 23-18

Section 23-1 Recovering Additional Energy

1. Technology can recover more oil and natural gas from old wells and mines and from a new source, oil shale.
2. Technology can make coal cleaner and more convenient to use than it is at present.
3. Breeder reactors can greatly extend the useful life of uranium fuel.

Section 23-2 New Sources of Energy

1. Renewable sources of energy like fresh water, wind, sunshine, and plants are recycled by natural processes within a relatively brief span of time.
2. The sun is the primary source of energy for the earth.
3. The energy of the sun can be captured and used to provide heat and hot water for homes, offices, and factories.
4. Solar energy can also be converted to electrical energy. This can be done by using thermal conversion and solar cells.
5. Biomass and animal wastes can be burned to produce heat or can be converted into biofuels.

Section 23-3 Additional Sources of Energy

1. Wind energy can be used to produce electrical energy.
2. The use of any nonrenewable or renewable energy source has advantages and disadvantages associated with it.
3. Geothermal, tidal, fuel-cell, and micro-hydroelectric sources can provide significant amounts of energy.

4. Fusion energy may become the world's primary energy source in the future.

Write the term from the list that best matches each statement.

biofuels, fuel cell, geothermal, oil shale, self-sustaining fusion reaction, solar energy, solvent refining, tide

1. rocks that form oil when heated
2. removes sulfur and ash from coal
3. sunshine
4. heat energy within the earth
5. water movement caused by moon's pull
6. electricity from H_2 without burning
7. When energy released is equal to or more than the energy consumed
8. energy sources from biomass

Explain the difference between the terms in each pair.

9. gasification, liquefaction
10. passive solar heating system, active solar heating system
11. Ocean Thermal Energy Conversion, photovoltaic conversion
12. biomass, gasohol

Write the missing terms for each sentence.

1. In recovering natural gas, explosives are used to ___.

2. Burning coal pollutes because of ___.
3. The fuel used up in breeder reactors is ___, and the extra fuel created in them is ___.
4. A field that is 50 feet by 50 feet gets enough sunlight to equal the energy in ___ of gasoline every day.
5. During the day, windows should allow ___, and during the night they should be ___ for effective solar heating.
6. In ocean thermal energy conversion (OTEC), water boils into steam at only 80° F because the boiling occurs under ___.
7. A ___ is placed on a roof in an active solar-heating system.
8. Plants and waste materials are ___ when used as fuel or ___ into other fuels.
9. A fuel cell uses ___ in order to generate ___.

INTERPRETING INFORMATION

The map in Figure 23-19 shows the locations of geothermal hot spots on the United States mainland. Each dot represents an area of promise for finding useful geothermal energy.

1. Estimate how many locations are shown having useful geothermal energy. In how many states are they found?
2. Which part of the United States has most of the hot spots? Which part has the least?
3. Using the map, find which states east of Chicago have geothermal hot spots.
4. Geothermal locations occur in clusters or in a line. Refer to another map to see what is special about the land where these clusters

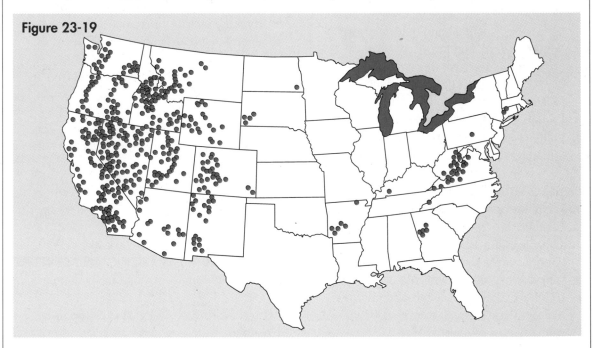

Figure 23-19

are located. What hypothesis can you make about where the hot spots are located?

THINK AND DISCUSS

Use the section numbers in parentheses to help you find each answer. Write your answers in complete sentences.

1. The energy of the sun is responsible for many of the alternative sources of energy. Show how this is true. (23-2, 23-3)
2. Compare the workings of a solar cell and a fuel cell. (23-2, 23-3)
3. Make an outline of the ways fossil fuels can be recovered or made more environmentally safe and convenient. (23-1)
4. What are the benefits and dangers of breeder reactors? (23-1)
5. What technological improvements will make solar cells more practical for homes and industry than they are now? (23-2)
6. What condition is necessary for tidal energy to be captured and made useful? (23-3)
7. List the technological problems that must be overcome before fusion of hydrogen can provide energy for society. (23-3)

GOING FURTHER

1. Obtain more information on one of the alternative sources of energy discussed in this chapter. Write and present a report in class.
2. Make a scrapbook of articles or of notes taken about TV programs on new energy sources. Highlight the key ideas or interesting things you learned from each.

COMPETENCY REVIEW

1. Which energy source is *not* renewable?
   a. oil   b. wind
   c. sunlight   d. plants
2. A great source of renewable energy is
   a. the sun.   b. oil.
   c. coal.   d. uranium.
3. What device is used for photovoltaic conversion?
   a. fuel cell   b. solar cell
   c. solar collector d. generator
4. Plants cannot be used for
   a. collecting solar energy.
   b. burning as a fuel.
   c. breeder reactors.
   d. making biofuels such as alcohol.
5. What helps a wind generator to produce electricity efficiently?
   a. wind of ten to 25 miles per hour
   b. large and long blades
   c. efficient design with ball bearing
   d. all the above
6. Which energy source has disadvantages and problems?
   a. only nonrenewable
   b. only renewable
   c. both nonrenewable and renewable
   d. neither nonrenewable nor renewable
7. The fuel made and also used within breeder reactors is
   a. uranium.   b. plutonium.
   c. hydrogen.   d. helium.
8. The fuel used in the fusion reaction is
   a. uranium.   b. plutonium.
   c. hydrogen.   d. helium.

CAREERS IN PHYSICAL SCIENCE

Perhaps, as a result of your studies, you have become interested in the career possibilities in physical science. Remember that there are two major goals of physical science:

1. to gain knowledge about the structure, laws, and composition of the world
2. to use this knowledge to improve life and society.

Scientists who focus on the first goal are part of basic science, or research. Scientists who focus on the second goal are part of applied science or technology. Usually, the work of physical scientists involves both goals.

Careers in Research

Physicist A physicist studies forces and matter, hoping to uncover their laws and behaviors. Today, physicists are exploring new frontiers. In their explorations, physicists try to answer questions such as these: Why does electricity flow without any resistance through some materials when they are very cold? Of what are protons, neutrons, electrons, and other subatomic particles in matter made? How can the nuclear energy locked in an atom be released by better fission and fusion reactions? What are the properties of laser light? What forces made the universe?

Chemist A chemist studies the physical and chemical properties of matter. Part of a chemist's research is to relate these properties to the structure of the molecules in the matter. Chemists can then develop new materials that have special properties. Chemists study the details of chemical reactions so that they can make new materials. Many chemists are involved with testing products for quality and safety.

Laboratory Assistant This person helps scientists by performing many research tasks, like collecting data, and by taking care of laboratory equipment. Often a laboratory assistant is more expert about a complicated device or procedure than the scientist for which they work.

Careers in Technology

Engineer Engineers apply science to design and make new technological devices and materials. Engineers build and operate the electric, water, highway, and sewer systems on which society depends. Here are some examples of questions engineers ask. How can tons, not just grams, of a new material be produced and shipped? How can cars and buildings be improved by design changes? How can big equipment like generators, oil wells, jet engines, and energy systems be made more efficient?

Engineering is divided into four specialties: chemical, electrical, mechanical, and civil. Each specialty has many subspecialties, like aerospace engineering or sanitation engineering. Architects usually have engineering training so they can design beautiful, useful, and strong buildings.

Technician This person may work with engineers to design or operate large equipment and factories. A technician may also work independently to install and repair complex equipment, or to give advice to engineering companies about making better designs and products.

Computer Specialist Computers are everywhere today. Businesses, banks, government offices, and science laboratories depend on computers. Small computer chips are in car engines, telephones, and TVs. Computer specialists create computer hardware, which consists of the electrical circuits, the chips, and other devices. They also create software. Software includes the programs, or instructions, that run the computer.

Careers in Society
People who study physical science do not necessarily become scientists. However, their background often helps them achieve success in other fields. For example, a lawyer with a background in physical science is very valuable in technical cases. Likewise, business administrators who know physical science are very valuable to technology companies. Physical scientists often work for the government as experts and policy advisors on technology. Many scientists are writers and teachers, helping students and society understand recent scientific discoveries and issues.

Education Needed
A good place to start a career in physical science is in high school. If possible, take courses in biology, chemistry, and physics.

Three or four years of mathematics are also recommended. Part of those math studies might include computer math and computer programming.

Usually college students take four years to obtain a bachelor of science (B.S.) degree in one of the main fields, like physics, or mechanical engineering. Many students then want to get a job in industry, research institutes, or government in order to become acquainted with "real life," or, the practical aspects of science as a job. However, many employers expect new employees to continue studying, perhaps at night or full time, in a two- or three-year master of science (M.S.) program.

A smaller percentage of people then may go on to get a doctor of philosophy degree (Ph.D.). Earning this degree involves research in some field of physical science and engineering. Doctorate degrees are usually expected of college teachers and research scientists.

Obtaining Information
Use your school or local library to find books about careers in science. If you are interested in a specific career, find out what kind of training, as well as how much training, you will need. The librarian can also help you locate professional societies that can send you information about career opportunities in their fields. For example, the American Chemical Society, the American Physical Society, and the American Society for Engineering Education are just three of the dozens of agencies to which you can write.

Scientific Notation

Scientific notation is a convenient way to write the very large and very small numbers used in science. In scientific notation, every number is written as a product of two special kinds of numbers. Here are two numbers written in scientific notation:

$$418,000,000 = 4.18 \times 10^8$$
$$0.00077 = 7.7 \times 10^{-4}$$

coefficient   power of ten

The first special kind of number is called the coefficient. The coefficient can be any number between 1 and 10, except 10 itself. Both coefficients shown above, 4.18 and 7.7, are between 1 and 10.

The second special number is a power of 10. The numbers 10^8 and 10^{-4} are powers of 10. Table 1-A shows powers of 10.

To write a number in scientific notation, place a caret mark (^) after the first *nonzero* digit in the number. Count the number of places between the caret and the decimal point, whether the decimal point is printed or understood. If the decimal point falls to the right of the caret, then the number of places indicates a *positive* exponent of 10. If the decimal point is to the left of the caret, then the number of places indicates a *negative* exponent of 10. The coefficient is obtained by placing a decimal point where the caret was inserted. Ending zeros may be dropped from the coefficient. Look at the following examples:

Table 1-A

10^x	Definition	Value
10^4	$10 \times 10 \times 10 \times 10$	10,000
10^3	$10 \times 10 \times 10$	1,000
10^2	10×10	100
10^1	10	10
10^0	1	1
10^{-1}	1/10	0.1
10^{-2}	$1/(10 \times 10)$	0.01
10^{-3}	$1/(10 \times 10 \times 10)$	0.001
10^{-4}	$1/(10 \times 10 \times 10 \times 10)$	0.0001

$$98,000 = 9\mathbin{^}8\mathbin{^}0\mathbin{^}0\mathbin{^}0. \qquad 9.8 \times 10^4$$
$$123 = 1\mathbin{^}2\mathbin{^}3. \qquad 1.23 \times 10^2$$
$$67.9 = 6\mathbin{^}7.9 \qquad 6.79 \times 10^1$$
$$0.11 = 0.1\mathbin{^}1 \qquad 1.1 \times 10^{-1}$$
$$0.0064 = 0.0\mathbin{^}0\mathbin{^}6\mathbin{^}4 \qquad 6.4 \times 10^{-3}$$

Practice

Change these numbers into scientific notation:

1. 186,000
2. 93,000,000
3. 45.82
4. 0.661
5. 0.0000053
6. 80 billion

Using Units in Calculations

A measurement consists of a number and a unit. In calculations that use measurements, the units follow regular mathematical rules. For example:

    a. 2 m + 3 m + 4 m = 9 m
    b. 2 m × 3 m × 4 m = 24 m³
    c. 18.2 pints − 5.3 pints = 12.9 pints
    d. 30 miles/50 min = 0.6 miles/min

Notice that in example a all the units are meters. On the other hand, in example b the unit of the answer is cubic meters rather than meters.

Units can be multiplied or cancelled. Multiplying and canceling units helps scientists obtain the correct unit when using formulas. To help you cancel units, write fractions like km/hr with horizontal fraction bars: $\frac{km}{hr}$. Here are examples:

distance = speed × time

$$\text{distance} = 60\ \frac{km}{\cancel{hr}} \times 1.5\ \cancel{hr} = 90\ km$$

mass = density × volume

mass = density × (length × width × height)

$$\text{mass} = \frac{5\ g}{cm^3} \times 4\ \cancel{cm} \times 3\ \cancel{cm} \times 2\ \cancel{cm}$$
$$= 120\ g$$

acceleration = change in speed/time

$$\text{acceleration} = \frac{72\ \frac{m}{sec}}{8\ sec} = \frac{9\ \frac{m}{sec}}{sec} = 9\ \frac{m}{sec^2}$$

cost = electricity price × power × time

$$\text{cost} = 12\ \frac{¢}{\cancel{kW}\ \cancel{hr}} \times 2\ \cancel{kW} \times 6\ \cancel{hr}$$
$$= 144¢ = \$1.44$$

Scientists often need to change the units on a measurement. For example, an object may be measured as 20 inches long, but what would that length be in centimeters or in meters? To make the change, multiply the original measurement by *conversion factors*, like 2.54 cm/1 inch and 1 m/100 cm.

$$\text{length} = 20\ \cancel{inches} \times \frac{2.54\ cm}{1\ \cancel{inch}} = 50.8\ cm$$

$$= 50.8\ \cancel{cm} \times \frac{1\ m}{100\ \cancel{cm}} = 0.508\ m$$

Conversion factors are obtained from *unit equalities*, which state how two units are related. For example, 1 inch = 2.54 cm is a unit equality. From this unit equality, two conversion factors can be obtained: 1 inch/2.54 cm and 2.54 cm/1 inch.

Note that conversion factors can be made for units not directly related to science like bagels, and even for nonsense units like zigs and zags.

When changing units, start with the known measurement. Use whichever conversion factors will permit unwanted units to cancel in the numerators and denominators. Here are examples:

How many kilograms are in 450 grams?

$$450\ \cancel{g} \times \frac{1\ kg}{1000\ \cancel{g}} = \frac{450 \times 1\ kg}{1000} = 0.45\ kg$$

Unit equality	Conversion factors obtained
1 hr = 3600 sec	$\dfrac{1\ hr}{3600\ sec}$ and $\dfrac{3600\ sec}{1\ hr}$
1000 cm³ = 1 L	$\dfrac{1000\ cm^3}{1\ L}$ and $\dfrac{1\ L}{1000\ cm^3}$
1 mile = 5280 ft	$\dfrac{1\ mile}{5280\ ft}$ and $\dfrac{5280\ ft}{1\ mile}$
13 bagels = $1.50	$\dfrac{13\ bagels}{\$1.50}$ and $\dfrac{\$1.50}{13\ bagels}$
12 months = 365 days	$\dfrac{12\ months}{365\ days}$ and $\dfrac{365\ days}{12\ months}$
3.4 zigs = 17.5 zags	$\dfrac{3.4\ zigs}{17.5\ zags}$ and $\dfrac{17.5\ zags}{3.4\ zigs}$

How many centimeters are in 3 kilometers?

$$3\ \cancel{km} \times \frac{1000\ \cancel{m}}{1\ \cancel{km}} \times \frac{100\ cm}{1\ \cancel{m}}$$

$$= \frac{3 \times 1000 \times 100\ cm}{1}$$

$$= 300,000\ cm$$

How fast is 60 miles/hr when changed into feet/second?

$$60\ \frac{\cancel{miles}}{\cancel{hr}} \times \frac{5280\ ft}{1\ \cancel{mile}} \times \frac{1\ \cancel{hr}}{3600\ sec}$$

$$= \frac{60 \times 5280\ ft \times 1}{1 \times 3600\ sec} = 88\ \frac{ft}{sec}$$

How many yards are in 1 meter?

$$1\ \cancel{m} \times \frac{100\ \cancel{cm}}{1\ \cancel{m}} \times \frac{1\ \cancel{in}}{2.54\ \cancel{cm}} \times \frac{1\ yd}{36\ \cancel{in}}$$

$$= \frac{1 \times 100 \times 1 \times 1\ yd}{1 \times 2.54 \times 36} = \frac{100\ yd}{91.44}$$

$$= 1.09\ yards$$

How many kilograms are in 3 tons, if 1 ton = 2000 lbs and 454 g = 1 lb?

$$3\ \cancel{tons} \times \frac{2000\ \cancel{lbs}}{1\ \cancel{ton}} \times \frac{454\ \cancel{g}}{1\ \cancel{lb}} \times \frac{1\ kg}{1000\ \cancel{g}}$$

$$= \frac{3 \times 2000 \times 454 \times 1\ kg}{1 \times 1 \times 1000} = 2724\ kg$$

Practice

1. What is the answer to this multiplication?

$$120\ \frac{ft}{min} \times 66\ \frac{lbs}{ft^3} \times 10\ ft^2 \times \frac{1\ min}{60\ sec} = ?$$

2. What conversion factors are obtained from 1 fortnight = 14 days and 1 week = 7 days? Use them to calculate, by the method shown, how many fortnights are in five weeks.

3. What is the speed of 200 cm/min when changed to km/hr? to mm/sec?

4. If 5 inches = 8 zags and 3.4 zigs = 17.5 zags, how many zigs will be in one foot? Use conversion factors.

Specific Gravity and Density

The **specific gravity** of a material is the ratio of its density to the density of water. The specific gravity of water is 1 and of copper is 8.9. Notice that specific gravities have no units. You can use specific gravity to find the density of a material in a large variety of units. To obtain the density of a material, multiply its specific gravity by the density of water in the units you desire. The formula is

$$\text{density} = \text{specific gravity} \times \text{density of water}$$

Here are some examples:

Find the density of copper in g/cm³.
The density of water in g/cm³ is 1 g/cm³.

$$\text{density} = 8.9 \times 1.00 \text{ g/cm}^3 = 8.9 \text{ g/cm}^3$$

Find the density of alcohol in pounds per gallon. From the table at right, the density of water is 8.3 lb/gal.

$$\text{density} = 0.79 \times 8.3 \text{ lbs/gallon} = 6.6 \text{ lbs/gallon}$$

Practice

1. What is the density of air in lbs/ft³?
Density (air) = 0.00121×62.4 lbs/ft³
Density (air) = 0.0755 lbs/ft³
Show that this means that in a 15 ft × 25 ft × 10 ft classroom, the air weighs about 280 pounds!
2. Calculate the densities of the following:
a. lead in g/cm³   b. sugar in kg/L
c. milk in lb/quart d. hydrogen in kg/m³

The density of a material in g/cm³ or g/mL is equal to its specific gravity.

Density of water in various units	
1.00 g/cm³	62.4 lb/ft³
1.00 g/mL	8.3 lb/gallon
1.00 kg/L	2.1 lb/quart
1000 kg/m³	0.43 lb/in³

Specific gravities of liquids			
alcohol	0.8	oil	0.91
gasoline	0.66	sulfuric acid	1.83
mercury	13.6	pure water	1.00
milk	1.05	sea water	1.03

Specific gravities of solids			
aluminum	2.7	plastic	0.9 – 2.0
beeswax	0.96	quartz	2.6
brass	8.0 – 8.7	rubber	0.9 – 1.2
copper	8.9	salt	2.2
gold	19.3	silver	10.5
human fat	0.91	steel	7.8
ice	0.92	sugar	1.6
iron	7.9	oak wood	0.7
lead	11.3	balsa wood	0.2

Specific gravities of gases	
(All at 20°C and atmospheric pressure)	
air 0.00121	helium 0.00017
ammonia 0.00073	nitrogen 0.00117
carbon dioxide 0.00185	oxygen 0.00133
hydrogen 0.00008	

Alphabetical Table of Elements

Name	Symbol	Atomic number	Atomic weight*	Mass number**
Actinium	Ac	89	227.028	227
Aluminum	Al	13	26.98	27
Americium	Am	95	243	243
Antimony	Sb	51	121.75	121
Argon	Ar	18	39.948	40
Arsenic	As	33	74.922	75
Astatine	At	85	(210)	(210)
Barium	Ba	56	137.33	138
Berkelium	Bk	97	(243)	(243)
Beryllium	Be	4	9.0122	9
Bismuth	Bi	83	208.98	209
Boron	B	5	10.81	11
Bromine	Br	35	79.904	79
Cadmium	Cd	48	112.41	114
Calcium	Ca	20	40.08	40
Californium	Cf	98	(251)	(251)
Carbon	C	6	12.011	12
Cerium	Ce	58	141.12	140
Cesium	Cs	55	132.91	133
Chlorine	Cl	17	35.453	35
Chromium	Cr	24	51.996	52
Cobalt	Co	27	58.9332	59
Copper	Cu	29	63.546	63
Curium	Cm	96	(247)	(247)
Dysprosium	Dy	66	162.50	164
Einsteinium	Es	99	(252)	(252)
Erbium	Er	68	167.26	168
Europium	Eu	63	151.96	153
Fermium	Fm	100	(257)	(257)
Fluorine	F	9	18.998	19
Francium	Fr	87	(223)	(223)
Gadolinium	Gd	64	157.25	158
Gallium	Ga	31	69.72	69
Germanium	Ge	32	72.59	74
Gold	Au	79	196.967	197
Hafnium	Hf	72	178.49	180
Helium	He	2	4.003	4
Holmium	Ho	67	164.93	165
Hydrogen	H	1	1.00794	1
Indium	In	49	114.82	115
Iodine	I	53	126.905	127
Iridium	Ir	77	192.22	193
Iron	Fe	26	55.47	56
Krypton	Kr	36	83.80	84
Lanthanum	La	57	138.906	139
Lawrencium	Lr	103	(260)	(260)
Lead	Pb	82	207.2	208
Lithium	Li	3	6.941	7
Lutetium	Lu	71	174.967	175
Magnesium	Mg	12	24.305	24
Manganese	Mn	25	54.938	55
Mendelevium	Md	101	(258)	(258)
Mercury	Hg	80	200.59	202
Molybdenum	Mo	42	95.94	98

*Parentheses give the mass number of the most stable isotope.
**Of the most abundant isotope on Earth.

Alphabetical Table of Elements

Name	Symbol	Atomic number	Atomic weight*	Mass number**
Neodymium	Nd	60	144.24	144
Neon	Ne	10	20.179	20
Neptunium	Np	93	237.048	237
Nickel	Ni	28	58.69	58
Niobium	Nb	41	92.91	93
Nitrogen	N	7	14.007	14
Nobelium	No	102	(259)	(259)
Osmium	Os	76	190.2	192
Oxygen	O	8	15.999	15
Palladium	Pd	46	106.42	106
Phosphorus	P	15	30.974	31
Platinum	Pt	78	195.08	195
Plutonium	Pu	94	(244)	(244)
Polonium	Po	84	(209)	(209)
Potassium	K	19	39.098	39
Praseodymium	Pr	59	140.908	141
Promethium	Pm	61	(146)	(145)
Protactinium	Pa	91	231.036	231
Radium	Ra	88	226.025	226
Radon	Rn	86	(222)	(222)
Rhenium	Re	75	186.207	187
Rhodium	Rh	45	102.906	103
Rubidium	Rb	37	85.468	85
Ruthenium	Ru	44	101.07	102
Samatium	Sm	62	150.36	152
Scandium	Sc	21	44.956	45
Selenium	Se	34	78.96	80
Silicon	Si	14	28.086	28
Silver	Ag	47	107.816	107
Sodium	Na	11	22.990	23
Strontium	Sr	38	87.62	88
Sulfur	S	16	32.06	32
Tantalum	Ta	73	180.95	181
Technetium	Tc	43	(98)	(98)
Tellurium	Te	52	127.60	130
Terbium	Tb	65	158.925	159
Thallium	Tl	81	204.383	205
Thorium	Th	90	232.038	232
Thulium	Tm	69	168.934	169
Tin	Sn	50	118.71	120
Tatanium	Ti	22	47.88	48
Tungsten	W	74	183.85	184
Unnilhexium	Unh	106	(263)	(263)
Unniloctium	Uno	108	(265)	(265)
Unnilpentium	Unp	105	(262)	(262)
Unnilquadium	Unq	104	(261)	(261)
Unnilseptium	Uns	107	(262)	(262)
Uranium	U	92	238.029	238
Vanadium	V	23	50.94	51
Xenon	Xe	54	131.29	132
Ytterbium	Yb	70	173.04	174
Yttrium	Y	39	88.9059	89
Zine	Zn	30	65.39	64
Zirconium	Zr	40	91.224	90

*Parentheses give the mass number of the most stable isotope.
**Of the most abundant isotope on Earth.

Common Units of Measurement

Scientists use the metric system of units. However, the system of units commonly used in the United States is the English system.

Length

In the English system, length is based on a unit of one inch (1 in). Most rulers divide the inch into spaces of halves, fourths, eighths, and sixteenths of an inch. The ruler shown in Figure 5-A is divided into halves, fourths, and eighths of an inch. Other units of length can be derived from the inch:

$$12 \text{ inches} = 1 \text{ foot (ft)}$$
$$3 \text{ feet} = 1 \text{ yard (yd)}$$
$$20 \text{ feet} = 1 \text{ fathom (f)}$$
$$5280 \text{ feet} = 1 \text{ mile (mi)}$$

Figure 5-A

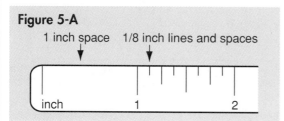

Sometimes it is useful to change a length in fractions of an inch into decimal form. For example, $2\,^3/_8$ inch can be written as 2.375 inch.

Helpful equalities

$^1/_8 = 0.125$	$^5/_8 = 0.625$
$^2/_8 = 0.25$	$^6/_8 = 0.75$
$^3/_8 = 0.375$	$^7/_8 = 0.875$
$^4/_8 = 0.5$	$^8/_8 = 1$

Area and Volume

Area can be measured in square inches (in^2) and square feet (ft^2). Volume can be measured in cubic inches (in^3) and cubic feet (ft^3). Figure 5-B shows the area and volume measurements of a cube.

Figure 5-B

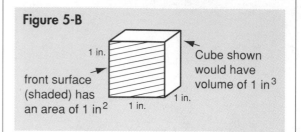

front surface (shaded) has an area of 1 in^2

Cube shown would have volume of 1 in^3

There are other common units that measure area and volume. The unit equalities for these are given in the table below.

Units for area	Units for volume
$1\ ft^2 = 144\ in^2$	$1\ ft^3 = 1728\ in^3$
$1\ acre = 43,560\ ft^2$	$1\ pint = 28\,^7/_8\ in^3$
	$1\ quart = 2\ pints$
	$1\ gallon = 4\ quarts$

Force, Weight, and Mass

The unit used to measure both force and weight is one pound (1 lb). Other units are one ton (2000 lbs) and one ounce ($^1/_{16}$ lb). The unit for mass in the English system is not well known; it is called one slug. If you divide your weight in lbs by 32, you will obtain your mass in slugs. On the moon, your mass would remain the same but your weight will be divided by about 6 because the moon's gravity is less than

Earth's gravity. For example, a person weighing 160 lbs on Earth has a mass of five slugs. On the moon, that person will have a weight of about 27 lbs, but will still have a mass of five slugs.

Density

Common units for density are lbs/in³, lbs/ft³, and lbs/gallon. Density values can be obtained from the specific gravities listed in Appendix 3. For example, the density of sea water is its specific gravity multiplied by the density of water in any desired unit, like lbs/gallon.

$$1.03 \times 8.3 \ \frac{lbs}{gallon} = 8.55 \ \frac{lbs}{gallon}$$

Speed

Common units for speed are ft/sec and miles/hr, which is abbreviated as *mph*.

Work, Energy, and Power

Work and energy are measured by the same unit, a foot-pound, which is written as ft-lb:

work = force × distance
work = 150 lb × 6 ft = 900 ft-lb

If this work is done in nine seconds, then a new unit for power is obtained, ft-lb/sec:

power = work/time
$$power = \frac{900 \text{ ft-lb}}{9 \text{ sec}} = 100 \ \frac{\text{ft-lb}}{\text{sec}}$$

A well-known unit for power is one horsepower (1 hp) which equals 550 ft-lb/sec. Horsepower measures the rate of energy use in car engines, motors, and large appliances. The answer to the equation above can be changed to horsepower by dividing it by 550.

$$power = 100 \ \frac{\text{ft-lb}}{\text{sec}} \times \frac{1 \text{ hp}}{550 \ \frac{\text{ft-lb}}{\text{sec}}}$$

power = 0.182 hp

Heat energy is measured in BTUs, British Thermal Units. One BTU of heat raises the temperature of 1 lb of water by 1°F.

Metric and English

The table below gives some key unit equalities between the metric system and the English system.

Measurement	Unit equality
length	1 inch = 2.54 cm
	1 mile = 1.61 km
volume	1 quart = 0.95 L
mass,	1 slug = 14.6 kg
weight	1 lb = 454 g (on Earth)
	1 lb = 4.5 N
speed	60 mi/hr = 96.6 km/hr
energy	1 ft-lb = 1.4 N-m, or J
heat	1 BTU = 252 cal
power	1 ft-lb = 1.4 watt
	1 horsepower = 746 watt
temperature	F = 1.8 (C) + 32
	C = 5/9 (F − 32)

GLOSSARY

Pronunciation Key
When difficult terms appear, they are respelled to aid pronunciation. A syllable in CAPITAL LETTERS receives the most stress. The table below lists the letters used for respelling. It includes examples of words using each sound and shows how the words would be respelled.

Symbol	Example	Respelling
ah	hypothesis	(hy-PAHTH-uh-sis)
aw	chlorophyll	(KLAWR-uh-fil)
ay	deceleration	(de-sel-uh-RAY-shun)
e	cryogenics	(kry-uh-JEN-iks)
ee	variable	(VER-ee-uh-bul)
ew	nucleic	(new-KLEE-ik)
f	chlorophyll	(KLAWR-uh-fil)
ih	electromagnet	(ih-lek-troh-MAG-net)
i	energy	(EN-uhr-gee)
k	aerodynamic	(ar-oh-dy-NAM-ik)
ks	mixture	(MIKS-ohur)
oh	thermodynamics	(thur-moh-dy-NAM-iks)
oo	supersonic	(soo-puhr-SAHN-ik)
or	thesaurus	(thuh-SOR-us)
oy	colloid	(kahl-OYD)
uh	ion	(EYE-un)
y	science	(SY-uns)
yoo	formula	(FOR-myoo-lah)
z	dissolve	(diz-AHLV)
zh	measure	(MEZH-ur)

absolute zero: coldest possible temperature; temperature at which all molecular motion stops; $-273°$ C, or $-469°$ F (p. 297)

absorbed light: see absorption

absorption: the loss of energy when a sound or light strikes a surface; absorption reduces loudness of the sound and brightness of light (pp. 257, 271)

acceleration (ik-sel-uh-RAY-shun): change in the speed or direction of an object's motion (p. 196)

acid: forms hydrogen ($+1$) ions when dissolved in water (p. 155)

action and reaction: see law of action and reaction

activation energy: energy needed to help a chemical reaction to begin, as with a match or spark plug (p. 81)

active solar heating system: system in which water or air heated by the sun is sent to where it is needed or stored using pumps, fans, and tanks (p. 422)

addictive: producing a compulsive physical need; habit forming (p. 183)

aerodynamic design (ar-oh-dy-NAM-ik): shapes that move through water and air with less friction (p. 234)

alcohol: organic compound with OH group (p. 178)

alkalis: strong bases with OH groups joined to metals from Group 1, such as NaOH (p. 157)

alloy: mixture of two or more metals with special properties and uses (p. 161)

alternating current: current in which electrons rapidly reverse their direction of flow; abbreviated AC (p. 330)

amino acids: organic compounds with COOH and NH_2 groups; form the proteins and tissues of living things; obtained from food (p. 180)

ammonium ion: NH_4, a many-atom ion; forms compounds as if it were a metal (p. 134)

ampere (AM-pir): unit of electric current, based on the number of electrons that flow per second (p. 326)

amplitude: height of a wave, measured from center line; loudness of a sound wave (p. 255)

analyze: look for patterns in data (p. 16)

antibiotics: drugs that kill germs (p. 183)

area: the number of unit squares that would fit inside a region; for a rectangle, area = length × width (p. 48)

atom: basic particle of matter; smallest particle of an element that still has the chemical properties of the element (pp. 16, 91)

atomic mass unit: symbol is u; compares masses of different atoms; based on carbon-12 atom being 12 u; 1 u = 1.661×10^{-24} g (p. 91)

atomic number: number assigned to each element; the number of protons in nucleus (p. 96)

atomic weight: average of the masses of all isotopes of an element, calculated according to their percentages found on Earth (p. 107)

average kinetic energy: determined from the masses and speeds of all the particles in an object; sets temperature of the object (p. 284)

Avogadro's number (av-uh-GAD-roz): 6×10^{23}; number of atoms in 1 gram-atomic mass of an element and number of molecules in 1 gram-formula mass of a compound (p. 126)

balanced equation: chemical equation with the same numbers of each kind of atom on both sides (p. 122)

ball bearings: small balls that roll between two surfaces, thereby reducing friction (p. 233)

base: compound which forms OH ions in water (p. 156)

benefits and burdens: advantages and disadvantages of a technology and of its impact on society (p. 22)

benzene ring: hexagonal pattern of six carbon atoms found in many organic compounds (p. 181)

bimetallic strip: object made of two thin metal strips welded back to back; bends when heated; used in thermostats and oven thermometers (p. 293)

biomass: plants grown for their use as a fuel or as a starting material to make fuels, not for their food value (p. 424)

boiling point: temperature at which a liquid rapidly changes into a gas, instead of just slowly evaporating (p. 73)

breeder reactor: nuclear reactor that produces energy and plutonium from uranium; the purified plutonium is then used as a fuel for making energy; the reactor "breeds" more fuel (p. 419)

BTU: British Thermal Unit; measures heat energy in English system (p. 212)

Bunsen burner: laboratory burner that uses natural gas to heat objects (p. 36)

burning: chemical reaction in which heat and light are produced quickly; rapid oxidation (p. 135)

by-products: materials produced in an industrial process in addition to the main product; medicines and plastics are by-products (p. 378)

calorie: unit to measure heat energy in metric system; 1 cal raises temperature of 1 g of water by 1° C (p. 287)

carbon-14 dating: method of estimating the age of objects produced since 50,000 BC using radioactive carbon-14 (p. 363)

catalyst: material which makes a chemical reaction happen or go faster but which is not changed by the reaction (p. 84)

Celsius: metric temperature scale represented by the symbol °C; ice melts at 0° C, water boils at 100° C (p. 50)

chain reaction: the fission of uranium atoms keeps going because the neutrons produced cause more fissions to occur (p. 359)

char: to turn black when heated; many organic materials char (p. 173)

chemical activity: element's ability to replace another element in a compound; a list that ranks elements by this ability (p. 143)

chemical bond: energy that holds atoms together in a molecule (p. 102)

chemical change: change in which one or more old substances disappear and one or more new substances appear (p. 75)

chemical energy: present in chemicals; stored in the bonds between atoms (p. 217)

chemical equation: a symbolic statement that gives the formulas of the materials which go into and come out of a chemical reaction (p. 121)

chemical formula: symbols that show what elements are in a compound, and how many atoms of each are in one molecule of the compound; for example, Na_3PO_4 (p. 115)

chemical property: ability of a material to combine with other materials or to undergo chemical changes by itself (p. 81)

chemical reaction: chemical change, starts with reactants and ends with products (p. 81)

chemicals: materials in the world and in living things (p. 28)

chemistry: branch of science that studies what things are made of and how they change (p. 9)

chlorophyll (KLAWR-uh-fil): green-colored organic molecule which permits plants to grow using the energy of sunlight; needed for photosynthesis (p. 182)

Clean Air Act: new federal law that sets standards regarding air pollution (p. 392)

coefficient: number placed in front of a formula to indicate how many molecules are being considered; in 5 Na_3PO_4 the number 5 is a coefficient (p. 116)

cogeneration: process of using the heat produced by a factory for heating buildings or water, or making electricity (p. 406)

colloid: mixture of a material with water that has some properties of solutions and some properties of suspensions; for example, milk (p. 166)

completed outer shell: 2 electrons in K shell or 8 electrons in all other shells, if they are the outer shell of an atom; stable arrangement of electrons which atoms move towards during chemical changes (p. 99)

complex machine: machine made up of two or more simple machines that interact to perform some task; for example, a car engine or clock (p. 243)

compound: substance made from two or more elements that are chemically combined (p. 69)

concave: mirror or lense that curves inward at the middle (p. 275)

concentrated solution: solutions containing a large amount of a chemical in a small amount of water (p. 34)

conclusion: findings or results of an investigation which support or reject the original hypotheses (p. 12)

condensation point: temperature at which a substance changes from a gas to a liquid (p. 73)

conduction: (1) movement of heat through a material, as from one end of a piece of metal to the other end (p. 289); (2) movement of electricity through a wire; the transferring of a static charge by direct contact (p. 320)

conductor: material able to let electricity pass through it; metals are conductors (p. 324)

conserve: save or preserve energy sources by less use and by wise use of energy (p. 400)

conservation of energy: law of science; in energy transformations, the total amount of energy remains a constant; no energy is lost or unaccounted for (p. 219)

contraction: decrease in length or volume as temperature decreases (p. 283)

controlled experiment: experiment in which one variable is tested at a time, keeping all other variables the same (p. 11)

controlled self-sustaining fusion reaction: a fusion reaction that produces enough energy to keep itself going at a steady rate (p. 430)

convection: movement of heat as warmer material rises or is forced to swirl in cooler material; for example, hot air rising from a stove (p. 289)

convex: mirror or lens that curves outward in the middle (p. 275)

corrosive: able to eat away or damage skin, clothing, and other surfaces (p. 29)

covalent bond: chemical bond formed between atoms by the sharing of electrons (p. 103)

criss-cross method: way to get the formula for certain compounds by using ion charges of the atoms involved (p. 116)

cryogenics (kry-uh-JEN-iks): science that studies events at low temperatures; for example, freezing of cells and organisms, superconductivity of electricity (p. 296)

current: see electrical current

curtailment: cutting back on energy use, like driving a car less or setting thermostat at 65° F instead of 70° F in winter (p. 401)

data: observations and measurements made during an experiment; facts and measurements given in tables and graphs (pp. 11, 53)

decanting: pouring off the liquid on top of insoluble solids which settled down to the bottom of a container (p. 77)

deceleration (de-sel-uh-RAY-shun): a slowing down in speed; a negative acceleration (p. 196)

decibels: unit for loudness of a sound (p. 256)

decomposition reaction: chemical change in which a compound breaks apart into its elements or into simpler compounds (p. 138)

density: mass of a material in a standard volume, such as grams per cubic centimeter. Given by the formula, density = mass/volume (p. 52)

dependent variable: variable which is not set or controlled by the researcher, but which emerges from the experiment (p. 58)

depressants: drugs which slow down chemical reactions in a person's body and mind; known as "downers" (p. 183)

diatomic molecule: molecule with two identical atoms, as O_2, H_2, and N_2 gases. (p. 122)

diluting: making a concentrated solution weaker by adding water (p. 34)

direct current: current in which electrons flow in one direction only; abbreviated DC (p. 330)

directly proportional: when variable x is multiplied by a certain amount, then variable y is also multiplied by that amount (p. 200)

direct-union reaction: chemical change in which two elements join directly to form a compound (p. 135)

dissolve: mix into water (p. 76)

Doppler effect: change in pitch of a sound because of movement; the pitch of a sound increases when the source of the sound moves towards the listener at a high speed; the pitch decreases when the source moves away (p. 259)

double-replacement reaction: chemical change in which two compounds exchange their metal ions thereby forming two new compounds; also known as ionic-exchange reaction (pp. 145–146)

efficiency: measure of how well a machine uses energy given to it; equals output work/input work (p. 233)

efficient: property of a machine with a high efficiency rating, such as 90%; a large part of energy goes to do useful work, and only a small part wasted as heat and friction (p. 234)

effort: work or energy put into a machine; equals effort force multiplied by an effort distance (p. 229)

electrical circuit (SUR-kut): complete path that electricity can follow along wires, from the source to electrical devices, back to the source again (p. 325)

electrical conduction: see conduction

electrical current: flow of electrons through a wire (p. 324)

electrical induction: process of producing a charge in a neutral object by the nearness of a charged object without touching (p. 319)

electrolysis: process of decomposing, or breaking apart, a compound using electricity (p. 139)

electromagnet (ih-lek-troh-MAG-net): temporary magnet produced by electric current; found in doorbells, tape recorders, and junkyards (p. 303)

electromagnetic energy (ih-lek-troh-mag-NET-ik): in electromagnetic waves like light, radio, and X-rays (p. 217)

electromagnetic waves: waves caused by rapid fluctuations of the electric and magnetic fields present everywhere, even in a vacuum; light is an example (p. 265)

electromagnetic spectrum: families of electromagnetic waves which are arranged and grouped together by their frequencies (p. 268)

electron shells: orbits in atoms within which electrons move (p. 95)

electrons: particles of negative charge which surround the nucleus of an atom; electricity consists of moving electrons (p. 94)

electroscope: device that detects electrical charges (p. 321)

element: material that is made up of one kind of atom; an element cannot be broken into any simpler material by ordinary means; one of the 108 materials from which all other substances are formed (p. 69)

energy: ability to do work (p. 211)

energy efficient: quality of machines that give more useful work and less heat for the energy put into them (p. 402)

energy-intensive: quality of a device or process which uses a lot of energy (p. 389)

energy transformations: change of energy from one form into another form, as electrical to heat in a toaster (p. 219)

environmental impact statement: statement required by law on all projects which may affect the environment, the ecology, the water resources, and the communities in an area; must list harmful effects, and how these will be minimized or removed (p. 392)

equilibrium (ee-kwuh-LIB-ree-um): situation in which forces acting on an object balance out, or cancel, each other so there is no resulting force (p. 192)

estimate: educated guess about a quantity that was not or could not be measured directly (p. 15)

endothermic: quality of a reaction which requires energy, usually as heat (p. 81)

exothermic: quality of a reaction which releases energy, usually as heat (p. 81)

expansion: increase in length or volume as temperature increases (p. 283)

family of elements: set of elements with similar chemical and physical properties; located in groups, or columns, on Periodic Table (p. 103)

filtering: separating dissolved and undissolved materials in a water mixture by passing it through fine paper (p. 78)

fission: nuclear reaction in which large atoms break apart into smaller atoms, releasing much energy; occurs in atomic bomb and nuclear reactors (p. 358)

fissionable: elements which can undergo fission; for example, uranium-235 and plutonium-239 are fissionable (p. 380)

flammable: easily set on fire (p. 29)

foot-pounds: unit for work and energy in English system; abbreviated ft-lb (p. 205)

force: a push or a pull; measured in newtons (N) or pounds (lbs) (p. 188)

forms of energy: ways in which energy can exist, such as chemical and light (p. 216)

formula mass: mass of one molecule obtained by adding masses of its atoms; also known as molecular weight (p. 125)

fossil fuels: oil, natural gas, and coal; formed by the decay of animals and plants (p. 376)

freezing point: temperature at which a substance changes from liquid to solid (p. 72)

frequency: number of waves that are made or pass by each second; unit is waves per second (p. 255)

friction: force created by rubbing; friction acts in the direction opposite to the motion of an object (p. 194)

fuel cell: device that produces electricity directly from the chemical energy in fuels without heat or burning (p. 428)

fulcrum: point on which a lever turns (p. 237)

fumes: gas that forms by evaporation from a liquid; vapor (p. 29)

functional group: group of atoms attached to an organic molecule which determine what kind of compound it is; OH, for example, is the functional group in all alcohols (p. 177)

fuse: safety device which prevents too much electricity from flowing in wires as more things are turned on; similar in purpose to a circuit breaker (p. 342)

fusion: nuclear reaction in which small atoms merge to make larger atoms, releasing much energy; occurs in the sun and stars (p. 359)

galvanometer: device to measure very small electric currents or voltages (p. 339)

gas: phase of matter which has no definite volume and no definite shape; gases fill the entire volume and shape of the container they are in (p. 72)

gasification: converting coal into gas fuels like methane, hydrogen, carbon monoxide (p. 416)

gasohol: mixture of gasoline and alcohol; the alcohol is a biofuel made from plants; see biomass (p. 424)

Geiger counter: device which measures amount of radiation emitted by a radioactive substance (p. 357)

generator: device for making electricity (p. 341)

geothermal energy: energy contained in the hot rock, lava, water, and steam found under the earth's surface; can be used to make electricity (p. 427)

global effect: worldwide, international impact of a technology (p. 21)

graduated cylinder: glass or plastic cylinder with volume markings placed along its side for measuring volumes of liquid in the laboratory (p. 50)

gram-atomic mass: sample of an element which equals its mass number weighed out in grams;

abbreviated G.A.M. (p. 125)

gram-formula mass: sample of a compound which equals its formula mass weighed out in grams; abbreviated G.F.M. (p. 126)

gravity: downwards force of Earth upon an object; force of attraction between any objects, created by their masses (p. 189)

greenhouse effect: gradual warming of the atmosphere due to increased amounts of carbon dioxide, methane, and other gases; more heat is trapped (p. 295)

grounding: wire that connects the outside body of electrical devices and appliances to underground pipes; prevents accidental shock by sending electrons into the ground; the third plug in home wires is usually for grounding (p. 343)

group: elements located in a column on Periodic Table; see "family of elements" (p. 103)

hard water: water that does not form soap suds or clean well due to dissolved calcium ions (p. 160)

half-life: time its takes for 50% of the radioactive element in a sample to decay into other elements; ranges from fractions of a second to millions of years for different elements (p. 355)

hazard: situation or substance that might harm people or the environment (p. 33)

heat: see heat energy

heat energy: energy created by motion of molecules in an object; referred to as "heat" (p. 217)

heat-resistant glass: glass used for laboratory tubing, test tubes and other glassware because it does not crack when heated (p. 39)

hemoglobin: red-colored organic molecule in blood which carries oxygen to cells (p. 182)

household chemicals: chemicals used for special purposes at home, such as medicines, cleaning fluids, polishes, and snow salt (p. 28)

hydrocarbon: organic compound containing just carbon and hydrogen atoms (p. 177)

hydroelectric energy: electricity produced by using water flowing in a river or from a dam to turn generators (p. 376)

hypothesis (hy-POTH-uh-sis): educated guess that predicts results of an experiment (p. 10)

ideal machine: imaginary machine in which there is no friction, input and output work are equal, and efficiency is 100% (p. 233)

illumination: brightness of a surface; depends upon brightness of and distance to the light source; see "intensity" (p. 272)

inclined plane: flat, sloping surface like a ramp; one kind of simple machine (p. 238)

independent variable: variable in an experiment which is set and controlled by the researcher; it is adjusted to see what its effect is upon dependent variable (p. 57)

indicators: chemicals which change color to show if a solution is acid or base; some show approximate pH (p. 157)

inefficient: property of a machine with a low efficiency rating, as less than 60%; only a part of the energy put in goes to do useful work and a large part is wasted on heat and friction (p. 234)

inert elements: elements that do not react with other elements; also called Noble Gases (p. 100)

inertia (in-UR-shuh): resistance of objects to changes in their motions; tendency for moving objects to keep moving and for objects at rest to stay at rest; inertia increases as mass increases (p. 198)

infrared wave: electromagnetic wave that has a large heating effect (p. 268)

input work: work put into a machine; see effort (p. 230)

insoluble: cannot dissolve in water (p. 77)

insulators: materials which do not conduct heat or electricity well; wood is a good insulator for heat and for electricity (pp. 289, 325)

intensity: brightness of a light source, like a bulb or a star (p. 272)

internal combustion engine: engine in which the pistons move by the pressure of expanding gas made by exploding the fuel inside the cylinders; used in cars and trucks today (p. 374)

inversely proportional: if variable x is multiplied by a certain amount, then variable y is divided by that amount (p. 200)

ion: atom with an electric charge due to electrons gained or lost (p. 100)

ionic bond: chemical bond formed by the transfer

of electrons from one atom to the another, creating oppositely charged ions which attract (p. 102)

ionic compound: compound held together by ionic bonds (p. 144)

ionic-exchange reaction: see double-replacement reaction

ionic solution: results when ionic compounds are dissolved in water causing them to form positive and negative ions in the solution (p. 144)

irregular solid: complicated shape that has no easy formula for calculating its volume, such as a stone, key, or hammer (p. 53)

isomers (EYE-su-murz): compounds with the same chemical formula but different structural formulas (p. 175)

isotopes: atoms of the same element having different mass numbers because they have a different number of neutrons; however, they must have the same number of protons (p. 97)

joule (JOOL): metric unit of work and energy; abbreviated J; 1 J = 1 newton-meter (p. 205)

kilogram: unit for measuring mass in the metric system; on the surface of the earth, 1 kilogram weighs about 9.8 newtons or 2.2 pounds (p. 50)

kilowatt-hour (KIL-uh-watt): unit for measuring electrical energy and calculating electric bills (pp. 212, 337)

kinetic energy: type of energy that is the result of the motion of an object; depends on mass and speed (p. 213)

laboratory chemicals: concentrated and purified substances used in laboratory work (p. 33)

laser: device which makes light that is very pure (single frequency) and coherent (waves in step) and carries large amounts of energy (p. 272)

law of action and reaction: Forces always come as equal but opposite pairs. If object x exerts a force on object y, then object y exerts an equal force back on object x. (p. 200)

law of definite composition: Elements join together in definite ratios of mass to make compounds. (p. 81)

law of machines: Machines cannot create work or energy; the work given out is, at best, equal to the work put in or less than the work put in. (p. 230)

law of reflection: A light ray bounces off a flat surface at an angle that equals the angle the ray made as it hit the surface. (p. 273)

law of refraction: A light ray bends towards the normal line as it moves from air into a denser medium like water or glass. (p. 275)

law of science: major conclusion that is supported by thousands of experiments and never found to be incorrect (p. 12)

lever: firm rod or board that rotates around a fixed spot, like a see-saw or nut cracker; a simple machine (p. 237)

liquefaction: process of converting coal into liquid fuels like gasoline, diesel fuel, and heating oil (p. 417)

liquid: phase of matter which has a definite volume but no definite shape (p. 72)

liter: metric unit for measuring the volume of liquids and gases; equal to exactly 1000 cm^3; abbreviated L; slightly larger than 1 quart (p. 49)

lodestone (LOHD-stohn): naturally magnetic rock (p. 303)

longitudinal wave: wave in which the particles of the medium vibrate parallel to the motion of the wave itself; sound is a longitudinal wave (p. 253)

lubricate: put oil between two surfaces so they slide with less friction (p. 233)

machine: device that transfers force and mechanical energy from one place to another place (p. 229)

magnetic energy: energy stored in magnets or created in electromagnets; creates force that attracts certain materials (pp. 218, 309)

magnetic field: pattern of forces around a magnet; can be seen using iron filings (p. 306)

magnetic poles: ends of a magnet; each is a different type of magnetism, designated as the north and south poles (p. 303)

magnetism: force of attraction for iron and other magnetic materials that exists in magnets and electromagnets (p. 303)

many-atom ion: group of atoms which are often found together and behave as if they were a single ion, such as NH_4, which has a charge of $+1$, and CO_3, which has a charge of -2 (p. 118)

mass: amount of matter in an object; measured in grams or kilograms in metric system (p. 50)

mass number: sum of neutrons and protons in the nucleus of an atom (p. 96)

matter: anything which occupies space and has mass (p. 68)

measure: use an instrument to find the exact amount of something (p. 14)

mechanical advantage: comparison of the force produced by a machine to the force used to operate it; MA = resistance force/effort force. The ideal mechanical advantage is for a frictionless machine. The actual mechanical advantage is for a machine with friction (p. 234)

mechanical energy: sum of the kinetic energy of an object and its potential energy due to height (p. 217)

medium: material through which a wave moves; air is a medium for sound waves (p. 251)

melting point: temperature at which a solid becomes a liquid (p. 72)

metals: elements which conduct electricity and have other properties in common; metals give away electrons to form positive ions (p. 100)

metalloids: elements which have properties between those of metals and nonmetals (p. 100)

microwave: electromagnetic wave used for cooking food and sending wireless telephone signals (p. 268)

miscible drive: process which uses organic liquids under high pressure to recover leftover petroleum from old wells (p. 415)

mixture: matter made up of two or more materials mixed together in which each retains its own properties; the percent of each material can vary (p. 69)

model: description or representation of an event or object. Models need not look like the real thing, they only need to behave like the real thing. (p. 16)

molecule: smallest particle of a compound that still has the chemical properties of the compound; made of atoms joined together (p. 91)

motion: change in position compared to a reference point (p. 195)

narcotics: drugs which dull the senses (p. 183)

neutralization: reaction of an acid and a base to form a salt and water (p. 158)

neutrons: particles found in the nucleus of an atom which have no electric charge and have a mass of approximately 1 u (p. 95)

newton: unit of force and weight in the metric system; abbreviated N (p. 190)

nonmetals: elements which do not conduct electricity and have other properties in common; nonmetals take in electrons to form negative ions (p. 100)

nonrenewable energy sources: an energy source that is not replaced once it is used; such as fossil fuels and uranium (p. 373)

nuclear energy (NU-klee-ur): energy stored in the nucleus of every atom; holds protons and neutrons together (p. 218)

nuclear reaction: process in which nuclei break apart or join together; produces enormous amounts of energy (358)

nucleus (pl. nuclei): tiny center of an atom which contains nearly all the atom's mass; consists of neutrons and protons held together by nuclear energy (p. 93)

observe: use senses to gather information (p. 14)

ocean thermal energy conversion: process which uses difference in temperature between layers of ocean water to produce electricity (p. 423)

ohm: unit for measuring the resistance of an object to electrical current (p. 328)

oil shale: rocks containing organic matter which can be decomposed into an oil similar to petroleum which contains various fuels (p. 415)

OPEC: Organization of Petroleum Exporting Countries; represents some of the countries that produce oil (p. 395)

ore: rocks containing compounds which are decomposed to produce metals (p. 161)

organic acids: organic compounds containing a

COOH group; in water the hydrogen atom comes off as a hydrogen ion, thereby making an acid solution (p. 180)

organic chemistry: branch of chemistry that studies compounds containing carbon (p. 172)

oscilloscope: device which shows a sound wave on a small TV screen; it is connected to a microphone (p. 259)

output work: work given out by a machine; see resistance (p. 230)

oxidation: (1) direct-union reaction in which an element joins with oxygen to make a compound, known as an oxide (p. 135); (2) the addition of oxygen atoms to a molecule, as in making alcohols from hydrocarbons (p. 178)

oxidizing flame: outer pale-blue cone of a Bunsen burner flame which can heat and give oxygen to materials; surrounds the inner reducing flame (p. 37)

parallel circuit: circuit in which there is more than one path for the current to flow in (p. 329)

passive solar heating system: system in which a building's design and materials collect the sun's energy for home heating in an effective way; does not use fans or pumps to move warmed air or water around the home or office (p. 421)

period: horizontal rows in Periodic Table (p. 106)

Periodic Table: table of all elements arranged according to increasing atomic number and positioned in families of like elements (p. 103)

permanent magnet: magnet which retains magnetism for a long time (p. 303)

petroleum: liquid mixture of many hydrocarbons found underground; source of many products; also known as oil or crude oil (p. 177)

pH: measurement of how acidic or basic a solution is: 0 is strongly acid, 7 is neutral, 14 is strongly basic (p. 157)

phase: state in which matter can exist: gas, liquid, or solid (p. 72)

photons: particles of light (p. 265)

photovoltaic conversion: process which occurs in solar cells; converts light energy directly to electricity (p. 424)

physical change: change in the form of a substance but not in its composition, such as melting, tearing, and dissolving (p. 75)

physical properties: characteristics of a material such as color and boiling point (p. 71)

physical science: the study of the materials, forces, and behaviors of objects (p. 9)

physics: branch of science that investigates the forces and motions of objects (p. 9)

potential energy: type of energy that is a result of chemical composition or of position, such as height above ground (p. 213)

pound: unit of force and weight in the English system; abbreviated lb (p. 190)

power: rate of using energy or doing work; the energy or work done per second or per hour; measured in watts and kilowatts; in electricity, power = voltage $\times$ amperes (p. 337)

precipitate: insoluble material which forms as a product in a reaction (p. 146)

pressurized container: spray cans containing a gas or liquid under high pressure; used for deodorants and paints; known as aerosols (p. 31)

products: materials made by a chemical reaction, written on the right side of the equation for the reaction (p. 121)

protons: positive particles in a nucleus with a mass of 1 u (p. 94)

psi: pressure in pounds per square inch (p. 405)

puncture: pierce or break open; it is dangerous to puncture a pressurized container (p. 31)

qualitative observations: observations which describe how an object or an event appears to the senses using words (p. 14)

quantitative observations: observations which describe an object or an event using numbers taken by measurement (p. 14)

radiation: (1) transfer of heat as a hot object gives off infrared electromagnetic waves, as in the glowing wires of a toaster or a red hot piece of iron that slowly cools off (p. 290); (2) invisible particles and bursts of energy that come out from radioactive elements, caused by the nuclei breaking apart (p. 353)

radioactive: property of materials which gives out

radiation from the nuclei of their atoms (p. 352)

radioactive decay: changes that occur when a radioactive element emits radiation from its nuclei (p. 354)

radioactive isotopes: those isotopes of an element which are radioactive; $C - 14$ is a radioactive isotope of carbon while $C - 12$ is not (p. 354)

rads: unit that measures how much radiation comes out of a radioactive material (p. 365)

rate of reaction: speed of a reaction; amount of product made in a certain time (p. 82)

raw data: measurements and observations made during an experiment without adjustments or further calculations (p. 56)

rays: straight lines of light (p. 265)

reactants: materials consumed in a chemical reaction, written on the left side of the equation of the reaction (p. 120)

reactor core: center of a nuclear reactor where the intense heat from the fission of uranium is produced (p. 381)

real machine: an actual machine in which there is friction, the output work is always less than the input work, and the efficiency is less than 100% (p. 233)

recoil: backwards movement resulting from a large reaction force made by a large action force; for example, the backwards jerk of a gun when a bullet is fired (p. 201)

recover: obtain more fuel or energy from a source which was considered used up (p. 403)

reducing flame: inner darker blue cone of a Bunsen burner flame which heats and removes oxygen from materials; surrounded by outer oxidizing flame (p. 37)

reference point: spot that is used to measure the motion of objects (p. 195)

reflected light: see reflection

reflection: part of a sound or light wave that bounces off a surface, such as a wall or mirror (pp. 257, 271)

refraction: the bending of light rays as they pass from one medium to another, such as from air into glass (p. 275)

regular solid: shape with an easy formula for calculating its volume; a cube or box is a regular solid with the formula, volume = length $\times$ width $\times$ height (p. 53)

rems: unit that measures how harmful to human beings is the radiation coming out of a radioactive material (p. 365)

renewable energy source: energy source that is replaced by natural processes within a short period of time after it is used; such as sunlight, wind, hydroelectric energy, plants, and trees (p. 372)

reserves: available supply of a source of energy, like coal, in a certain country, or in an area of a country (p. 391)

resistance: (1) work that is done by a machine; equals resistance force multiplied by a resistance distance (p. 229); (2) measure of how difficult it is to move electricity through an object; electrical resistance is measured in ohms (p. 328)

resonance: increased strength of vibrating that occurs when two objects have identical vibrating frequencies (p. 258)

retrofitting: putting new, advanced devices on an old system; for example, installing computer controls on old elevators in a building (p. 407)

salt: compound containing a metal and nonmetal part; a product made by the neutralization of an acid and a base (p. 133)

saturated solution (SACH-uh-rayt-ud): solution that cannot dissolve any more material; maximum amount already dissolved (p. 164)

saturated compounds: organic compounds in which all carbon atoms are joined by single bonds, by sharing one pair of electrons (p. 174)

scale: (1) numbers placed along the axes of a graph (p. 56); (2) the numbers used in reading a measuring instrument (p. 327)

science: facts, ideas, and investigations about the world (p. 8)

science process skills: special skills used to carry out scientific research, such as observing, organizing, and predicting (p. 14)

scientific method: process used by researchers in stating problems and finding solutions (p. 10)

sectors: parts of the economy; the industrial/

commercial, residential, and transportation sectors need energy for different purposes (p. 387)

series circuit: circuit in which there is only one path for the current to flow; electrical devices linked in a line (p. 329)

simple machines: devices which are used as machines themselves and which form the working parts of more complicated machines; includes lever, pulley, gears, etc. (p. 229)

single-replacement reaction: reaction in which a more active element replaces a less active element in a compound (p. 141)

solar energy: sunlight used as an energy source (p. 420)

solid: phase of matter which has a definite volume and shape (p. 72)

soluble: able to dissolve in water (p. 77)

solubility: maximum number of grams of a material which can dissolve in 100 g of water at a given temperature (p. 164)

solubility curve: graph that shows how solubility changes with temperature (p. 165)

solution: mixture made when a material dissolves in water; solutions are clear and material does not settle (p. 77)

solvent refining: process using special liquids to remove the ash and sulfur found in coal; makes coal more useful (p. 416)

sound energy: energy contained in vibrations of air molecules in a sound wave (p. 217)

specific gravity: ratio of the density of a material to the density of water; useful for finding densities in any desired units by using the formula: density = specific gravity × density of water (p. 441)

spectra: lines of color emitted by each element when heated; spectra reveal the electron shell structure of atom (p. 95)

speed: distance travelled in a unit of time; speed = distance/time; some units are miles/hr and m/sec (p. 196)

standard of living: quantity and quality of key aspects of living; includes food, jobs, education, health, housing; and leisure-time; a country's standard of living is supported by its energy sources and consumption (p. 395)

static electricity: charges, either negative or positive, that remain in one place; the negative charges are electrons (p. 316)

stimulants: drugs which speed up the chemical reactions in a person's body and mind; also known as "uppers" (p. 183)

structural formula: diagram which shows the arrangement of atoms in a molecule (p. 173)

subscript: numbers placed within a formula that tell how many atoms of an element are in a molecule; in Na_3PO_4 the subscripts are "3" for Na, an unwritten "1" for P, and "4" for O (p. 115)

supersaturated solution: solution with more than the maximum amount of material that normally dissolves; supersaturated solutions are unstable and easily release the excess material as a precipitate (p. 164)

supersonic: faster than the speed of sound, which is about 330 m/sec or 740 miles/hr (p. 252)

suspension: mixture made when a material is stirred with water but does not dissolve; suspensions are cloudy and material settles to the bottom (p. 77)

technological assessment: process of analyzing and predicting the consequences for business and society of a new system or device (p. 244)

technology: devices or procedures that apply science for improving life and society (p. 19)

theory: idea that helps explain the results of many experiments or observations (p. 12)

thermal pollution: heat dumped into the air or rivers by factories, electric generating plants, and vehicles (p. 295)

thermodynamics (thur-moh-dy-NAM-iks): study of heat (p. 220)

thermostat: device that keeps a room at a desired temperature by turning the furnace or air conditioner on and off (p. 293)

tide: flow of ocean water caused by gravity of the moon; may be used for making electricity in certain places (p. 427)

toxic: poisonous (p. 29)

tracer: radioactive chemical used to follow the flow of liquids or solids in a patient, or to follow atoms in a chemical reaction; important for medical and chemical research (p. 362)

trade-off: compromise between technological choices that considers the advantages and disadvantages of each choice (p. 23)

transformer: device that changes the voltage of AC electricity (p. 331)

transmitted light: light which passes through a medium, such as the sunlight that passes through a window (p. 271)

transverse wave: wave in which the particles of the medium vibrate perpendicular to the motion of the wave itself; water waves are transverse waves (p. 253)

turbine: device with blades arranged on wheel, similar to a pinwheel; turns when air, water, or high-pressure steam flows through it; used to turn generators to make electricity (p. 376)

ultrasonic sound: sound with frequencies higher than human hearing can detect, more than 20,000 waves/s (p. 259)

ultraviolet wave: an electromagnetic wave that kills bacteria, causes sunburn, and makes certain paints and rocks glow with color (p. 268)

unbalanced forces: forces acting on an object which do not cancel each other; leads to a change in motion (p. 193)

units: standard amount upon which a measurement is based; in "40 minutes," the unit is minutes (p. 45)

universal solvent: water, because it can dissolve so many materials (p. 163)

unsaturated solution: solution containing less than maximum amount of material that can be dissolved; able to dissolve more (p. 164)

unsaturated compound: organic compound in which some carbon atoms are joined together by double or triple bonds, by sharing two or three pairs of electrons (p. 175)

uranium-238 dating: process that uses radioactive U-238 to estimate the age of rocks (p. 363)

useful energy: forms of energy that are easily and almost completely converted into useful work, like chemical energy in gasoline; heat is not a useful form of energy (p. 223)

vacuum: absence of air or any other material; empty space (p. 12)

valence electrons: electrons in the outer shell of an atom; valence electrons determine the chemical behavior of an atom (p. 100)

vapors: see fumes (p. 29)

variable (VER-ee-uh-bul): aspect, factor, or condition in an experiment that might affect the results of the experiment (p. 11)

ventilated: place where fresh air flows (p. 29)

vibrate: shake rapidly back and forth around one spot (p. 249)

voltage: unit for measuring the pressure or force of electrons in a circuit; for example, most flashlight batteries are 6 volts (p. 326)

volume: measurement of the space within or occupied by an object (p. 47)

water softeners: materials which change hard water to soft water by removing dissolved calcium ions (p. 160)

water-displacement method: method for finding the volume of an object; an object is submerged under water in a graduated cylinder. The object's volume is the difference in the water level readings on the cylinder. (p. 53)

wasted energy: part of the energy put into a machine which turns into heat instead of performing useful work (p. 233)

wave: a series of disturbances that move in a repeating pattern (p. 250)

wavelength: distance between the peaks of a wave (p. 255)

weight: pulling force of Earth upon an object; measured in newtons (N) and pounds (lbs) (p. 190)

work: action in which a force moves an object for a distance; work = force × distance; units are newton-meters, joules, foot-pounds (p. 204)

Page numbers printed in **bold** type have illustrations or tables

M

Mach 1, 251
machines, 210–11, 228, 230–32, 243
  efficiency of, 233–34, 241
  effort/resistance, 229
  mechanical advantage, 234–36,
  237–40
  real/ideal, 233
  simple, **229**
magnesium, 82, 101, 108, 121, 136, 156
magnets, 76, 137, 213, 218, 302–04,
  305, 308–10
magnetic field, 306, 307, 338, 339, 430
mass, 50, 188
  acceleration, 199–200
  gravity, 189–90
  inertia, 199
  kinetic energy, 216
mass number, 96, 125, **442–43**
matter, 12, 68–70, 71, 190, 283
  chemical properties/changes, 81
  phases, or states, of, 72, 78
  physical properties/changes, 71–79
measurement, 14, 47, 327, 444-45
  of force and weight, 190
  of length, 47
  of mass, 50, 189–90
  of speed, 196
  of time, 47
  of volume, 48
mechanical advantage,
  actual/ideal, 234–36
  inclined plane, 238–40, 241
  lever, 237–38
mechanical energy, 217, 221, 228–29
melting point, 72, 285
Mendeleev, Dimitri, 103
mercury, 100, 138
metals, 100–01, 133, 289
  alkali, 107
  alloys, 161–62, 234, 303, 307
  in periodic table, 108–09
metalloids, 100, 108
meters, electric,
  galvanometer, 339
  accuracy in reading, 327

metric system, 46–51
  and common (English) units, 444–45
microwaves, 267–69
mirrors, 273–75
mixture, 69, 75–6
  separation of, 76–7
  solutions as mixtures, 163
models, 16, 92, 98, 119, 123
  of atom, **18**, 93–8
  of magnets, 308–09
molecular mass. *See* formula mass
molecule, 90–2, 111–14
motion, 195, 258–59
  acceleration, 196
  force, 188, 192–95, 206
  inertia, 198
  kinetic energy, 213
  laws of 194, 198–203
  speed, 196, 197
motor. *See* electric motor
music, 248, 250, 255, 259

N

natural gas, 373, 374, 376, 378–79, 416
neon, 161
neutralization, 158–59
neutron, 95, 110
  fission, 358–59, 381
  isotopes, 96, 363
  radioactivity, 353–54
Newton, Isaac, 269
newton-meter, 205, 212
nickel, 69, 161, 303
nitric acid, 137, 154
nitrogen, 69, 119, 137, 161
noble gases. *See* inert elements
noise pollution, 259–60
nonmetal, 100, 102, 133
  in periodic table, 108–09
nuclear magnetic resonance, NMR,
  311
nuclear power plant. *See* nuclear reactor
nuclear reaction, 358, 359
nuclear reactor, 22, 218, 367, 375, 381,
  419

nuclear waste, 367, 368, 375, 393
nucleus, 93, 94, 96–9, 110, 218,
  352–54, 358–60

O

observation, 14
ocean thermal energy, 423
Ohm's law, 328
oil. *See* petroleum
oil shale, 415
ores, 161
organic chemistry, 172–75, 182–83
  functional groups, 177–82
  hydrocarbons, 177–78
oscilloscope, 259
output. *See* systems analysis
oxidation, 135, 150
oxidation number, 117
oxygen, 69, 80–81, 91, 97, 103, 106,
  107, 115, 122, 133, 161, 182
  preparation of, 138
ozone, 184

P

parallel circuit, 329–30
peat, 379
period, 106
periodic table of elements, 103–109,
  104–105
petroleum, 81, 373, 378, 394–95, 403,
  415
pH, 157
phases, 72–4, 78, 284–85
phenolphthalein, 157, 158
photon, 265, 266
photosynthesis, 81, 182
photovoltaic conversion, 424
physical changes, 75–9
physical properties, 71–4
physical science, 8–9
  careers, 436–37
physics, 9
pitch, 255
plastics, 168